Biology of Plagues:
Evidence from Historical Populations

The threat of unstoppable plagues, such as AIDS and Ebola, is always with us. In Europe, the most devastating plagues were those from the Black Death pandemic in the 1300s to the Great Plague of London in 1665. For the last 100 years, it has been accepted that *Yersinia pestis*, the infective agent of bubonic plague, was responsible for these epidemics. This book combines modern concepts of epidemiology and molecular biology with computer modelling. Applying these to the analysis of historical epidemics, the authors show that they were not, in fact, outbreaks of bubonic plague. *Biology of Plagues* offers a completely new interdisciplinary interpretation of the plagues of Europe and establishes them within a geographical, historical and demographic framework. This fascinating detective work will be of interest to readers in the biological and social sciences, and lessons learnt will underline the implications of historical plagues for modern-day epidemiology.

SUSAN SCOTT is a research worker in historical demography in the School of Biological Sciences at the University of Liverpool.

CHRISTOPHER J. DUNCAN is Emeritus Professor of Zoology also in the School of Biological Sciences at the University of Liverpool.

They previously co-authored *Human Demography and Disease* (1998) and have published many papers on population dynamics and historical epidemiology.

Biology of Plagues: Evidence from Historical Populations

SUSAN SCOTT AND CHRISTOPHER J. DUNCAN

School of Biological Sciences
University of Liverpool

CAMBRIDGE UNIVERSITY PRESS
Cambridge, New York, Melbourne, Madrid, Cape Town, Singapore, São Paulo

Cambridge University Press
The Edinburgh Building, Cambridge CB2 2RU, UK

Published in the United States of America by Cambridge University Press, New York

www.cambridge.org
Information on this title: www.cambridge.org/9780521801508

First published 2001
This digitally printed first paperback version 2005

A catalogue record for this publication is available from the British Library

Library of Congress Cataloguing in Publication data

Scott, Susan, 1953–
Biology of plagues/Susan Scott and Christopher J. Duncan
p. cm.
Includes bibliograhical references and index.
ISBN 0 521 80150 8
1. Epidemics. 2. Epidemics – Europe – History – 16th century.
3. Epidemics – Europe – History – 17th century. 4. Black death – Europe.
5. Plague. I. Duncan, C. J. (Christoher John) II. Title.
RA650.6.A1 S36 2001
614.4′94 – dc21 00–063066

ISBN-13 978-0-521-80150-8 hardback
ISBN-10 0-521-80150-8 hardback

ISBN-13 978-0-521-01776-3 paperback
ISBN-10 0-521-01776-9 paperback

Contents

Preface

When studying the population dynamics of northwest England for our earlier book *Human Demography and Disease* (also published by Cambridge University Press) we became interested in the biology of the plagues that beset Europe after the Black Death. A plague struck this part of England and spread rapidly in 1597–98 and it was obvious from a basic training in zoology that this was not an outbreak of bubonic plague. By making a full family reconstitution study of the community at Penrith in Cumbria (where some 40% of the population died) it was possible to trace the spread of the disease between named individuals in the same family and between households. From this starting-point, we have made an interdisciplinary study of the epidemiology and biology of the plagues that have afflicted western Europe, concentrating on the outbreaks from the Black Death, which began in 1347, to the Great Plague of London. We have combined modern epidemiological concepts, computer modelling of epidemics, recent molecular biology studies, spatial analysis techniques, time-series analysis of the epidemics and the careful analysis of the sequence of infections in selected epidemics. We hope that our monograph will be of interest to a wide variety of readers who will come to look at historical plagues with different eyes.

Once again we thank Dr S. R. Duncan, of the University of Oxford, for introducing us to the intricacies of time-series analysis and for developing the mathematical models that we have used.

We are grateful to members of the Cheshire Parish Register Transcription project, in particular Mr and Mrs C. D. Leeming and Mr and Mrs J. R. Fothergill, for providing unpublished data. Mrs J. J. Duncan also provided invaluable assistance in reading documents in Secretary Hand and in translating original articles.

We salute the pioneering work of Dr G. Twigg and gratefully acknowledge his generous help in the early stages of this project.

S.S.
C.J.D.

Conversion table for imperial to metric units

Imperial unit	Metric equivalent
1 inch	25.4 millimetres
1 foot	0.3048 metre
1 yard	0.9144 metre
1 mile	1.609 kilometres
1 acre	0.405 hectare
1 square mile	259 hectares

1
Introduction

We first became interested in plagues when studying the demography of northwest England (Duncan *et al.*, 1992; Scott & Duncan, 1998), where an epidemic in the town of Penrith in 1597–98 killed some 40% of the population and initiated endogenous oscillations in the annual numbers of births and deaths. In this way, its effects persisted for 150 years. The outbreak spread rapidly, travelling 20 to 30 miles in 2 or 3 days and it was obvious that it was a biological impossibility that this was an outbreak of bubonic plague. We initially thought that this must have been an isolated outbreak of an unknown and unique infectious disease (Scott *et al.*, 1996) but further study convinced us that this regional epidemic had many points in common with other outbreaks in England that were believed to be bubonic plague.

In this book, we have attempted an objective (though not exhaustive) study of the plagues that have ravaged humankind for hundreds of years, giving the biological, demographic and epidemiological viewpoints of the available historical evidence. Obviously, the difficulties faced are vastly greater than those of a modern epidemiologist investigating a new outbreak of an unknown disease today. He or she has an array of techniques available from microbiology and molecular biology, can take biopsy and autopsy samples with the back-up of a pathology laboratory, can make on-site investigations of the ecology and epidemiology of the disease, and can discover the clinical features and mode of transmission of the infection. Even so, some features of present-day outbreaks of Ebola, such as the elucidation of the reservoir host, are not yet established with certainty. Where the disease has a complex biology, as in bubonic plague, it took years of painstaking study before all the details were elucidated.

1.1 What is a plague?

The Bible uses the word 'plague' to describe an affliction that was regarded as a sign of divine displeasure or as an affliction of humankind such as the plague of locusts. Nowadays, it is a term used to describe a deadly epidemic or pestilence and *The Wordsworth Encyclopedia of Plague and Pestilence* (Kohn, 1995) lists a seemingly endless catalogue of historical epidemics from all over the world, including smallpox, cholera, typhus and malaria. They were all infectious and potentially lethal, caused high mortality and were serious historic events. The influenza pandemic of 1917–19, with a final death toll worldwide estimated at more than 20 million, is a good example. The incubation period was often less than 2 days so that its worldwide spread was dependent on 20th century means of rapid travel that could move people in bulk, namely steamtrains and steamships of which the troop ships of the First World War are a good example.

However, the basic etiology of these diseases is now usually well understood and, in spite of the terrible death toll, the percentage mortality of the affected populations was not relatively high (Langmuir *et al.*, 1985). In this book, we are mainly, but not exclusively, concerned with the Black Death, arguably the most awful epidemic ever to have struck, which raged in Europe from 1347 to 1350, and the unremitting succession of plagues that followed it for 300 years. These reached their peak in continental Europe during 1625–31 and in England in 1665–66, but they then disappeared completely after about 1670. When these plagues struck a naive population, where we have reasonably accurate data available in Italy and England, mortality could reach about 50%; we do not know what was the causative agent in these terrible epidemics.

1.2 Four ages of plague

From whence did the Black Death come? The probable answer to this question is that it originated in the Levant, but Europe had suffered from a series of mysterious plagues for many years before 1347 and it is possible to identify tentatively and arbitrarily four historic ages of plague.

1.2.1 Plague at Athens, 430–427 BC

The epidemic that struck Athens in 430 BC remains one of the great medical mysteries of antiquity and has been discussed by a number of scholars (Morens & Littman, 1972, 1994; Poole & Holladay, 1979;

Longrigg, 1980; Langmuir *et al*., 1985; Kohn, 1995; Olson *et al*., 1996; Retief & Cilliers, 1998), but the first vivid description was given by Thucydides, himself a victim who survived the outbreak (see Page, 1953). It is sometimes termed the Thucydides syndrome because of his evocative narrative.

People were stricken suddenly with severe headaches, inflamed eyes, and bleeding in their mouths and throats. The next symptoms were coughing, sneezing, and chest pains followed by stomach cramps, intensive vomiting and diarrhoea, and unquenchable thirst. The skin was flushed, livid and broken with small blisters and open sores. The patients burned with fever so extreme that they could not tolerate being covered, choosing rather to go naked. Their desire was to cast themselves into cold water, and many of those who were unsupervised did throw themselves into public cisterns, consumed as they were by unceasing thirst. Many became delirious and death usually came on the seventh or eighth day of the illness, although those who survived the first phase often died from the weakness brought on by constant diarrhoea. Many who recovered had lost their eyesight, their memory, or the use of their extremities.

This plague is believed to have originated in Ethiopia and travelled through Egypt and the eastern Mediterranean before reaching Athens. The first cases appeared in Piraeus, the Athenian port and base for many travellers and merchants who probably contracted the disease in their journeys abroad. It spread rapidly to the upper city and whole households were left empty. Mortality among doctors, as among other attendants of the sick, was especially high. Fearful of an attack by the Spartans, the Athenian leader Pericles ordered the inhabitants of the surrounding countryside to move inside the city, where they could be protected by the army and the fortified walls. Many country dwellers, coming to an already overpopulated city, had no place to live except in poorly ventilated shacks and tents. This mass of people, crowded together in the hot summer, created a situation that was ideal for the rapid transmission of the disease. Though there were many dead bodies lying unburied, there was said to be a complete disappearance of birds of prey and dogs. Apparently it was rare to catch the disease twice, or if someone did, the second attack was never fatal. A peak case rate was reached during the Spartan siege, which lasted 40 days, after which the crowded refugees dispersed. The disease remained at a low level through 429 BC (when Pericles died of it) and returned in force in the summer of 428 BC at the time of another Spartan siege. The disease was quiescent, or even absent, from the winter of 428 BC until the summer of 427 BC, but broke out again in the autumn or early winter of

427 BC. This epidemic lasted no less than a year, but there is no further mention of the disease. The total number of Athenians who died is not recorded but, over the 3-year period, of 13 000 enrolled hoplites (soldiers), 4400 died – a mortality rate of 33%. Hagnon took the fleet and sailed to Potidaea carrying the plague there also and this made dreadful havoc among the Athenian troops. Even those who had been there previously and had been in good health caught the infection and so 1050 men out of 4000 were lost in about 40 days.

There have been several identifications of the causative agent of the plague at Athens, including smallpox (Littman & Littman, 1969), scarlet fever, measles and typhus (Shrewsbury, 1950; Page, 1953) but these are all now discredited. Langmuir *et al.* (1985) concluded that the clinical descriptions clearly indicated the involvement of specific organ systems and that there was an obvious inflammatory condition of the eyes and respiratory tract; this acute respiratory infection was severe and probably necrotising; the initiation with vomiting followed by empty retching and later by 'watery diarrhoea' strongly suggested a gastroenteropathy mediated by the central nervous system rather than a local inflammatory process. Langmuir *et al.* (1985) believed that the skin lesions were suggestive of bullous impetigo. They did not suggest that the Thucydides syndrome was identical with the modern toxic shock syndrome but believed that the same basic pathogenic mechanisms were involved, in that there was infection in predisposed hosts by a possibly non-invasive *Staphylococcus* sp. that was capable of producing an exotoxin similar to toxin-1 of the toxic shock syndrome (Rasheed *et al.*, 1985). This toxin may have differed from toxin-1 in that it produced predominantly enterotoxic effects and less profound circulatory collapse, and had only moderate or no erythrogenic potential.

Morens & Littman (1992, 1994) have approached the plague at Athens from a different viewpoint and have arrived at a conclusion that is strongly opposed to that of Langmuir *et al.* (1985). We describe their hypothesis briefly because they use mathematical modelling techniques that we shall also employ to elucidate the epidemiological parameters of later plagues. They have reduced the reliance on clinical symptoms in favour of the epidemiology of the disease because pre-modern descriptions, which lack detailed information on serology and accurate accounts of rashes and other clinical features, always retain a high degree of uncertainty. Use of the Reed and Frost mathematical model (section 2.5) led them to conclude that, under any conditions of crowding that probably prevailed in Athens in 430 BC, an epidemic of influenza would have died out rapidly in a few weeks. They excluded all common diseases and most respiratory diseases and concluded that the cause of the Athenian epidemic could be limited to either a reservoir disease (zoonotic or vector-borne) or one of the few respiratory diseases that are associated with an unusual means of persistence: either environmental/fomite persistence, or adaptation to indolent transmission among dispersed rural populations. They suggested that the diseases in the first category include typhus, arboviral diseases and bubonic plague and, in the second category, smallpox. Retief & Cilliers (1998) also reviewed the epidemiological evidence and agreed that the only possibilities are epidemic typhus, bubonic plague, arboviral disease and smallpox.

Other workers have suggested that the plague at Athens was an early manifestation of Ebola (sections 1.3 and 13.15). Olson *et al.* (1996) stated that a modern case definition of Ebola virus infection records sudden onset, fever, headache, pharyngitis followed by cough, vomiting, diarrhoea, maculopapular rash, and haemorrhagic diathesis, with a case fatality rate of 50% to 90%, death typically occurring in the second week of the disease. In a review of the 1995 Ebola outbreak in Zaire, the Centre for Disease Control and Prevention reported that the most frequent initial symptoms were fever (94%), diarrhoea (80%), and severe weakness (74%), with dysphagia and clinical signs of bleeding also frequently present. Symptomatic hiccups were also reported in 15% of patients. Olson *et al.* (1996) concluded that the profile of the plague at Athens was remarkably similar to that of the recent outbreaks of Ebola in Sudan and Zaire. Certainly, as we shall see, this devastating epidemic had features in common with later plagues in Europe (section 1.3).

1.2.2 The plague of Justinian

Procopius, the Greek historian, believed that this epidemic (like the plague of Athens) originated near Ethiopia. The pandemic began in Egypt in AD 541 and it then moved through Asia Minor, Africa and Europe, arriving in Constantinople, the capital of the Byzantine Empire, in the late spring and summer of AD 542. Merchant ships and troops then carried it through the known western world and it flared up repeatedly over the next 50 years, causing an enormous mortality, perhaps aided by wars, famines, floods and earthquakes. The plague raged in Constantinople for 4 months in AD 542, with the death toll rising from 5000 to 10 000 per day and even higher during the three most virulent months. The Byzantine emperor Justinian fell ill and recovered, but 300 000 people were said to have died in Constantinople alone in the first year, although Russell (1968) and Twigg

(1984) believed these figures to be greatly exaggerated. The officials were completely overwhelmed by the task of disposing of the dead bodies (Kohn, 1995).

Procopius recorded that people (understandably) were terrified, knowing that they could be struck without warning. The first symptoms were a mild fever which did not seem to be alarming, but bubonic swellings followed within the next few days. Once the swellings appeared, most sufferers either went into a deep coma or became violently delirious, sometimes paranoid and suicidal. It was difficult to feed and care for them properly, although mere contact with the sick did not seem to increase the chances of contracting the disease. Most victims died within a few days, but recovery seemed certain for those whose buboes filled with pus. Black blisters were a sure sign of immediate death; otherwise, doctors often could not predict the course of the disease or the success of various treatments. Autopsies revealed unusual carbuncles inside the swellings and these clinical features led to the conclusion that the Justinian plague was a pandemic of bubonic plague (Kohn, 1995). Shrewsbury (1970) agreed, but suggested that other serious diseases, such as smallpox, diphtheria, cholera and epidemic influenza were also present. It is not possible to be certain from the evidence available, but the rapid spread over great distances, the heavy mortality and other biological features of the pandemic suggest that bubonic plague was not the major component, but that some other infectious disease, spread person-to-person, was responsible.

1.2.3 The Great Age of plagues: the Black Death and thereafter

The Black Death erupted in Sicily in 1347 and the pandemic spread through Europe during the next 3 years, reaching Norway (where two-thirds of the population died; Carmichael, 1997) and Sweden and crossing to England (and thence to mainland Scotland, the Hebrides, Orkney and the Shetland Islands) and to Ireland (Biraben, 1975) and, possibly, to Iceland and Greenland (Kohn, 1995). Its arrival presaged a continuous succession of epidemics in Europe for the next 300 years before it disappeared completely around 1670. The enormous mortality of the Black Death had a major impact on the demography of Europe, and the population of England did not fully recover for 150 years. Events during 1347–50 are described in Chapter 4 and the demographic consequences of a major mortality crisis on a population are discussed further in section 13.17. Were the multiplicity of plagues throughout Europe from 1350 to 1670 all the result of the same causative agent as that responsible for the Black Death, albeit with some minor mutations during that time? It is not possible to answer this question with certainty, but we believe that the most probable explanation of the etiological and epidemiological details is that it was so.

During the second half of the 14th century, the epidemics in England and continental Europe were less virulent but the infection gradually regained its ferocity, reaching its peak around 1630 in France and 1665–66 in England.

1.2.4 Bubonic plague in the 20th century

The details of the complex biology of bubonic plague were finally unravelled around 1900: it is a bacterial disease of rodents depending on the rat flea for its spread and on a reservoir of resistant rodent species for the maintenance of the disease. Only occasionally does the infection spread to humans when one is bitten by a rat flea but, in the days before antibiotics and modern medicine, this was usually fatal and serious epidemics could be established.

Unfortunately, historians of Europe in the 20th century, almost universally, have concluded that all plagues in the Middle Ages were bubonic, in spite of the fact that the people at that time saw clearly that it spread person-to-person and, even in the 14th century, had already instituted specific quarantine periods. A major objective of this book is to examine the historical facts dispassionately, eschewing any preconceived notions about the behaviour of rats and fleas and to determine the nature of the epidemics of the Middle Ages in Europe. Normally, we refer to these as plagues, but where there is possible confusion with bubonic plague, we designate the former as haemorrhagic plague. To distinguish between haemorrhagic and bubonic plagues we begin in Chapter 3 with a detailed account of the complex biology, etiology and epidemiology of bubonic plague and explain how, in consequence, the spread and maintenance of the disease in rodents and humans is strictly constrained.

Although we have described bubonic plague from 1900 to the present day as the fourth age of plague, the disease has been identified (presumably correctly) and recorded in detail from China since AD 37 (Wu, 1926; Wu *et al.*, 1936) and it is likely that it has been present in India and in a gigantic swathe across central Asia for hundreds of years. It probably extended westwards to the Levant and the north African coast and may have erupted sporadically in the warm Mediterranean coastal regions in the 6th century AD (Twigg, 1984) and it was certainly present there in the Middle Ages and continued with occasional epidemics in the 18th century (see Chapter 12).

Where climatic conditions were suitable and reservoir rodent species were present locally, endemic bubonic plague could be established. Where only the climate was suitable, as in the coasts of France, Spain and Italy, epidemics of bubonic plague could potentially break out, having been brought into the ports by sea, but these terminated once the local rats had died. During the third age of plagues, 1347–1670, therefore, Europe probably experienced minor outbreaks of bubonic plague along the Mediterranean coasts of Italy, Spain and France in addition to the major epidemics of haemorrhagic plague.

Finally, endemic bubonic plague erupted in a series of epidemics in India at the very end of the 19th century and spread across southeast Asia, and so began the fourth age of plagues. As we show in Chapter 3, the arrival of steamships then allowed rats and their fleas carrying bubonic plague to be rapidly transported from the grain stores on the docks of China to subtropical regions wherever suitable indigenous rodents occurred.

1.3 The dangers of emerging plagues

From whence did the plague of Athens, the Justinian plague and the Black Death come? How did they emerge with such sudden ferocity? We have seen in the 20th century the emergence of a number of new deadly diseases that are largely resistant to medical science: scientists have identified more than 28 new disease-causing microbes in 1973 (Olshansky *et al.*, 1997). Indeed, it has been suggested that the history of our time will be marked by recurrent eruptions of newly discovered diseases (e.g. Hantavirus in the American West), epidemics of diseases migrating to new areas (e.g. cholera in Latin America), diseases that become important through human technologies (water cooling towers provided an opportunity for legionnaires' disease) and diseases that spring from insects and animals because of human-engendered disruptions in local habitats. Two of the terrors that haunt are the fears that new, unstoppable infectious diseases will emerge and that antibiotics will be rendered powerless. To some extent, these processes have been occurring throughout history. What is new, however, is the increased potential that at least some of these diseases will generate large-scale pandemics, such as a resurgence of the 1918 influenza pandemic; the global epidemic of human immunodeficiency virus (HIV) is the most powerful and recent example. Yet the acquired immune deficiency syndrome (AIDS) does not stand alone; it may well be just the first of the modern, large-scale epidemics of infectious diseases. The world has rapidly become much more vulnerable to the eruption and, most critically, to the widespread and even global spread of both new and old infectious diseases. This new and heightened vulnerability is not mysterious; the dramatic increases in the worldwide movement of people, goods, and ideas is the driving force behind the globalisation of disease because not only do people travel increasingly, they also travel much more rapidly, and go to many more places than ever before. The lesson is clear: a health problem in any part of the world can rapidly become a widespread health threat (Mann, 1995).

Most emergent viruses are zoonotic, with natural animal reservoirs a more usual source of new viruses than is the spontaneous evolution of a new entity. Human behaviour increases the probability of the transfer of viruses from their endogenous animal hosts to humans. The original source of the AIDS pandemic has been traced back to a subspecies of chimpanzee that has been used for food in West Central Africa, the hunters being exposed to infected blood during the killing and dressing. The virus has probably been living harmlessly in chimpanzees for hundreds of years and may have been transferred to humans throughout history, but the socio-economic changes in Africa provided the particular circumstances leading to the spread of HIV and AIDS.

An outbreak of encephalitis in Malaysia in 1999, which killed 76 people may have been caused by a more deadly version of the Hendra virus, which was first identified in Australia 5 years previously. The difference is that, whereas the virus in the earlier outbreak did not spread easily between animals, the Malaysian version apparently did: all the Malaysian victims were connected with pig rearing. Health officials in Asia now fear that a dangerous new human pathogen has emerged that has spread from fruit bats via pigs and consequently more than 300 000 pigs were slaughtered in southern Malaysia as an initial precautionary measure. The spinal fluid taken from five patients contained a paramyovirus (named 'Nipah') and analysis of the amino acid and RNA sequences confirmed that it is related to the deadly Hendra virus. Why has the virus suddenly begun to kill pigs and people when the bats may have harboured it safety for centuries?

The Ebola virus, a member of the Filoviridae, burst from obscurity with outbreaks of severe haemorrhagic fever. It was first associated with an outbreak of 318 cases and a case fatality rate of 90% in Zaire and caused 150 deaths among 250 cases in Sudan. Smaller outbreaks continue to appear periodically, particularly in East, Central and southern Africa. In 1989, a haemorrhagic disease was recognised among cynomolgus macaques imported into the USA from the Philippines; strains of Ebola virus were isolated and serologic studies indicated that the virus is a prevalent cause of infection among macaques. Epidemics have resulted from person-to-person transmission, nosocomial spread and laboratory infections but it must be emphasised that the mode of primary infection and the natural ecology of these viruses are unknown. The possible role of the Ebola virus as the causative agent in haemorrhagic plague is discussed in section 13.15.

A mysterious epidemic of Marburg virus (related to Ebola virus) broke out in a remote area of the Democratic Republic of Congo, Central Africa, in December 1998. At least 72 miners suffered from fever, pain, rash and bleeding and 52 had died by May 1999. The victims had spent time in caves and bats are considered to be the leading contender for an animal reservoir of the virus; monkeys die too quickly from the virus for them to be considered for this role.

A virulent influenza pandemic struck from 1917 to 1919, with a final worldwide estimated death toll of more than 20 million lives (Kohn, 1995). It has been termed Spanish influenza (dryly known as 'the Spanish Lady') because this was believed to be the first serious point of attack, with 8 million Spaniards falling ill in 1917–18. It then struck at military bases throughout Europe and death rates mounted ominously in 1918. At the same time (beginning in March 1918) acute respiratory infections were reported at military installations in the USA and by October some US army camps were reporting a death every hour; Britain was then counting 2000 deaths per week, with London at about 300 deaths per week. Country after country felt the ravages of the disease. The weak, the young and the old usually suffer worst in epidemics, but the age group 21 to 29 years proved to be the most vulnerable in this outbreak of Spanish influenza. While manifesting the ordinary symptoms of influenza (headache, severe cold, fever, chills, aching bones and muscles), the Spanish form also generated complications such as severe pneumonia (with purplish lips and ears and a pallid face), purulent bronchitis, mastoid abscess and heart problems. The frightening disease subsided after the end of the First World War and later vanished completely but, by then, it had attacked every country in the world, particularly China, India, Persia, South Africa, Britain, France, Spain, Germany, Mexico, Canada, the USA and Australia.

A radical genetic mutation, called antigenic shift, accounts for the appearance of new viral subtypes capable of engendering influenza pandemics. New viral types originate in ducks, chickens, pigs and other animals, in which reservoirs of influenza viruses change genetically and are then passed into the environment, and to human beings. The strain that caused the 1918 epidemic, H1N1, was found inside pigs and there is always the fear that this strain may resurface, perhaps in as virulent a form as in 1918. Many pandemics originate in Asia, notably China, where enormous numbers of ducks, pigs, and other virus-producing animals live in close proximity to human beings (Kohn, 1995). Avian influenza A (H5N1) virus has recently been shown to be transmitted from patients to healthcare workers in Hong Kong and this finding may portend 'a novel influenza virus with pandemic potential' (Bridges *et al.*, 2000).

Fragments of the virus responsible for Spanish influenza were found in 1998 in the lungs of a woman who died in the 1918 epidemic and whose body was preserved by huge layers of fat and the frost of Alaska and it is hoped that it will be possible soon to map the RNA of the virus to identify the gene that made it so deadly. Preliminary work has produced the complete sequence of one key gene and the existing strain to which the 1918 sequences are most closely related is A/Sw/Iowa/30, the oldest classical swine influenza strain. More recently, influenza 1918 RNA has been found in respiratory tissue and the brains of Spitzbergen coal miners who died in the epidemic and Oxford (2000) suggested that this could be a piece in the jigsaw linking pandemic influenza to the ensuing outbreak of the sleeping disease encephalitis lethargica.

A virulent and drug-resistant form of typhoid caused by the pathogen *Salmonella typhi*, which kills 600 000 people a year, has now emerged in Vietnam. The study of its genome is now almost complete: the nucleus contains three separate pieces of DNA, a massive coil some 4.5 million bases long and two plasmids, smaller loops of genetic data. One of the plasmids contains an array of offensive and defensive genes, which probably explain the potency of this strain of typhoid. It came as a great surprise when it was discovered that the other plasmid contained a sequence of 50–60 genes that are found in *Yersinia pestis*, the bacterium of bubonic plague and thus the Vietnamese microbe appears to be fortified with the genes of other pathogens (Farrar, 2000).

There is a seemingly endless catalogue of lethal infectious diseases that have emerged. Some of these have been described in a very lively manner by Garrett (1995): Lassa fever, Bolivian haemorrhagic fever, Marburg virus, the Brazilian meningitis epidemic and the Hantaviruses. Health officials in New York City reported, in August 1999, an outbreak of what appeared to be St Louis encephalitis, a disease that can spread to humans from birds via mosquitoes. However, it has now been discovered that the infectious agent is West Nile virus, which is normally found in Africa and Asia and is also transmitted by mosquitoes. Helicopters sprayed entire neighbourhoods in Queens, New York, after the disease killed horses, thousands of birds and several people; there has been a fresh outbreak in New York City in summer 2000 and what really alarms American health officials is the danger of the disease establishing itself permanently in the country. It remains an open question as to how the virus reached the USA (Boyce, 1999). So, it should not surprise us that the classical pandemics of historical times emerged and it is probable that they originated as viral zoonoses. Viruses have a great capacity for mutating and are opportunistic parasites; the worrying thought is (as suggested above) where and when will they next strike?

1.4 Populations and metapopulations

The study of how disease affects groups, or populations, of people is known as epidemiology; the discipline began when doctors wanted to study outbreaks of infectious diseases such as cholera and bubonic plague. Epidemiological studies today gather such data as age, race, sex and even social class, together with the incidence of the disease (the number of new cases appearing in a given time period) and its prevalence (the number of sufferers at any one time). The information can then be used to establish patterns in the disease and thus pinpoint aggravating factors.

Epidemiology can be defined in a number of different ways as, for example, 'the science of the infective diseases – their prime causes, propagation and prevention. More especially it deals with their epidemic manifestations' (LeRiche & Milner, 1971). This definition can then be extended because, if a communicable disease conforms to biological laws, epidemiological processes could be interpreted in terms of medical ecology (Gordon & LeRiche, 1950). Thus what we are studying in this book are the health and diseases of populations and groups and, in contrast to clinical medicine, the unit of study in epidemiology is the population and not the individual (Morris, 1957).

We investigate firstly the epidemiology of plagues in towns, large and small, treating them as circumscribed populations that have an identity but, of course, are not completely closed – infectives will have come into the population and a proportion of the inhabitants may have fled when an epidemic has been recognised. The temporal and spatial spread of the plague within the community (or unit) is governed by the household infection rate and by the ways in which it can spread to other households and thence to other streets and the results may then be compared with other populations. If the spread of the plague is density dependent, the pattern of the epidemic would be expected to be different in communities of different sizes. The city of London, as we shall see, was a complex population; a very large number of individuals crammed together but with subsets delineated by class and parish, with partial intercourse. The population was freely open, with many immigrants, travellers and merchants arriving daily by land and sea.

The spread of an outbreak of plague may also be studied at a higher population level, i.e. throughout a geographically defined area that might be the size of a country or even part of a continent. Examples are island Britain, and the Iberian peninsula, which was effectively separated from continental Europe by the Pyrenees and, in both, plague epidemics had to enter from the sea via the ports. These may be called metapopulations, a term used by ecologists to describe a population of populations. The study of metapopulation dynamics in biology is normally concerned with the behaviour of a single species over time; there are no static populations and likewise there is no such thing as a static metapopulation. The metapopulation concept in ecology is closely linked with the processes of population turnover, extinction and the establishment of new populations. Ecological metapopulation theory, with one important exception, has not been applied to human populations; indeed, as originally defined, it is not strictly applicable because it deals with extinctions and recolonisations and makes the simplifying assumption that each ecological site is regarded as being in one of two alternative states, either empty or filled at their local carrying capacity, characteristics that were rarely found in England during the age of plagues. The exception concerns studies of spatial heterogeneity and the epidemic spread of infectious diseases through a human metapopulation where individuals can be either infected or uninfected, an example of the interaction between demography and disease.

The spread of epidemics is an important part of modern Geography and we have used such techniques as disease centroids to trace the spatial movements of the plague in a metapopulation where it was endemic (see section 2.12). It becomes evident that the Black Death had a different pattern of spread from subsequent plague epidemics which, in turn, exhibited a range of sharply differing characteristics. The Black Death recognised no boundaries, either natural or human engendered, and spread in a wave-like movement all across Europe and to off-shore islands in about 3 years before disappearing. We can regard its territory for this brief period as a 'supermetapopulation'. Bubonic plague as a disease of rodents, and secondarily of humans, was certainly established as endemic across a huge subtropical area by 1900, from the Levant across to China and Southeast Asia. It has persisted for many years and we can regard this also as a 'supermetapopulation'. Haemorrhagic plague slowly established itself in Europe after the Black Death, with France as its endemic centre. England, Spain and Italy experienced epidemics of differing frequency, being separated from France by various geographical features, but France expanded from a metapopulation into a 'supermetapopulation' in the Middle Ages, which was composed of present-day Germany, the Benelux Countries and France. Here, plague was maintained as endemic, there being a handful of widespread epidemics somewhere maintained by long-distance travelling infectives.

1.5 A cautionary note

It is sometimes difficult to determine whether a marked increase in deaths in a year (a mortality crisis) was really the consequence of a plague epidemic. The health authorities in the city states of northern Italy in the 14th century went to great lengths to distinguish between minor (which they disregarded) and major (which were very serious) 'pests' (as they were called) by examining the victims personally, but historians rarely have such direct evidence on which to base their conclusions. Livi-Bacci (1977) relied on the size of the crisis and wrote

> For several parts of Tuscany between 1340 and 1400 I have calculated that on average a serious mortality crisis – defined as an increase in deaths at least three times the normal – occurred every 11 years; the average increase in deaths was at least sevenfold. In the period 1400–50 these crises occurred on average every 13 years and deaths increased fivefold. In the following half century (1450–1500) the average frequency declined to 37 years and the average increase to fourfold.

Shrewsbury (1970) considered that 'When more than 66% of the total annual burials occurs in the three months of July to September inclusive, the record is almost certainly indicative of an outbreak of bubonic plague' and this led him to conclude, for example, that there were multiple plague epidemics in northwest England in 1623. However, this area was living on the margins of subsistence at this time and mortality was sensitive to a 5- to 6-year cycle in grain prices (Scott & Duncan, 1998) and the constituent communities suffered major mortalities not only in 1623, but in 1587–88 and 1596–97 also. We have shown that in these years the peak of wheat prices coincided with a low in the 12-year cycle of wool prices (Scott & Duncan, 1997, 1998). Those populations that depended on both commodities suffered severely whereas those that depended on only one for their livelihood escaped unscathed. They did not suffer from plague in 1623. It is evident that by the end of the 16th century all towns could recognise plague when it struck their community and, if an outbreak is not recorded in the parish registers, a rise in mortality should be assumed to be the consequence of a plague epidemic only with extreme caution.

1.6 Pioneers in the study of plagues

We all owe a debt of gratitude to Yersin and his co-workers, to Wu and to the Plague Commission of India for the way in which they slowly and meticulously unravelled the complex biology and epidemiology of bubonic plague. A splendid piece of detective work. However, in this book we also wish to acknowledge the work and writings of a number of people who have influenced us and on whom we have relied heavily:

(i) First, Charles Creighton, 1847–1927, is the doyen of epidemiologists whose *History of Epidemics in Britain* was published in two volumes in 1891 and 1894. He graduated from Aberdeen University MA in 1867 having studied Latin, Greek, Mathematics, English, Logic, Moral Philosophy, Natural Philosophy and Natural History, and as Bachelor of Medicine and Master of Surgery in 1871. His approach in his classic work, which was to provide a chronicle of death and disease in the life and people of England, was that of a professional historian and he worked with great care on his sources (Eversley, 1965). We have relied heavily on his data series in our earlier work on lethal infectious diseases (Duncan *et al.*, 1993a,b, 1994a,b, 1996a,b; Scott & Duncan, 1998). He probably knew something about the biology of bubonic plague when he was writing in 1891 because this was being elucidated at the time but he does not seem to assume that this was necessarily related to the plagues in England that he was describing and, consequently, his descriptions are not modified to fit within the life histories of the rat and flea. In later life, he spent 3 months in India at the end of 1904 and reported about rats living in the mud walls of houses and of dead rats being found in a house where the inmates had died of bubonic plague (Underwood, 1965).

(ii) In the preface to the first edition of his book *Infectious Diseases: Epidemiology and Clinical Practice*, published in 1969, A. B. Christie wrote 'A good book, it has been said, should be opened with expectation and closed with profit . . .'. His treatise not only lives up to these high standards, but it is read with pleasure: he makes even dull topics interesting, spicing his account with classical allusions, gentle humour and personal anecdotes. He writes authoritatively and clearly on every infectious disease and this is particularly apparent when he deals with bubonic plague. His clinical experience across continents is revealed when he says that he believes that he had patients in Libya with bubonic plague who were infected by contact with a camel that had been ailing before slaughter and had a swelling in its neck.

(iii) Professor J. F. D. Shrewsbury, a microbiologist, has given us a great work of scholarship in his *A History of Bubonic Plague in the British Isles*, published in 1970, in which 'He has ransacked virtually all published local histories and parish records and he has read very widely in contemporary chronicles and memoirs' (Morris, 1977). Although we have relied heavily on his studies as a data source, we have not attempted to repeat the details of his findings and we suggest that readers who require more information about plagues in England should refer to this basic source book. It is a little dull and confusing in places but is occasionally illuminated by his dry humour:

'John Toy ascribed the visitation to God's punitive anger, because He had already twice warned the people of Worcester of their sins by inflicting slighter outbreaks of the disease upon the city; but it apparently never occurred to him that the Almighty would not thus degrade the Infinite to single Worcester out for such irrational punishment, for Worcester was certainly no more sinful than Lincoln, Salisbury, Canterbury, or any other English episcopal centre. It certainly never drew part of its revenue from brothels like the see of Winchester . . .'

'[T]he parish of St Giles, Cripplegate, where a parishioner was summoned in April to appear at the next sessions to answer "for receivinge people into his house sick of the plague brought from other parts to the prejudice of the parish" and for having "at the same tyme another sick of the French pockes [who] liveth incontynently with one Fayth Langley". Was he running the seventeenth-century equivalent of a nursing home?'

'In 1610 the churchwardens of St Margaret's, Westminster, paid 6*d*. to "Goodwife Wells for salt to destroy the fleas in the churchwardens' pew". Evidently the Anglican worshippers of the seventeenth century were as tormented by this ectoparasite as the monks had been in Salimbene's day. Most of the fleas, which undoubtedly were equally devout and attentive in most English parishes, were the human flea . . .'

As his title suggests, Shrewsbury believed whole-heartedly that bubonic plague was responsible for the majority of the plague epidemics in England and Scotland and yet, as a trained medical microbiologist, he saw that the facts on many occasions, made this a biological impossibility. He was therefore frequently forced to adapt his conclusions. When plague was reported in the months December to February he stated that it must have been a mild winter. When other facts about an epidemic did not fit bubonic plague he frequently declared it to be an outbreak of typhus, even when plague is recorded in the registers. He invented what he called 'trailer epidemics' to circumvent other difficult events. He was well aware that the mortality levels in many of the epidemics were much higher than would be expected in bubonic plague, particularly in the Black Death, and he reluctantly concluded that the sources from which he had quoted had overestimated the death toll.

Nevertheless, Shrewsbury steadfastly maintained that bubonic plague was the cause of most of the plague epidemics in the British Isles and it is most unfair that he should have twice been attacked, apparently for daring to suggest that there might be weaknesses in the story.

Gottfried (1978) wrote

'Herein lies one of the book's major shortcomings – Shrewsbury's failure to investigate any but printed and easily accessible chronicles and letters. No effort is made to search more obscure printed and manuscript sources: and even when original data are searched, it is done in an extremely uncritical manner. Often, the validity of the records is denied on the basis of uncorroborated value judgements and twentieth century medical information . . . One of his major premises is that epidemic bubonic plague has not changed in character "during the period of recorded history". This is contrary to what other epidemiologists have written. Shrewsbury diminishes the significance of the effects of pneumonic plague in fifteenth century England, saying that it cannot "occur in the absence of the bubonic form". This too seems to run contrary to the evidence . . . Also, interregional travel was far more common in the Middle Ages than Shrewsbury indicates, and was by no means restricted solely to merchants. Thus, both bubonic and pneumonic plague could survive in sparsely populated regions.'

Morris (1977) made a longer and more vigorous attack, particularly because Shrewsbury, who was a medical microbiologist, refused to allow the pneumonic form of bubonic plague to have a role in the epidemics:

'for some reason he has chosen to turn a blind eye to any evidence of pneumonic plague. He does not notice how often the victims are said to have succumbed in three days and if he meets with any reference to plague in cold weather he jumps to the conclusion that the disease must have been something else, preferably typhus . . . But there is much evidence, all of it ignored by Shrewsbury, that the Great Pestilence of 1348–50 contained a high percentage of pneumonic cases and indeed that in many places the plague first appeared in its pneumonic form. This would easily account for the high mortality which Shrewsbury is anxious to whittle down.'

Morris also attacks because of Shrewsbury's statement that bubonic plague has an unvarying relationship with rodent enzootics:

> 'Shrewsbury's main contention is that the country would have had to be constantly re-infected by fresh importations of plague-bearing rats. He has not thought of the possibility that England might well have become an enzootic area in which some rats at any given time are diseased. This is odd since he knows very well that in other parts of the world plague has taken permanent root and produced notorious enzootic or endemic centres. Indeed he argues, mistakenly as it happens, that India has always been one such centre from which Europe has drawn its periodic re-infections . . . Besides, if England became, as obviously it did, a permanently enzootic area in the seventeenth century, why should it not have done so two centuries earlier? That plague was endemic, or at least enzootic in London, needing no imported re-infections, for more than half a century before 1665 is abundantly clear from the annual mortality bills.'

(iv) Biraben (1975, 1976) in his two-volume work *Les hommes et la peste en France et dans les pays européens et mediterranéens* has assembled an impressive set of data on plague epidemics in Europe after the Black Death. He has combed the literature extensively and his bibliography runs to over 225 pages. We have used these data-sets for analysis in Chapters 11 and 12.

(v) Graham Twigg is a zoologist who has specialised in the biology of rodents and who has discussed with Dr D. E. Davis the status of rats in the Middle Ages. In 1984 he wrote *The Black Death: A Biological Reappraisal*, in which he carefully develops the evidence that shows that bubonic plague was not the cause of this great pandemic. He summarised his seminal work in the conclusion 'The logistics of the epidemic in England support the hypothesis of an air-borne organism of high infectivity and virulence, having a short incubation period and being spread by respiratory means' (Twigg, 1989). All students of plague should read his work.

1.7 Objectives

An epidemiologist must, by definition, be an historian, even if only in the short term. We present a new analysis of the plagues that scourged Europe from the 14th to the 17th centuries, approaching from biological, ecological and epidemiological viewpoints. We analyse the historical data (hopefully objectively) using modern techniques of theoretical epidemiology, clinical molecular biology, computer-based modelling and the spatial models of epidemic spread that have been developed by geographers.

There is a substantial literature on the Black Death and the Great Plague of London in 1665 but, during the intervening 300 years, Europe suffered from repeated outbreaks of the pestilence and these epidemics have received less attention. Were these epidemics all the result of the same infectious agent? What were its epidemiological characteristics? Were all, some or none the result of bubonic plague? What determines the dynamics of plague epidemics?

To answer these questions we begin by defining epidemiological concepts, such as transmission probability and basic reproductive number, and then explain how the epidemics of some infectious diseases can be modelled with the aid of computer-driven simulations. Once the basic parameters of a disease in historic times have been determined or estimated it is possible to construct models of the epidemics from which the underlying etiology can be suggested.

We have also tried to include the human story of the epidemics in England, showing how each population responded to the outbreak, how the disease spread through the community, how the members responded and how they made their wills. We give detailed case studies of the epidemics at Penrith (Chapter 5) and Eyam (Chapter 10) and have devised a new method of analysing and displaying the spread of the infection in each family group using family reconstitution techniques; this is the only means by which the epidemiological characteristics (e.g. incubation, latent and infectious periods, contact rates and transmission probability) can be determined.

We begin the story of the age of plagues in Chapter 3 with an account of the Black Death and the subsequent outbreaks in the 14th and 15th centuries, but it is not until the 16th century, when parish registers started in England, that firm and detailed information becomes available. The epidemiological characteristics of plague can be deduced therefrom and, the key feature that emerges is the lengthy incubation period of this infectious disease. When one is armed with this information, the reasons for the spread and behaviour of haemorrhagic plague in continental Europe over a period of 300 years (described in Chapter 11) become clear: the key to understanding its epidemiology is the endemic status of the pestilence in France.

Chambers (1972) suggested that long-term demographic trends may have often been caused, not by Malthusian fluctuations in the balance between population levels and food supplies, but by independent biological changes in the virulence of disease and by the rise and fall of the great epidemic scourges, which were not economic in origin. Slack (1977a) therefore concluded that the epidemiology of plague is a subject that bears on some of the central issues of demographic and social history. Historians have long been puzzled by the paradoxical rapid recovery of the population of England after the undoubted heavy mortality of the Black Death and in section 13.17 we examine some of the demographic consequences of a major mortality crisis in a single population, using the techniques of time-series analysis and computer modelling. We show that, although a population can apparently recover remarkably quickly, subtle demographic consequences could still be detected for over 100 years after the plague had disappeared.

2
Epidemiological concepts

Any serious attempt to elucidate the identities of the infectious agents in the plagues that struck at Europe over several centuries must begin with a scientific study of their biology and characteristics so far as we can discover them – every disease leaves its fingerprints on which the epidemiologist may work. Infectious agents may be viral, bacterial or protozoan, as well as larger animals such as nematodes or helminths, but they all have one feature in common, namely that humans are their ecological niche wherein they have shelter and food and can reproduce prodigiously. However, transmission from one host to another is fundamental to the survival strategy of all infectious diseases because a host will eventually clear the infection or die and hence the arrival of a disease in an individual depends crucially on the occurrence of that disease in other members of the population (Halloran, 1998). Transmission may be direct, person-to-person (as in measles or smallpox) or indirect, involving an intermediate host (e.g., the anopheline mosquito, which transmits the protozoan parasite *Plasmodium* of malaria). However, zoonoses are primarily diseases of animal hosts that are occasionally transmitted to humans. Examples are Lyme disease and bubonic plague; it is important to note that events are *not* critically dependent on the human population in these diseases.

We found the review by Halloran (1998) invaluable when we prepared this brief overview of modern epidemiological concepts; the works by Gisecke (1994), Lilienfeld & Stolley (1994) and Last (1995) are also recommended. The sequence of events during a simple, directly transmitted disease (such as influenza or chicken pox) is shown in Fig. 2.1; the infection of a susceptible person is followed by an incubation period – the interval between the entry of the agent and the appearance of the symptoms (Fig. 2.1A); it is not a fixed number of days and the actual period is often dependent on the infectious dose. It is generally thought of as the time

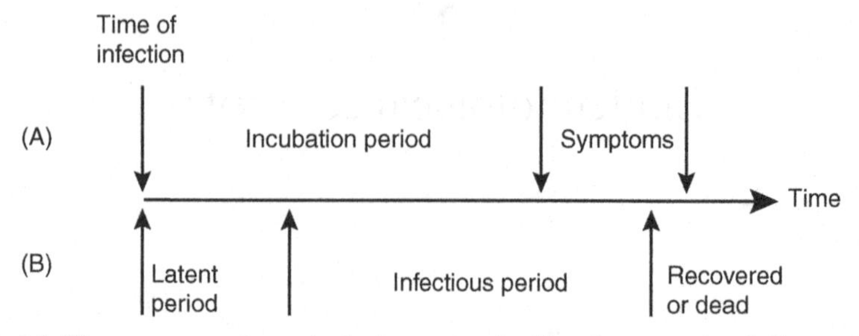

Fig. 2.1. The sequence of events during a simple, directly transmitted disease. The dynamics of the disease are shown in A and the dynamics of infectiousness in B. The victim may cease to be infectious before the end of the symptomatic period.

required for the multiplication of the microorganism within the host up to a threshold point where the pathogen population is large enough to produce symptoms in the host (Lilienfeld & Stolley, 1994). Thus each infectious disease has a characteristic incubation period that is largely dependent on the rate of growth of the organism in the host (Benenson, 1990) but, in addition to the effect of the dose, this is modulated by the immune response of the host, so that the incubation period varies among individuals. The disease eventually runs its course and for the infectious agent to persist it must have infected at least one other human. There are different outcomes for the host, who may have died (as usually happened in plagues) or recovered, when he or she may have an immunity of variable duration that protects from future exposure to the disease. Part B in Fig. 2.1 correlates the infectious period with the time-course of the disease; there is a latent period after infection before the host becomes infectious. Finally, the host passes to the non-infectious stage, often before the symptoms disappear, and transmission then becomes impossible.

The time-courses of parts A and B (Fig. 2.1) and their relationship to one another are specific for each infectious agent (Halloran, 1998); the relationship between the incubation and latent periods are of critical importance. In chickenpox (Varicella zoster), the latent period is shorter than the incubation period so that patients become infectious before the development of symptoms. Indeed, for many childhood diseases the period of greatest infectivity is just towards the end of the incubation period; these diseases are particularly difficult to control because an apparently healthy person will establish many contacts in everyday life and, consequently, isolation in quarantine when the symptoms have appeared is of limited value. The latent period of HIV is of the order of days to weeks whereas the median incubation period, before symptoms appear, is greater than 10 years, during which time a great many people may be infected (Halloran, 1998). We show in Chapter 5 that people in England infected with plague were infectious for some 22 days before the symptoms appeared, a critically important factor that ensured the establishment of an epidemic in spite of some of the public health measures that were enforced.

2.1 Transmission probability

The probability that the successful transfer of a parasite from an infective source to a susceptible host is defined as the transmission probability and its estimation is important for understanding the dynamics of the infection. It depends on the characteristics of the infective source, the parasite, the susceptible host and the type of contact. Considering the possible infections for the plagues, the infectious source could be another person (as in influenza), an insect vector (as in the rat flea in bubonic plague) or bacterially contaminated drinking water (as in cholera). The epidemiologist dealing with present-day epidemics has a battery of techniques to hand to determine the characteristics and biology of a disease but it must be remembered that, in spite of this, the remorseless spread of HIV still continues in sub-Saharan Africa. It is much more difficult to derive an estimate of the transmission probability of the plagues that occurred 400 years ago.

Halloran (1998) described two main methods for estimating transmission probability. In the first, infectious individuals are identified and the proportion of contacts that they make with susceptibles that result in transmission is determined (secondary attack rate). In the second method (the binomial model), susceptibles are identified and data are gathered on the number of contacts they make and their infection outcomes but this technique is less suitable for our purposes.

2.2 Secondary attack rate

In this case-contact approach to the estimation of transmission probability, infectious persons (primary or index cases) are identified and then the susceptibles who come in contact with them are determined. The conventional secondary attack rate (SAR) is strictly a ratio and is a measure of contagiousness. It is defined as the probability of the occurrence of the disease among susceptibles following contact with a primary case as follows:

$$\text{SAR} = \frac{\text{number of persons exposed who develop the disease}}{\text{total number of susceptibles}}$$

Using the data available for the plagues that we describe in detail in subsequent chapters, estimations can be made of the household SAR, which is concerned solely with infections within the house, although, even with a full family reconstitution, the exact number of resident susceptibles cannot be known; for example, some of the children may have left home. There is another caveat to bear in mind; the record of a burial in the registers is the only indication of an infection and we have no means of identifying individuals who contracted the plague and recovered. However, we have been encouraged by the consistency with which the pattern of the spread is replicated in each parish that we have analysed. The technique that we have used is shown diagrammatically in Fig. 2.2, where the point of infection of the index (primary) case is shown (P_1); this indicates the start of the infection in the household. The latent period of P_1 (see Fig. 2.1) follows this starting point and anyone else infected during this time is designated as a co-primary (P_2) because they cannot have been infected by P_1. Anyone infected during the infectious period of P_1 (Fig. 2.2) is designated a secondary (S_1) or co-secondary (S_2, S_3 etc.).

Tertiary and higher cases are those occurring after the maximum allowable time interval for the secondary cases. We assume that the infectious period is terminated in haemorrhagic plague by the death of the victim, who was probably displaying symptoms, on average, for the last 5 days of his life. The picture is complicated when co-primaries are identified because their infectious periods extend beyond the death of P_1. By working through the data for a large number of households derived by family reconstitution from the parish registers, it is possible to derive estimates of the latent, infectious and incubation periods for haemorrhagic plague, as shown in Fig. 2.2.

Thus, in assessing the household SAR, it is necessary to determine whether each case in each household is a co-primary, secondary, tertiary, or higher generation case. The estimated household SAR is the total number of secondary cases in all households divided by the total number of at-risk susceptibles in all households. Co-primary cases are excluded from the denominator; tertiary or higher cases are excluded from the numerator but included in the denominator. As we shall see, there was wide variation in the household SAR of the plague, apparently dependent on such factors as the season and the progress of an epidemic.

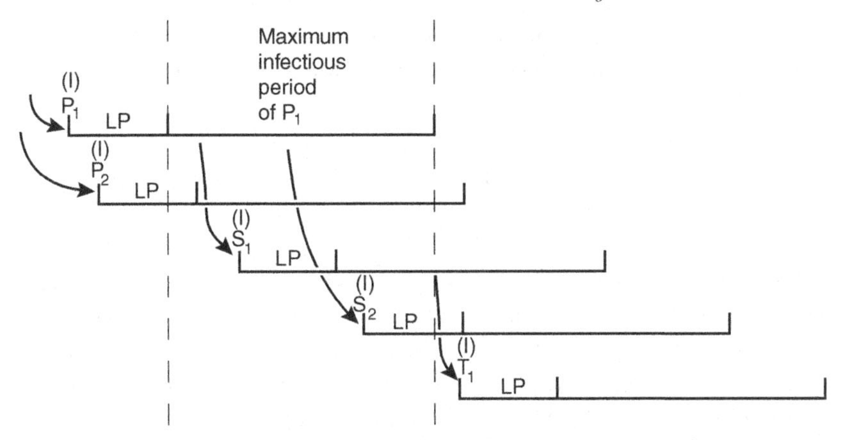

Fig. 2.2. Diagram to illustrate the time-course of the spread of an infectious disease such as haemorrhagic plague. The time between the point of infection (I) and the end of the infectious period (which may coincide with the death of the victim) is shown as a scaled horizontal line which is subdivided into latent (LP) and infectious periods. In this example, two primary cases (P_1 and P_2) are shown and the time limits when P_1 can infect secondary cases (P_1's infectious period) are shown by the vertical dashed lines. Two secondary cases (S_1 and S_2) are shown; they could have been infected by either P_1 or P_2. One tertiary case (T_1) is shown; I occurred after the end of the infectious periods of both P_1 and P_2 and so this could not be a secondary case.

2.3 Basic reproductive number, R_0

A second important parameter in the epidemiology of infectious diseases is the basic reproductive number, R_0. For diseases caused by viruses and bacteria, R_0 is the average number of persons directly infected by an infectious case during his or her entire infectious period when the infective enters a totally susceptible population. R_0 does not include the new cases produced by the secondary cases, or further down the chain (Halloran, 1998). R_0 is a dimensionless number. If R_0 equals 8 for a disease in a given population, then the introduction of one infective would be expected, on average, to produce 8 secondary infectious persons.

For microparasitic infections, R_0 is a composite of three important aspects of infectious diseases: the rate of contacts (c), the duration of infectiousness (d), and the transmission probability per potentially infective contact (p). The average number of contacts made by an infective during the infectious period is the product of the contact rate and the duration of infectiousness, cd (Halloran, 1998). The number of new infections produced by one infective during the infectious period is the product of the number of contacts in that time interval and the transmission probability per contact:

$$R_0 = \begin{matrix}\text{number of}\\ \text{contacts per}\\ \text{unit time}\end{matrix} \times \begin{matrix}\text{transmission}\\ \text{probability}\\ \text{per contact}\end{matrix} \times \begin{matrix}\text{duration of}\\ \text{infectiousness}\end{matrix} = cpd$$

An infectious disease does not have a specific value for R_0 that differs within particular host populations at particular times. For example, contact rates in rural areas will be lower than contact rates in urban areas, i.e. R_0 of smallpox would have been lower in rural England than in London (Scott & Duncan, 1998). As we shall see, R_0 for the plague in one locality apparently differed markedly at different seasons.

R_0 of *Yersinia pestis*, the bacterium of bubonic plague of rodents (see Chapter 3), in infections in a rat population may be high during the season of high flea density but low when the fleas are not active. This is an example of an indirectly transmitted disease, where the bacterium is transmitted between two different host populations, the rat and the flea; it is clearly a more complex situation than in person-to-person infection, as is illustrated in Fig. 2.3. R_0 for indirectly transmitted diseases depends on the product of the two components of transmission.

R_0 assumes that all contacts are with susceptibles and that none are already immune, unlike the situation with the regular lethal smallpox epidemics in rural towns in 17th century England where virtually the only susceptibles were those born since the last epidemic (Duncan *et al.*, 1993a; Scott & Duncan, 1993). The plague at Penrith in 1597–98 is described in detail in Chapter 5 and it was found that the mortality in the different age groups was indiscriminate. Some 45% of the population died but it seems certain that some individuals were exposed to the infection but survived; they may have contracted the disease but not have died of it, or they may have been resistant. We begin by assuming that all individuals in the plague at Penrith and certainly in the Black Death (but not necessarily in London in the 17th century) were potentially susceptible.

The parameter R_0 allows the comparison of diseases from the viewpoint of population biology – the aim of this book. When a disease is endemic (i.e. the disease lingers at about the same incidence for a long time) an infectious case produces, on average, one new infectious case, i.e. $R_0 = 1$. If $R_0 < 1$, the disease will disappear, whereas if $R_0 > 1$ an epidemic will be initiated. As an epidemic proceeds the number of susceptibles will inevitably fall and when $R_0 < 1$, the epidemic will die out.

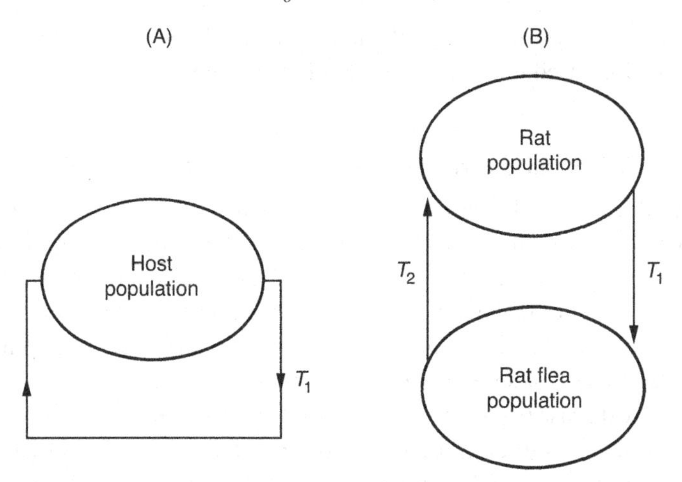

Fig. 2.3. Diagram to illustrate direct and indirect transmission. The quantities T_1 and T_2 represent a summation of the transmission parameters for the flow of the infectious agent. In direct, person-to-person transmission, as in measles (A), $R_0 = T_1$. In indirect transmission, as in bubonic plague where *Yersinia pestis* cycles between rodents and fleas (B), $R_0 = T_1 \times T_2$.

2.4 Virulence, R_0 and the case fatality ratio

Virulence is a measure of the speed with which a parasite kills an infected host (Halloran, 1998). Since R_0 is a function of the time spent in the infective state, R_0 could decrease as virulence increases. If the parasite is so highly virulent that it kills its host quickly, then R_0 could be <1 and the parasite will die out. This is an important point because, except in a few very large populations where it may have been endemic, plague epidemics in individual localities in Europe inevitably died out after 1 or 2 years, as we describe in section 13.7. However, plague was pseudo-endemic in France for some 250 years, i.e. the disease was always present somewhere in this vast metapopulation and it continued to cycle round the major towns, in each of which it usually persisted for only 1–2 years (see section 11.2). Viewed in this way, there is evolutionary pressure on parasites to become less virulent and to develop a more benign relationship with the host and it must be remembered that haemorrhagic plague, a particularly malignant disease, disappeared after about 1670.

The case fatality ratio is the probability of dying from a disease before recovering or dying of something else; as virulence increases, the case fatality ratio increases (Halloran, 1998). It is impossible to estimate accurately the case fatality ratio of the plague because of the inadequacy of the data but it was certainly very high indeed, although there is evidence that it may have decreased in the later stages of some epidemics.

2.5 Serial generation time: the Reed and Frost model

The unpublished mathematical model of epidemics developed by Reed and Frost has been described by Maia (1952). In a closed population of size N within which people intermingle fairly uniformly, it is assumed that, in a certain period of time t, every individual will have about the same number of contacts with other individuals. If the degree of intimacy of the contact is postulated to be sufficient for a patient with a certain contagious disease to transmit the disease to a susceptible person, this number of contacts, K, will be the average number of contacts for transmission of the disease per individual per time t. If t is made equal to the serial generation time, the individuals infected during one period will then be infectious during the next.

The serial generation time has been defined as the period between the appearance of symptoms in successive cases in a chain of infection that is spread person-to-person and, in present-day infections, is readily determined by observation of patients. However, where the latent and infectious periods for historical plagues can be determined (see section 2.2 and Figs. 2.1 and 2.2), the mean serial generation time can be estimated as the time between two successive infections as follows:

L is the duration of latent period (days);
I is the duration of infectious period, pre-symptoms (days);
S is the duration of period showing symptoms and presumably infectious (days).
Total infectious period $= I + S$

Mean time of transmitting the disease from the primary to the secondary case is assumed to be the mid-point of the infectious period $= (I + S)/2$. Therefore, the mean time at which the primary case infects a secondary case $= L + [(I + S)/2]$ days after the point of infection = the serial generation time, t.

If K is the average number of adequate contacts per individual per serial generation time, t, and the population size is N, then the probability of an adequate contact between any two given individuals during time t will be

$$p = \frac{K}{N-1} \tag{2.1}$$

and

$$q = 1 - p \tag{2.2}$$

will be the probability of any given individual avoiding adequate contact with any other given individual during time t.

Thus the population is at any time, t, composed of cases, C_t, susceptibles, S_t, and immunes, I_t, and the probability of any given individual avoiding contact with any of the cases will be q and, with all the C_t cases, will be

$$Q_t = q^{Ct} \tag{2.3}$$

and the probability of any given individual having at least one adequate contact with any of the cases will be

$$P_t = 1 - Q_t = 1 - q^{Ct} \tag{2.4}$$

In the transmission of disease we are interested only in the contacts between cases and susceptibles. Thus, in the next time period $(t + 1)$, we shall have

$$C_{t+1} = S_t(1 - q^{Ct}) \tag{2.5}$$

The theory presented by Reed and Frost rests on certain assumptions. (i) The infectivity of the organism is not altered during the course of the epidemic, i.e. p is constant. (ii) Immunisation of susceptibles through asymptomatic or subclinical infections does not take place. (iii) Although q is considered a constant as applied to a particular epidemic, the theory does not postulate that q is constant for a certain disease; it may change from one occasion to another in the same community and it may be different in different communities at the same time.

Consequently, the theory assumes that the rise and fall of epidemics, at least when evolving in a short period of time, will be dependent upon the numbers of susceptibles available and their depletion through infection and (usually in haemorrhagic plague) death, to a subliminal level or complete exhaustion.

However, the theory cannot explain diseases such as bubonic plague (see Chapter 3) and typhus with multiple hosts, such as insect vectors and animal reservoirs, where the situation is too complex for measurement of all the factors and the disease in humans does not follow Reed and Frost dynamics.

The results of modelling Equation 2.5 are shown in Figs. 2.4 and 2.5, using a population of 1200 and thereby replicating a rural town in Tudor times. The overall *shapes* of the graphs are similar, usually rising to a peak of cases more slowly than they decay, but changing the parameters makes a major difference to the *time-course* of an epidemic. It can be seen that in most epidemic profiles at least 90% of the susceptible population become infected.

Figure 2.4 illustrates the time-courses of the epidemics of different infections in which the mean number of contacts is standardised at 12 (i.e. $p = 0.01$). Serial generation times are as follows: A = 3 days (influenza); B = 10 days (measles); C = 15 days (chickenpox); D = 22 days (haemorrhagic plague). All the epidemics, with $p = 0.01$, were completed in 5 serial generation times and hence lasted from 15 days in influenza (A) to 110 days in plague (D).

Figure 2.5 illustrates the effects of changing the average number of contacts when the serial generation time is maintained constant at 22 days, i.e. equivalent to haemorrhagic plague. When the number of contacts was set at the low value of 3 (Fig. 2.5A) the epidemic lasted over 200 days (9 generation times) and there was a slow build-up to the peak of cases. Increasing the number of contacts (Fig. 2.5B and C) markedly reduced the duration of the epidemics until, with an average of 40 contacts, the epidemic exploded dramatically and was finished in 4 generation times (Fig. 2.5D).

Clearly, Equation 2.5 depends on the random mixing of infectives and susceptibles, which is not the case in some circumstances. Nevertheless, the modelling shows how each disease leaves its fingerprints on the epidemic, dependent on the serial generation time through which it can be identified by the epidemiologist. It also reveals how the time-course and pattern of an epidemic is modified by local factors such as the probability of an adequate contact between two individuals, which, in turn, is dependent on such factors as the density and social customs of the population, i.e. p will be much higher (i) at a local summer fair than when the community is snow-bound in the depths of winter and (ii) at the start of an epidemic when isolation and quarantine practices were not in force. We shall see that this is an important point because there were marked seasonal differences in the dynamics of plague, both in England and continental Europe.

2.6 Contact rates

Thus the contact patterns in a population play an important part in determining exposure and transmission. The rate of contact is governed by population density; people come into contact with an infective in an urban environment more frequently, even if infection was by air-borne transmission over many yards, than if they were less densely distributed, as in a rural environment. Population density therefore plays a role in determining the value of R_0 in diseases that are spread by casual contact.

The theory of regular epidemics of infectious diseases that are spread person-to-person has been widely studied (for detailed accounts, see Anderson & May, 1982a,b, 1985, 1991; Olsen & Schaffer, 1990; Tidd *et al.*, 1993; Bolker & Grenfell, 1993, 1995) and basic models have been presented that cover the spread of a virus in a population, a proportion of which is made up of non-immune, and hence susceptible, individuals who may be exposed to the disease and become infected. Of these, a proportion will die but some (depending on the nature of the disease) will recover and will then be immune. Studies of measles in the 20th century (Bolker & Grenfell, 1993; Fine, 1993) or smallpox in the 17th and 18th centuries (Duncan *et al.*, 1993a,b, 1994a,b) are good examples. These are termed SEIR (susceptibles-exposed-infectives-recovered) models and can be summarised as follows. The population, N, is assumed to remain constant where the net input of susceptibles (new births) equals the net mortality, μN (where μ is the overall death rate of the population, and hence the life expectancy of the population $= 1/\mu$). The population is divided into susceptibles (X), latents (infected, not yet infectious, H), infectious (Y) and recovered (and hence immune, Z); see Fig. 2.1. Thus $N = X + H + Y + Z$. It is assumed that the net rate at which infections occur is proportional to the number of encounters between susceptibles and infectives, βXY (where β is a transmission coefficient). Individuals move from latent to infectious at a per capita rate, σ, and recover, so becoming immune, at rate γ. The dynamics of the infection as it spreads through these classes are then described (see Anderson & May, 1982b) by the following equations:

$$\mathrm{d}X/\mathrm{d}t = \mu N - \mu X - \beta XY \tag{2.6}$$

$$\mathrm{d}H/\mathrm{d}t = \beta XY - (\mu + \sigma)H \tag{2.7}$$

$$\mathrm{d}Y/\mathrm{d}t = \sigma H - (\mu + \gamma)Y \tag{2.8}$$

$$\mathrm{d}Z/\mathrm{d}t = \gamma Y - \mu Z \tag{2.9}$$

The mathematical theory of epidemics is essentially concerned with the introduction of a 'seed' of infection into a largely susceptible population

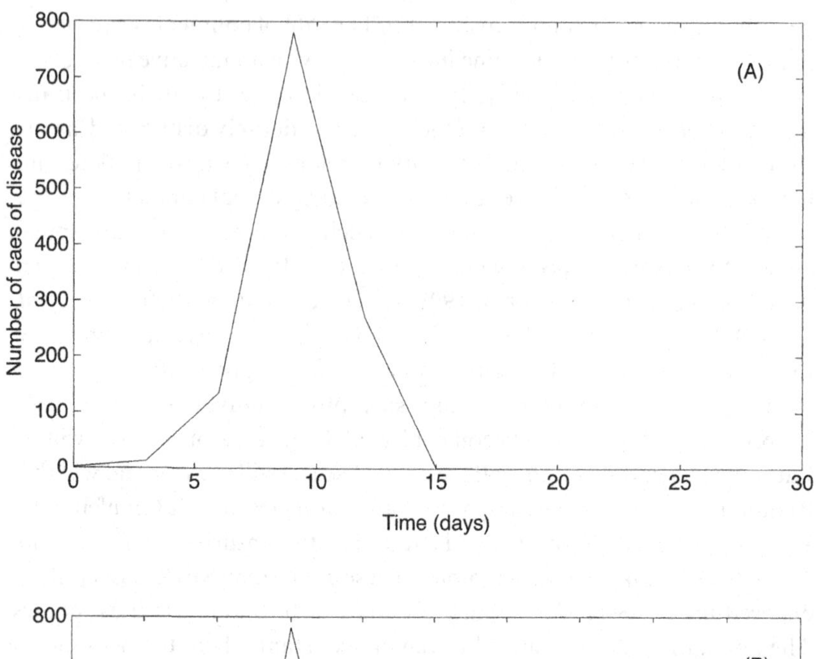

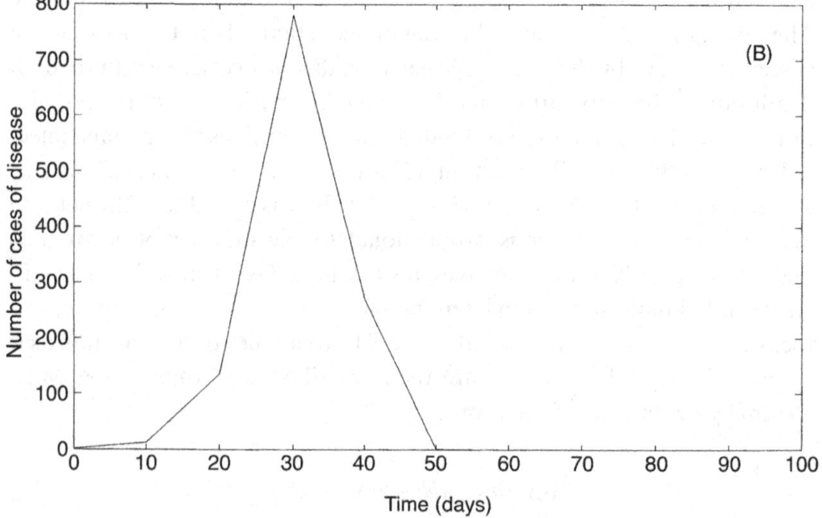

Fig. 2.4. Results of Reed and Frost modelling for a population where $N = 1200$ (representing a Tudor rural town) and the mean number of effective contacts = 12. Serial generation times (days): A = 3, B = 10, C = 15 and D = 22. The duration of the epidemic is dependent on the serial generation time. Note the different scales.

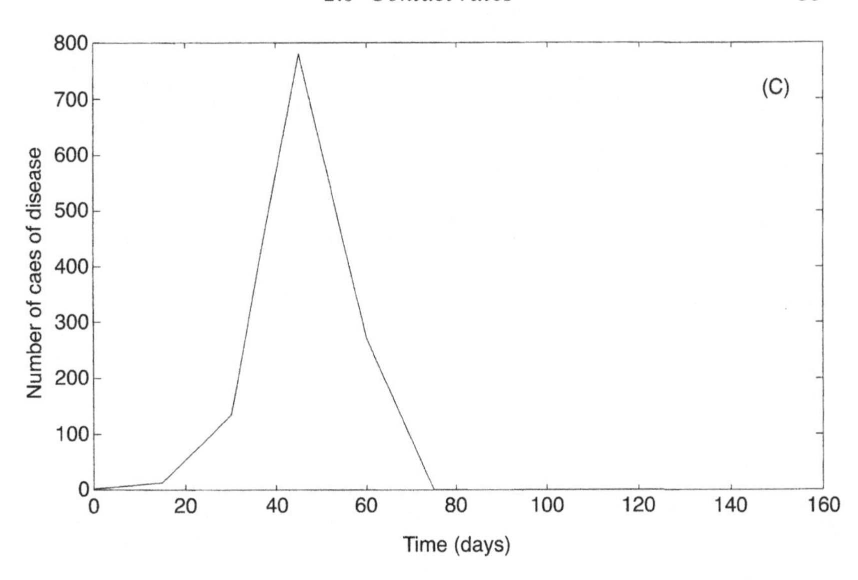
(C)
Number of caes of disease
Time (days)
0
100
200
300
400
500
600
700
800
0
20
40
60
80
100
120
140
160

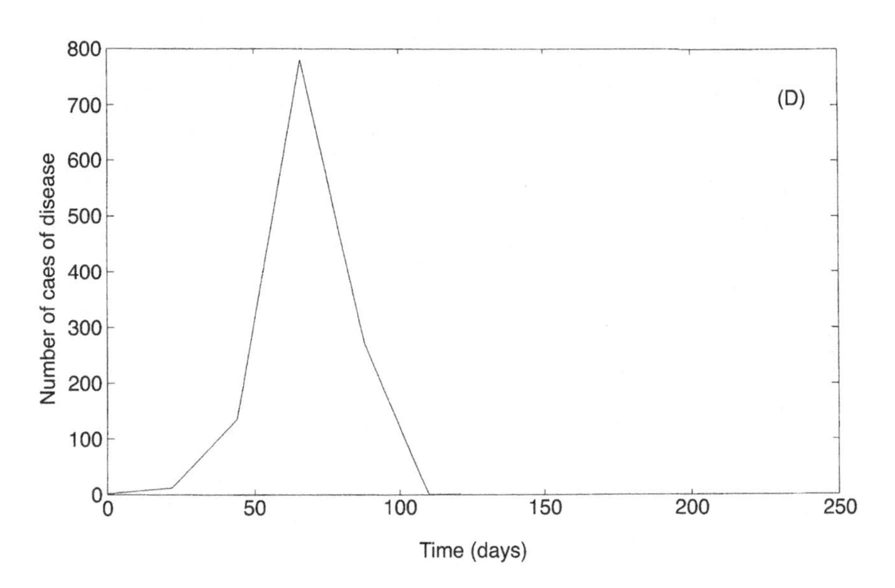
(D)
Number of caes of disease
Time (days)
0
100
200
300
400
500
600
700
800
0
50
100
150
200
250

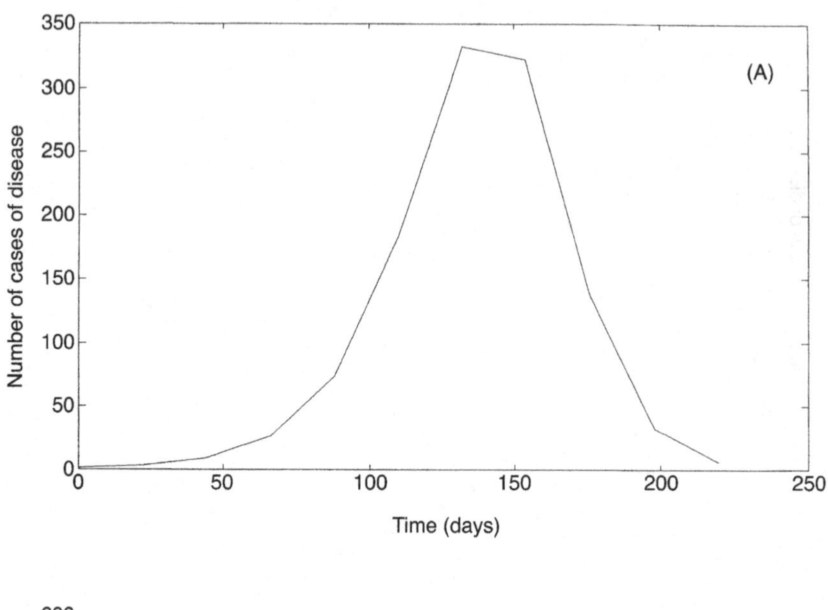

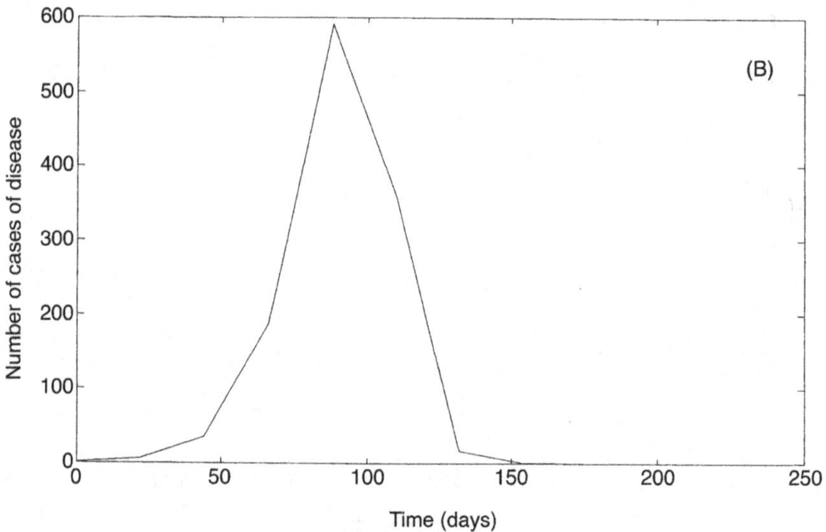

Fig. 2.5. Results of Reed and Frost modelling for a population where $N = 1200$ and the serial generation time = 22 days (replicating haemorrhagic plague). Mean number of effective contacts: A = 3, B = 6, C = 20 and D = 40.

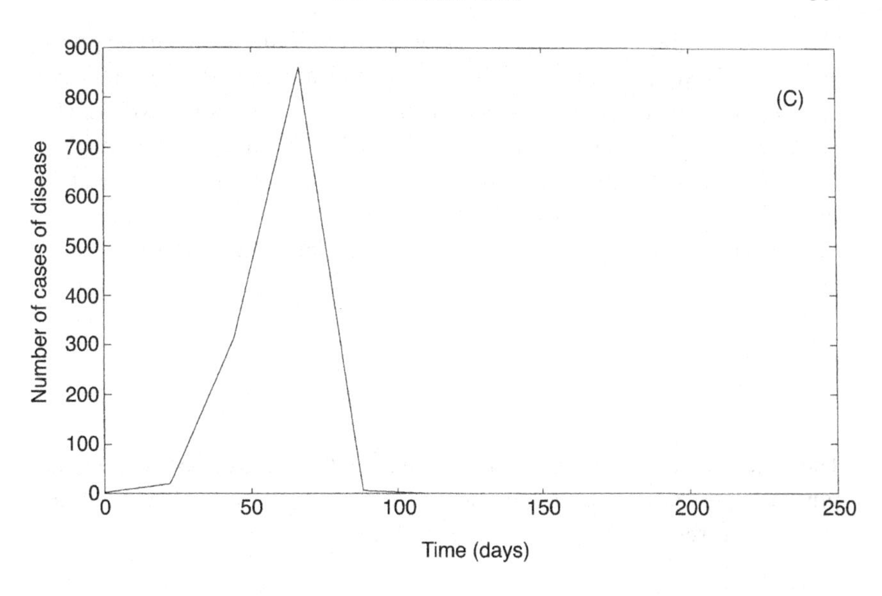
(C)
Number of cases of disease
Time (days)
0
100
200
300
400
500
600
700
800
900
0
50
100
150
200
250

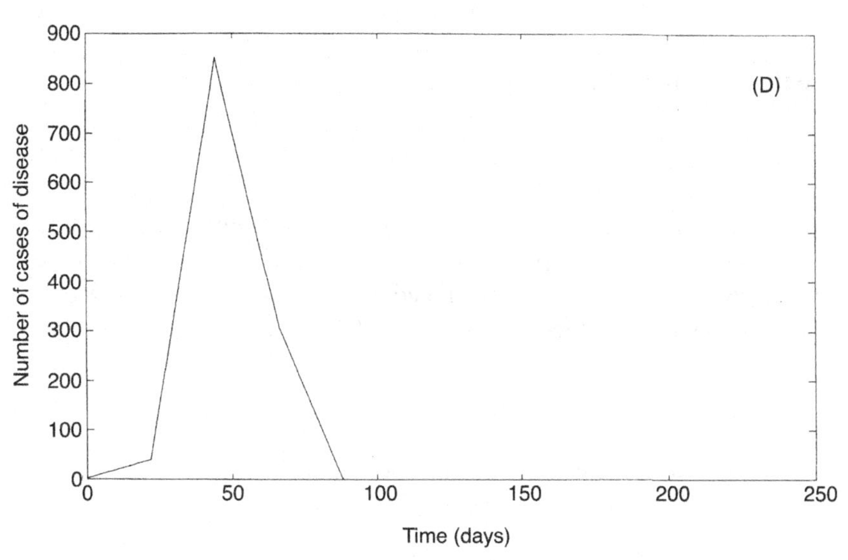
(D)
Number of cases of disease
Time (days)
0
100
200
300
400
500
600
700
800
900
0
50
100
150
200
250

(Anderson & May, 1991). The disease will maintain itself within the population provided that the reproductive rate of the infection, R, is greater than or equal to unity; R is the expected number of secondary cases produced by an infectious individual in a population of X susceptibles (see section 2.3). The value of R equals R_o in a disease-free population, which, as will be shown later, was the case in the Black Death and for many of the plagues in rural England in the 16th century. For a system defined by Equations 2.6 to 2.9,

$$R_o = \frac{\sigma\beta X}{(\sigma + \mu)(\gamma + \mu)} \tag{2.10}$$

The criterion $R_o > 1$ for the establishment of the disease (see section 2.3) can be expressed as the requirement that the population of susceptibles exceeds a 'threshold density', $X > N_T$, where

$$N_T = (\gamma + \mu)(\sigma + \mu)\beta\sigma \tag{2.11}$$

so that Equation 2.10 can be expressed as

$$R_o = X/N_T \tag{2.12}$$

For most infectious diseases (but not e.g. HIV), the duration of the latent ($1/\sigma$) and infectious ($1/\gamma$) periods are of the order of days to (at most) a few weeks, whereas life expectancy in Tudor and Stuart times ($1/\mu$) was approximately 25 years (Scott & Duncan, 1998). Under these circumstances,

$$\sigma \gg \mu \text{ and } \gamma \gg \mu$$

and Equations 2.10 and 2.11 may be approximated as

$$R_o = \beta X/\gamma \tag{2.13}$$

and

$$N_T = \gamma/\beta \tag{2.14}$$

If the disease can establish itself, then, at equilibrium, $R_o = 1$ and the density of susceptibles is equal to the threshold density, N_T.

Some of the parameters that determine R_0 are specific to the infectious agent; for example, the latent and infectious periods and the component of the transmission coefficient (β) that is related to the transmissibility of the disease (see section 2.3). Other components of R_0, such as the density of susceptibles (X) and the component of β that reflects the average frequency of contacts between individuals, in contrast, vary greatly in different localities, depending on environmental, demographic and social conditions (Anderson & May, 1982b; see section 2.3).

In many common infectious diseases, the density of susceptibles depends primarily on the birth rate in the community – most of the individuals alive will have experienced and survived previous epidemics and so will be immune. The build-up of a sufficient density of susceptibles would be achieved under these circumstances only by new births or by the immigration of individuals who had not previously been exposed. This situation is exemplified by the outbreaks of smallpox at Penrith, where regular epidemics were established at 5-yearly intervals because the population took 5 years to build up a sufficient density of susceptible children by new births (Duncan *et al.*, 1993a, 1994a; Scott & Duncan, 1993).

This regular pattern of epidemics of a lethal infectious disease is clearly different from the irregular outbreaks of plague in the 16th and 17th centuries in England and continental Europe, but the story of smallpox does underline the critical importance of the density of a susceptible population if a disease is to explode. Epidemics will be unable to develop in low-density rural communities (where $X < N_T$) and the disease will not persist in the absence of a continual inflow of infectives.

Many infectious diseases in the 17th, 18th and 19th centuries showed regular, predictable epidemics; examples, in addition to smallpox, include diphtheria (Scott & Duncan, 1998), whooping cough (Duncan *et al.*, 1996a, 1998), measles (Duncan *et al.*, 1997) and scarlet fever (Duncan *et al.*, 1996b). Many of these studies are of large populations, ranging from cities and the metropolis to the metapopulation of the whole of England and Wales, where the disease was endemic (i.e. with a substantial number of cases each year), but superimposed thereon were major, regular epidemics.

For diseases that are of short duration relative to the host life-span (i.e. the majority, including haemorrhagic plague visitations and the standard serious infectious diseases), the interepidemic period, T, is approximately described by the following equation

$$T = 2\pi[LD/R_0 - 1)]^{\frac{1}{2}} = 2\pi(AD)^{\frac{1}{2}} \tag{2.15}$$

where D is the sum of the duration of the latent and infectious periods, L

the human life expectancy of the population, and A the average age at infection, as above (Anderson & May, 1991). It can be shown (Scott & Duncan, 1998) that Equation 2.15 is formally equivalent to

$$T = \frac{2\pi}{\sqrt{\mu[N\beta - (\mu + v)]}} \tag{2.16}$$

where N is the number in the population, μ the death rate (= 1/life expectancy), and v the rate of recovery (= 1/infectious period). Hence, the interepidemic interval in any particular situation is determined by $N\beta$, the product of the transmission rate and the population size/density.

2.7 Decaying and driven epidemics

The deterministic model of Equations 2.6 to 2.9 of 'standard' infectious diseases, such as measles or smallpox, exhibits a damped oscillation, in which the epidemics gradually decay and the disease settles to a stable, steady-state (endemic) level. The boom and bust of the initial epidemic is followed by a period in which the pool of susceptibles is restocked by births and eventually a new, but less severe, epidemic is triggered. This second epidemic is less severe because there are now fewer susceptibles (the bulk of the population is now immune) and so the overshoot is less dramatic. Successive epidemics are increasingly mild (Anderson & May, 1991).

Clearly, the epidemics of the 'standard', lethal infections of history did not decay rapidly but persisted with undiminished vigour over many years (see Fig. 2.6). It has been suggested that stochastic effects could indefinitely perpetuate the oscillation of the system, thereby maintaining the epidemics. Alternatively, seasonality in transmission (a feature of many infectious diseases, including outbreaks of haemorrhagic plague) or weather (dry conditions had an effect on the epidemics of smallpox and scarlet fever; Scott & Duncan, 1998) could pump-up the decaying oscillation and lock the system into sustained cycles with periods that are determined by Equation 2.16 (London & Yorke, 1973; Yorke & London, 1973; Dietz, 1975; Yorke *et al.*, 1979).

We have suggested (Scott & Duncan, 1998) that epidemics could be maintained if the system were directly driven by an oscillation in the amplitude of the transmission coefficient, β, which, in practice, would be mainly an oscillation in susceptibility brought about by the regular variation of such external factors as seasonal weather conditions or nutritive levels.

2.8 Time-series analysis of data

The statistical technique of time-series analysis allows the investigation of continuous data over time and identifies and characterises the cycles therein. It is of particular use to the demographer when studying baptism and burial records and to the epidemiologist when analysing annual cases of an infectious disease. We have described time-series analysis in detail previously (Scott & Duncan, 1998) and have given worked examples. We have used the MATLAB program and that given by Shumway (1988); in brief, the techniques available include:

(i) *Spectral analysis*. The data-series (i.e. the number of births, deaths, or other events in each year) are fed into the computer program, which analyses the relative importance (or strength) of the different cycles contained within the series and identifies their wavelength or period (i.e., the number of years for a complete cycle or oscillation). The significance of these cycles can then be tested by the program.
(ii) *Filtering*. When one cycle or more has been identified by spectral analysis, this program designs a filter that removes noise and unwanted oscillations from the data-series and the resulting cycles can be displayed.
(iii) *Cross-correlation function* (*ccf*). This program compares two filtered data-series over a standard time period; it provides an estimate of the significance of the correlation between them and of the delay (or lag) between the two cycles.

Difficulties arise for the demographer when (as would be expected in human populations) the cycles detected are non-stationary and have a period that varies slightly (e.g. a 6-year cycle may fluctuate between 5 and 8 years).

2.9 Lethal smallpox epidemics in London, 1650–1900: a case study

In this section we look at smallpox deaths in London and show how time-series analysis can be used to elucidate the dynamics of the disease. The story that emerges from this example illustrates the theory of infectious diseases that has been described above; we try to bridge the gap between historical studies of infectious diseases and the current interest in the mathematical modelling of epidemics in the 20th century (where excellent and complete data-series are available), which is exemplified by the work of Bartlett (1957, 1960), Anderson & May (1991) and Grenfell (1992).

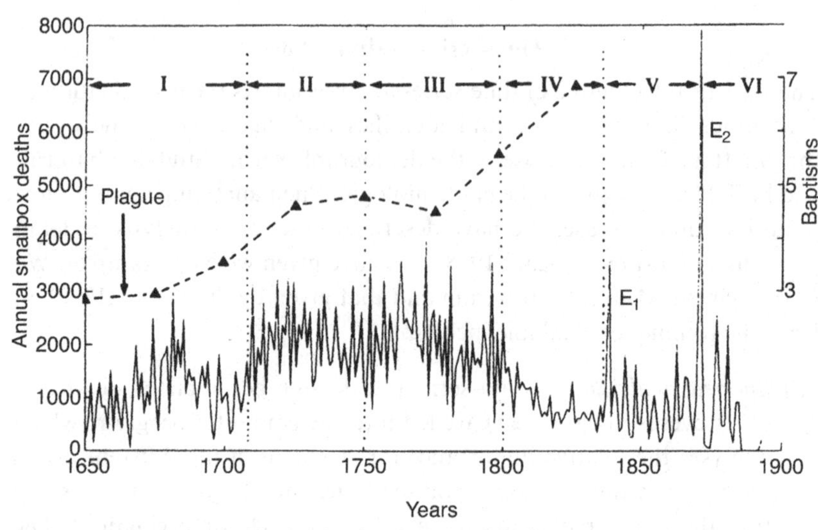

Fig. 2.6. Annual smallpox deaths (ordinate) in London over some 250 years, 1647–1893, divided into cohorts: I = 1647–1707; II = 1708–50; III = 1751–1800; IV = 1801–35; V = 1836–70; VI = 1871–93. The plague in 1665 and the major smallpox epidemics of 1838 (E_1) and 1871 (E_2) are indicated. The dashed line (closed triangles) gives the cumulative number of baptisms in the preceding 25 years (thousands), right-hand ordinate. Data sources: Bills of Mortality (Creighton, 1894) and Wrigley & Schofield (1981).

Annual smallpox deaths in London, 1647–1893, are displayed in Fig. 2.6, which shows the fluctuations and the trend in the basic endemic level, on which are superimposed clear epidemics of the disease in years when there was a sharp rise in mortality. The important features illustrated by this figure are as follows. (i) There was severe mortality from the Great Plague in London in 1665, which had an effect on the population dynamics (see section 13.17). (ii) Inoculation against smallpox was more widely administered after about 1750 and vaccination was introduced in 1796 and became compulsory for infants in 1853; these practices clearly modified the pattern of the epidemics after 1800 and eventually led to the disappearance of the disease in London at the end of the 19th century (effective $R_o < 1$). (iii) The late outbreaks of smallpox in the mid-19th century (shown as E_1 and E_2 on Fig. 2.6), particularly the epidemic in 1871, which triggered *decaying* epidemics and showed that the underlying dynamics of the disease had now changed.

The changing pattern of the epidemics has been studied by time-series analysis, and Fig. 2.6 can be divided into separate periods, each having a characteristic interepidemic interval (T):

(I) 1647–1707; T changing from 4 years to predominantly 3 years.
(II) 1708–1750; endemic level rising, T changing from 3 to 2 years.
(III) 1751–1800; T firmly established at 2 years.
(IV) 1801–1835; introduction of variolation and vaccination produced a reduction in the pool of susceptibles; endemic level falling steadily; epidemics greatly reduced in amplitude; $T = 2$ to 3 years.
(V) 1836–1870; epidemics reinitiated following a major outbreak in 1838; endemic level continuing to fall; $T = 4$ years. Non-driven SEIR dynamics.
(VI) 1871–1893; the major epidemic in 1871 triggered three further decaying epidemics before smallpox ceased to be a serious disease. Non-driven SEIR dynamics.

Since T is determined by $N\beta$, the product of population size and the transmission coefficient, see Equation 2.16, it is possible to suggest reasons for the changing value of the interepidemic interval. During 1647–1750 (cohorts I and II), N (measured by cumulative baptisms, Fig. 2.6) rose steadily and, concomitantly, T fell from 4/3 years (1647–1707) to 3/2 years (1708–1750). However, after 1750, when T clearly changed to 2 years (suggesting a *rise* in $N\beta$), surprisingly, baptisms were stationary; concomitantly wheat prices were rising sharply and we suggest that malnutrition caused an overall increase in susceptibility (β) to smallpox and consequently a rise in $N\beta$ and a fall in T to 2 years after 1750 (Scott & Duncan, 1998). Time-series analysis of the seasonal weather conditions shows that, during the first three cohorts, the epidemics were strongly correlated with low seasonal rainfall, and we conclude that the spread of smallpox was favoured by dry conditions that could act as the driver for maintaining the epidemics. In summary, the interepidemic interval was reduced in 1647–1750 mainly because of rising N, and in 1750–1800 mainly because of rising β. This case study illustrates how the dynamics of an infectious disease can be modified by population size, density, malnutrition, seasonal weather conditions and vaccination.

2.10 Mixing patterns

We have seen that population density plays a role in determining the values of R_0 and T in diseases that are spread through person-to-person contact such as measles, smallpox or influenza or, as we shall suggest, probably in haemorrhagic plague. The simplest assumption about the contact pattern in a population is that it occurs by random mixing, with every person

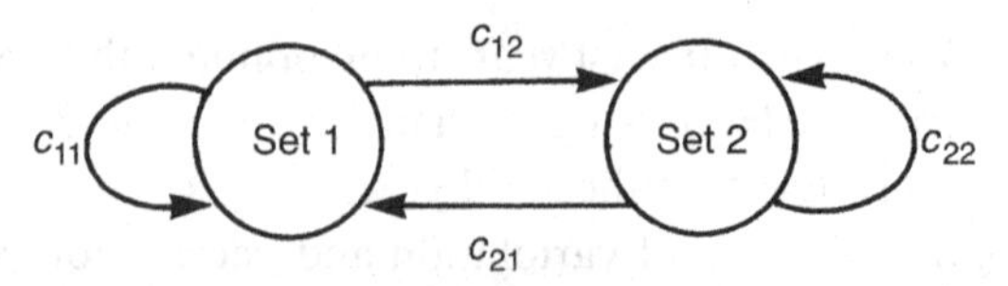

Fig. 2.7. Mixing patterns of two groups where c_{12} = contact rate of set 1 with set 2, c_{21} = contact rate of set 2 with set 1, and c_{11} and c_{22} = contact rates within sets 1 and 2, respectively. From Halloran (1998).

having an equal chance of making contact with, and being exposed to, infection by each other person. However, most populations do not mix randomly but have subgroups that mix more with their own members than with other groups. These subgroups in the community during a plague were clearly households where the household SAR was of paramount importance in determining the spread of the infection through the family (see the case studies at Penrith, Chapter 5, and Eyam, Chapter 10). The magistrates in London enforced quarantine:

> The misery of those families is not to be expressed; and it was generally in such houses that we heard the most dismal shrieks and outcries of the poor people, terrified and even frightened to death by the sight of the condition of their dearest relations, and by the terror of being imprisoned as they were.
>
> *(Defoe, 1722)*

We show later that, in rural towns, the pestilence spread rapidly within a family once it was introduced but transmission to other households was slower and more difficult and this is a key point – if the plague were to persist and spread in a population and not die out it must effect transmission to at least one other household.

The contact rate of individuals of group 1 with individuals of group 2 is denoted by c_{12} (Fig. 2.7) and the contact pattern is described by a mixing matrix that has the same number of rows and columns as the number of mixing groups. The entries in the matrix represent the rate of contacts of individuals within and between the groups. The mixing pattern of two groups is represented by the matrix:

$$C = \begin{bmatrix} c_{11} & c_{12} \\ c_{21} & c_{22} \end{bmatrix}$$

On the diagonals are the rates of contacts within groups, c_{11} and c_{22}. The off-diagonal entries, c_{12} and c_{21}, represent the rates of contacts between the groups corresponding to that row and column (Halloran, 1998).

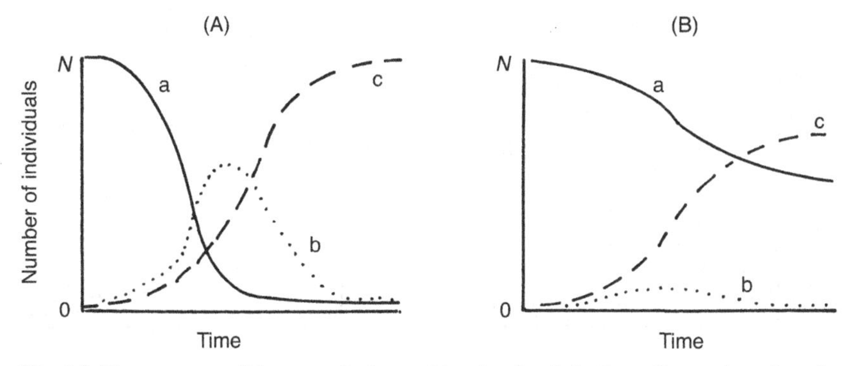

Fig. 2.8. Time-course of the spread of an epidemic of an infectious disease in a closed population of N susceptibles. Susceptible people become infected and then, after the latent period, become infectious; they then develop immunity. (A) Epidemic with a high R_0; everyone becomes infected during the epidemic which dies out because there are no susceptibles left. (B) Epidemic with a low R_0; the epidemic dies out before all the susceptibles become infected. a, susceptibles; b, infectives; c, recovered and immune. From Halloran (1998).

2.11 Open versus closed population dynamics

London, where haemorrhagic plague was probably endemic in the 17th century and where there was steady immigration (Landers, 1993), can be regarded as an open population. A rural town in England or continental Europe, on the other hand, can be regarded effectively as closed populations during a plague, with few births (on this time-scale) and no immigration.

We consider first a closed population of N initially susceptible people who are assumed to be mixing randomly with contact rate c; everyone in the fixed cohort is initially in the uninfected state, X, at time $t = 0$. Suppose a measles virus is introduced into the population so that one person enters the infectious state, Y. If $R_0 > 1$ the epidemic is expected to spread. The process in closed populations is illustrated in Fig. 2.8A and B. The infection spreads from the first infective to the average number, R_0, of susceptibles, depending on the rate of contact c, the transmission probability p, and how long the person is infectious. As the epidemic spreads, the number of susceptibles (X) decreases while the number of people with immunity (Z) increases. Incidence of infection will increase until the number of available susceptibles becomes a limiting factor, when the number of new cases begins to decrease until the parasite dies out. Thus, when R_0 is high in closed populations, an epidemic will explode (see Fig. 2.8A), comparable with the Reed and Frost model (Figs. 2.4 and 2.5), and with plague epidemics in rural towns. However, when R_0 is low in closed populations

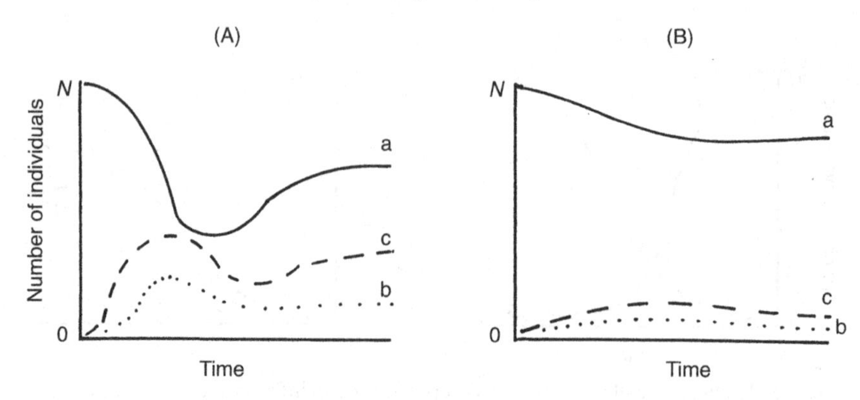

Fig. 2.9. Time-course of the spread of an infectious disease in an open population of N susceptibles where an epidemic is followed by endemic persistence. (A) High R_0. (B) Low R_0; the prevalence of susceptibles, infectives and immune people is in dynamic equilibrium; the infectious agent does not die out because of the supply of new susceptibles. a, susceptibles; b, infectives; c, recovered and immune. From Halloran (1998).

(see Fig. 2.8B), the epidemic does not explode but dies out before all the susceptibles become infected. Examples of the plague following this pattern are in communities of low density (e.g. in Cheshire, sections 9.1.2 and 9.6.1) and outbreaks in winter in rural towns in England before the major epidemic in the following summer (sections 5.5, 10.3 and 10.4).

If a population is open (e.g. London in the 17th century), with births and immigration providing a steady supply of susceptibles, the parasite may not die out but can persist and become endemic. Figure 2.9B illustrates an open population with a low R_0; after an epidemic the susceptibles, infectives and immunes remain in dynamic steady state, with the annual number of new incident cases remaining low and steady. When R_0 is high in the open, endemic situation, the level of infectives is higher and the number of susceptibles is correspondingly lower (Fig. 2.9A).

2.12 Spatial components of epidemic spread

We have summarised the temporal aspects of the dynamics of infectious diseases in sections 2.5 and 2.6 and have given examples of the use of modelling the basic equations in section 2.9; epidemiologists have elaborated these models by incorporating spatial components, particularly when a population is distributed non-uniformly in space in such a way that the rates of transmission are significantly higher in some places than in others. Variability in transmission rates can arise when some hosts live in dense aggregates in cities while others live in small or remote groups (Bailey, 1975; Cliff & Haggett, 1988; Anderson & May, 1991; Cliff, 1995). These models are of current interest in the design of immunisation programmes. However, relatively few studies have successfully incorporated explicit spatial components into epidemic models that are both mathematically tractable and geographically and epidemiologically plausible (Cliff, 1995). The work has concentrated on identifying the ways in which space determines the corridors by which epidemics move from one geographical area to another, as well as the velocity and direction of disease propagation both within a metapopulation and on a worldwide scale (Bailey, 1975; Cliff, 1995). These spatial corridors may change over time and our study of haemorrhagic plague suggests that they were major trade routes and river systems. The techniques used to analyse the spatial components of modern epidemics include (i) lag maps to identify the time–space ordering of epidemic spread, (ii) the treatment of maps as graphs where the nodes represent areas coded in some epidemiologically meaningful way and the edges of the graphs represent the corridors of spread, (iii) spatial autocorrelation techniques, (iv) assessment of the velocity of spread, (v) analysis of spatial scale and (vi) centroids (Cliff, 1995).

Current research confirms that geographical space behaves in a non-linear way in directing the spread of a disease under present-day conditions. The geographical spread of common transmissible diseases, such as measles, in urbanised societies today frequently exhibits two components: (i) relatively long-distance diffusion occurs by spread from town to town, often leap-frogging the intervening countryside between urban areas and (ii) in-filling of the space between towns by localised spread outwards from each infected urban centre into the surrounding countryside (Cliff, 1995). The spread of the Black Death illustrates these two components but subsequent plagues in Europe followed a different type of spread. We show in Chapter 4 that the Black Death spread as waves from its points of introduction in Italy and France to northern Europe and its average daily rate of spread through the metapopulation can be computed. In Chapter 7 we analyse the spread of the plague in northern England in 1597–98 and show how transmission was not wave like but was along well-defined corridors of communication and that it behaved like a typical infectious disease, exhibiting Reed and Frost dynamics within each effectively closed population.

Thus the spatial components of an epidemic leave clues for the epidemiologist trying to determine the nature of the infectious agent. The very slow spread of bubonic plague in India described in Chapter 3 reflects its complex underlying biology and is quite unlike the spread of 'typical' infections such as the Black Death. The key to understanding the dynamics of the spread of an infectious disease is the serial generation time; influenza has an infectious period of 2–3 days and during some of this time the patient may be prostrate with the illness but, nevertheless, has up to 48 hours to spread the disease over continents by modern air, rail and road travel or by steamtrains in the influenza pandemic of 1917–19. The situation was very different 600 years ago when rapid and widespread dissemination of influenza would have been much more difficult and, consequently, its spread and effects would have been less dramatic. In contrast, infectious diseases with very long latent and pre-symptomatic infectious periods (as we suggest for haemorrhagic plague) would allow widespread progressive, wave-like dissemination (see the Black Death, Chapter 4) when movement through the metapopulation was mainly on foot, or, in later centuries, more rapidly along the established communication corridors when travellers and carriers had time to move long distances across Europe by horse.

3
The biology of bubonic plague

Humans have been afflicted by bubonic plague for hundreds of years, particularly in central Asia and China, where it is endemic, and epidemics have flared up at sporadic intervals when the conditions are right. Localised outbreaks have been described in ports in England in the 20th century and there are a number of reports of ships coming from the East on which bubonic plague was diagnosed in crew members. Heavy mortality can be experienced *over long periods of time*: 12 million people died of bubonic plague in India between 1898 and 1957. Bubonic plague is a lethal infection of rats that is spread to humans via the rat flea when environmental and other conditions are suitable. The biology and epidemiology of the disease was only fully elucidated at the end of the 19th century and, because some of its clinical features were recorded in contemporaneous reports of the Black Death, 20th century accounts, almost without exception, give bubonic plague as the disease causing that pandemic. It is, therefore, important to describe the biology of bubonic plague in detail before accepting it as the cause of the Black Death and other subsequent plagues in England up to 1665. Christie (1969), who had wide experience of infectious diseases, has provided an admirable account of the epidemiology of modern bubonic plague and the following pages are very much based on his overview.

3.1 History and geographical distribution of bubonic plague

Wu *et al.* (1936) have listed the 232 occasions when there was an annual outbreak of pestilence in China from AD 37 to 1718 and it has persisted through to the early years of the 20th century. Very few clinical details are known of the early plagues but we must tentatively agree with Wu that these were also bubonic/pneumonic plague. Twigg (1984) suggested that bubonic plague had been present for hundreds of years in areas within or near the central Asiatic plateau and he considered this to be the home of the disease. Stemming therefrom were the following major foci of the disease:

(a) in the foothills of the Himalayas between India and China,
(b) in Central Africa in the region of the Great Lakes,
(c) scattered across the entire length of the Eurasian steppe from Manchuria to the Ukraine.

Hankin (1905) believed that Garhwal in India (focus (a), above) was the area where bubonic plague was endemic. It is a mountainous and somewhat inaccessible region that had little ordinary traffic with the rest of India. However, Hankin (1905) stated that the first recorded plague in Garhwal in humans was only in 1822; thereafter irregular outbreaks were recorded during the 19th century, usually confined to a few villages, some said to be severe, and 535 deaths were recorded in the epidemic of 1877. Only a handful of people died in most outbreaks and these are probably authentic cases of bubonic plague that led up to the pandemic in India which began in 1895 and continued for more than half a century.

Before the outbreak of plague in Garhwal in 1822, Hankin (1905) recorded only the following pestilences in India:

1344 Army of Sultan Mahommed Tughlak destroyed by pestilence probably near Deogiri, a town a short distance from Nassik.
1611 Plague said to have begun in Punjab. It lasted 7 years, and spread to Delhi, Agra, Cashmere and Kandahar.
1683 Confined to Western India; lasted for 8 years in Ahmedabad.
1812 Began in Gujerat and lasted 9 years.

East Africa (focus (b), above) is believed to have been affected by plague in the 6th century AD (Roberts, 1935) and Twigg (1984) recorded a plague in this region at Mombasa in 1697 that lasted almost 3 years. If this were bubonic plague, it probably spread from the caravan routes from the north to the east coast and it was known in the northeastern Congo and at Kisumu on Lake Victoria. Epidemics alleged to be plague occurred at Kisumu in Kenya throughout the 18th and 19th centuries.

McNeill (1977) has argued that focus (c) originated with the trade routes of the Mongul Empire, the Silk Road between Syria and China across the deserts of Central Asia. He suggested that the Mongol movements brought the causative agent of bubonic plague, the bacterium *Y. pestis*, to the rodents of the Eurasian steppes. This focus could well have been the launch pad for any epidemics of bubonic plague in the Mediterranean region after

1300, but the disease had probably occurred there from the 6th century onwards. Pollitzer (1954) considered that the first really satisfactory evidence of bubonic plague was a pandemic beginning in AD 542 and believed that it came from Ethiopia, suggesting a central African origin.

Twigg (1984) concluded that it was likely that bubonic plague was present in lands around the Mediterranean in some of the epidemics in the 6th century, probably in association with other diseases whose impact was equally severe and whose differentiation from bubonic plague was not easy to define at the time. McNeill (1977) considered that it may have come from an old focus in northeast India or in central Africa. Twigg (1984) assumed that the black rat (see section 3.5) must have existed in the Mediterranean ports and cities in the 6th century in order to sustain any epidemics of true bubonic plague. The black rat had not spread to northern Europe by the time of the plague of Justinian (Shrewsbury, 1970) and, consequently, the disease was confined to the Mediterranean coastlands (Twigg, 1984). Bubonic plague disappeared from southern Europe at the end of the 6th century and Twigg (1984) stated that nothing that was firmly identifiable as this disease was heard of in the Mediterranean region for the next 700 years.

Molecular biology techniques were used to resolve problems of historical etiology. DNA extracts were made from the pulp of teeth extracted from skeletons excavated from graves near Provence, southern France, which were dated: (i) between the 13th and late 14th centuries and so did not necessarily come from people who died during the Black Death as is presumed by Raoult *et al.* (2000); (ii) about 1590 (Drancourt *et al.*, 1998); and (iii) during the outbreak of bubonic plague in 1722 (Drancourt *et al.*, 1998), see Chapter 12. No skeleton showed macroscopic signs of disease. The incorporation of primers specific for *Y. pestis rpoB* (the RNA polymerase β-subunit-encoding gene) and the recognised virulence-associated *pla* (the plasminogen activator-encoding gene) repeatedly yielded products that had a nucleotide sequence similar to that of modern-day isolates of the bacterium. The specific *pla* sequence was obtained from six of the 12 plague skeleton teeth but none of the seven controls. Thus a nucleic acid-based identification of ancient bubonic plague was achieved which confirmed the presence of the disease during the 14th century and at the end of the 16th century on the Mediterranean coast.

Hankin (1905) described how sporadic bubonic plague rumbled on in India in the 19th century (see above) and it also became established in Yum-nan, China. In 1855, troops were sent to suppress a rebellion and plague spread further, probably as a consequence of the movement of refugees. It reached the provincial capital, Kunming, in 1866 and Canton and Hong Kong in 1894, a fairly slow rate of spread. Bubonic plague was then carried by rats and fleas from the rat-infested warehouses of the Chinese ports to many of the warmer parts of the world. India, North and East Africa and the Mediterranean coast had experienced bubonic plague before, as we have seen, and were largely free from the disease at this time, but now epidemics broke out again.

India was infected via Calcutta in 1895 and via Bombay in 1896 and the great plague pandemic of the 20th century (this time truly bubonic plague) had begun. Egypt had been free from plague for 55 years but it reappeared at Alexandria in 1899, moving slowly inland and reaching most parts of the Nile Valley as far as Aswan. Plague reappeared in Tunis in 1907 after being absent from the coastal areas of North Africa, west of Egypt, since 1822. Between 1895 and 1903, most of the major ports in the tropics became plague infected and then the disease became widespread across the continents, assisted by more modern means of transport (Twigg, 1984).

Indeed, the development of modern transport was the key to the widespread dispersal of the 20th century pandemic of bubonic plague: steamships replaced sail, so sharply reducing the time on intercontinental voyages and allowing bubonic plague to be transmitted by rats and fleas to distant ports, whereupon the railways rapidly carried the disease far inland. South Africa was invaded through its ports from 1899 to 1902 and epidemics broke out in Cape Town and Durban with 766 and 201 cases, respectively. After 10 years, bubonic plague had spread via gerbils and other rodents, eventually covering 50 000 square miles; there were 167 outbreaks *but with only 372 human cases* and these were confined to the villages.

Bubonic plague reached San Francisco in rats from the Orient between 1900 and 1904, infecting local rats but rapidly spreading into a variety of rodents. After 40 years, it had invaded 10 states in the USA, becoming the most extensive plague focus in the world, but it remained a wild, rural plague and never exploded in the cities. In the 60 years between 1908 and 1966, only 115 cases of bubonic plague with 65 deaths in humans were recorded in the USA (Christie, 1969), differing by many orders of magnitude from the terrible mortality of the Black Death and, for example, from that of the haemorrhagic plague epidemic at Lyons in 1628–29 when some 35 000 people died (section 11.2.2).

Shipping, with rats and fleas in the cargo, is the easiest way of introducing bubonic plague and so it is not surprising that the disease has been recorded in Britain in the 20th century. There were 82 cases (17 of them fatal) on 54 ships arriving in England. In two instances, plague was transferred to land and on one occasion a man died. Bubonic plague probably existed from 1900 to 1907 in a small area of Glasgow and outbreaks were also recorded in Liverpool, Hull, Bristol and London. These foci were all in ports and there is no evidence of any extensive spread; Twigg (1984) described in detail the only known occasion of authenticated bubonic plague in a rural situation in England. In summary, the outbreak occurred in 1910 in Suffolk, 4 miles from the port of Ipswich, which received grain boats. Three of the victims who died were in the same family and a woman who had nursed at their cottage and who lived a quarter of a mile away was the fourth to die. At this time, brown rats and hares were found dead in the fields and examination showed that these were infected with the bacilli of bubonic plague over quite a wide area. After a careful analysis of the evidence, Twigg (1984) concluded that an epizootic had become established in the brown rats and hares and that the first victim, a 9-year-old girl who had recently stayed on an isolated farm, had caught bubonic plague from an infected flea; the other three victims had died of pneumonic plague when nursing her.

3.2 *Yersinia pestis*

The causative agent of bubonic plague is a small ovoid, Gram-negative bacillus, *Yersinia pestis*, which is non-motile and non-sporing. The organism forms a capsule or envelope when grown on serum agar at 37 °C: this capsule contains an antigen that is distinct from the somatic antigen and may be concerned with the ability to resist phagocytosis and so with the virulence of the organism; bacilli without a capsule may occur in chronic lesions in rats with latent plague. *Yersinia pestis* is not a highly resistant organism, being killed in 5 minutes at 55 °C. It does not survive drying for more than 2 days, except in dry flea faeces. The bacterium remains alive in its host, vector or a burrow during long quiescent periods in the wild and is able to revert to full virulence when the environment becomes favourable (Christie, 1969).

Different varieties of *Y. pestis* have been distinguished, namely var. *orientalis*, var. *antigua* and var. *mediaevalis*, and different rodents harbour different yersiniae in different places. *Rattus rattus* and *R. norvegicus* usually harbour *orientalis*, as do ground-squirrels (*Citellus*) in America, hares in the Argentine, jack rabbits in California and bandicoots in India and Ceylon. *Antigua* is carried by marmots in Manchuria and Mongolia and by ground-squirrels and hamsters in the southeast of the former

USSR, whereas *mediaevalis* is the common parasite of gerbils in Kurdistan, Turkey and Iraq (Christie, 1969).

At least 18 antigenic components of *Y. pestis* have been identified and these are found in *all* strains isolated from various parts of the world. There is no qualitative difference in these components and this has made serological typing of *Y. pestis* difficult. There are two main antigenic complexes, one in the capsule is heat labile and the other, a somatic antigenic complex, is heat stable. The capsular antigen contains a polysaccharide protein component (Fraction 1) that is specific for *Y. pestis* and seems to be associated with resistance to phagocytosis. Such strains may be able to survive inside host phagocytes because of two somatic antigens (V and W) that are associated both with resistance to phagocytosis and also with the ability to survive and multiply *inside* host phagocytes. The temperature may be around 25 °C in warm climates inside the gut of a cold-blooded flea and at that temperature *Y. pestis* may be found apparently lacking both Fraction 1 and VW antigens: such bacilli can be ingested and destroyed by flea polymorphonuclear phagocytes. If they are transferred to a warm-blooded rodent host, they may still be ingested by host monocytes, but at 37 °C they seem to undergo a phenotypic change and emerge fully virulent, with Fraction 1 and VW antigens. Changes in virulence probably occur in the wild in various hosts: avirulent strains isolated from chronic lesions in rats regain virulence after passage through normal animals in the laboratory, and subtle changes in the environment probably affect the virulence of strains and so influence the epizootic pattern, quiescent or active, of plague foci. The virulence of a strain also varies with the host: a strain lacking Fraction 1 is not virulent for guinea pigs though it is for mice.

The 4.38 megabase-pair (Mb) genome of *Y. pestis* is currently being sequenced; it carries a 70 kilobase-pair (kb) plasmid that encodes an effector protein, Yop1, that enters human macrophages causing diminished immune defences (Rosqvist *et al.*, 1988; Guan & Dixon, 1990; Cornelis & Wold-Wulz, 1997; Mills *et al.*, 1997). One of the ways that cells can regulate the activity of an enzyme is by reversible covalent modifications when a phosphate group is added to a specific serine, threonine or tyrosine residue. An enzyme can then undergo conformational changes that either increase or decrease its activity. An essential virulence determinant of *Yersinia* has been shown to be the activity of a specific protein tyrosine phosphatase. Site-directed mutagenesis was used to show that the *Yersinia* phosphatase possesses an essential cysteine (Cys) residue required for catalysis and the amino acid residues surrounding it are highly conserved, as are other amino acid residues in the *Yersinia* and mammalian protein tyrosine phosphatases, suggesting that they use a common catalytic mechanism (Guan & Dixon, 1990; Schubert *et al.*, 1995; Fauman & Saper, 1996).

Scholars defending the view that *Y. pestis* was the causative agent during the age of plagues aver that little can be deduced from present-day studies of the disease because epidemic bubonic plague has changed in character (Gottfried, 1978), but Twigg (1993) has emphasised that there is still only one serotype of *Y. pestis* despite bubonic plague having spread to 200 rodent species, many species of flea and all ethnic groups. This suggests antigenic stability, a view confirmed by the identity of the nucleotide sequences of 16th century *Yersinia* from dental pulp with present-day isolates of the bacterium (section 3.1) and it enables us to relate the biology of modern plague to plague in the past with some confidence and thus to compare modern outbreaks with those of earlier centuries. As Carmichael (1986) has stated, 'unless there is very persuasive, unassailable evidence to the contrary we must begin from the position that infectious diseases, including *Y. pestis* in human communities of the European late Middle Ages are similar in both epidemiological and clinical presentation to analogous twentieth-century infections'.

3.3 The rodent host

As we have said, bubonic plague is a disease of rodents and it had long been recognised that rats or other rodents came out of their holes before an outbreak in humans. However, over 300 mammalian species are susceptible to the disease; mice, rats and guinea pigs are all readily infected, rabbits less so. Monkeys vary in susceptibility; dogs, although difficult to infect experimentally, often show serological evidence of infection in plague foci; cats, pigs, cattle, sheep, goats and horses are difficult to infect, although they may also sometimes be found infected in wild plague foci; birds, except sparrows, seem to be wholly resistant. Camels are difficult to infect experimentally, but can probably be infected in the wild and may spread the disease to humans. Monkeys have been infected with pneumonic plague from inhaling aerosols of plague cultures, as have rats, guinea pigs, mice and marmots.

This difference in rodent susceptibility is of great importance in the persistence of plague foci because the disease will die out in an area where the host (e.g. a rat) is highly susceptible, but it persists where there is a balance between susceptible and resistant hosts. In Siberia and Mongolia, for example, marmots, susliks and tarabagans are subject to recurrent, acute outbreaks that might eventually eliminate the plague focus through lack of hosts; but the local gerbils and voles are more resistant to *Y. pestis* and so they can serve to maintain the enzootic in the area because they do not die from the infection. In Kurdistan, Turkey, Iraq and Syria the hosts are gerbils: two of them, *Meriones vinogradivi* and *M. tristrami*, are highly susceptible, but *M. persicus* and *M. libycus*, are resistant, so that the plague exhibits cyclic epidemics in these rodents. In central Colorado, however, an isolated colony of prairie dogs (*Cynomys gunnisoni*) was wiped out by plague: they were highly susceptible and died, although the fleas remained alive in the burrows for at least a year after the last prairie dog had died.

Thus rodents are the true natural hosts of *Y. pestis*; other animals (including humans) are accidental hosts only. They inhabit a wide range of habitats (mountains, plains, steppes, deserts, cultivated fields and forests) in temperate and tropical areas. Rats injected subcutaneously with a small number of virulent *Y. pestis* die within 2–8 days. Necrosis and oedema are found at the site of inoculation and the regional lymph nodes are swollen and embedded in haemorrhagic subcutaneous tissue; the spleen is enlarged, the liver and lungs are hyperaemic and there is often a pleural effusion. Small necrotic foci occur in the liver and spleen of animals surviving for a week. Rats may be infected by a trace of infective material smeared over the conjunctiva that results in rapid death, with enlargement of cervical lymph nodes and spleen and haemorrhages in the stomach and duodenum. Wild rats have buboes in their necks because fleas most commonly bite them there (Christie, 1969).

3.4 Murine versus sylvatic phases

When bubonic plague is confined to rats, as it usually is when being carried by ship or when it is present in a port, it is said to be in the 'murine' phase. Frequently, it transfers from the rats to the indigenous endemic rodents, such as gerbils or ground-squirrels, whose populations cover vast areas of country, and it is then said to form a 'sylvatic' reservoir (Twigg, 1978). This is a permanent reservoir, a potential source of infection that may affect people, other rodent species or even return to rats. Such a plague focus may remain static for a number of years, *Yersinia* being passed back and forth between the rodents and fleas with only a seasonal variation in numbers. This is described as *enzootic plague* and it is maintained by the balance between resistant and susceptible hosts in the focus; it is present in the rodents at all times but causes only a limited number of deaths. However, sporadically, this balance is disturbed and bubonic plague then spreads widely and rapidly and causes many rodent deaths. This is described as

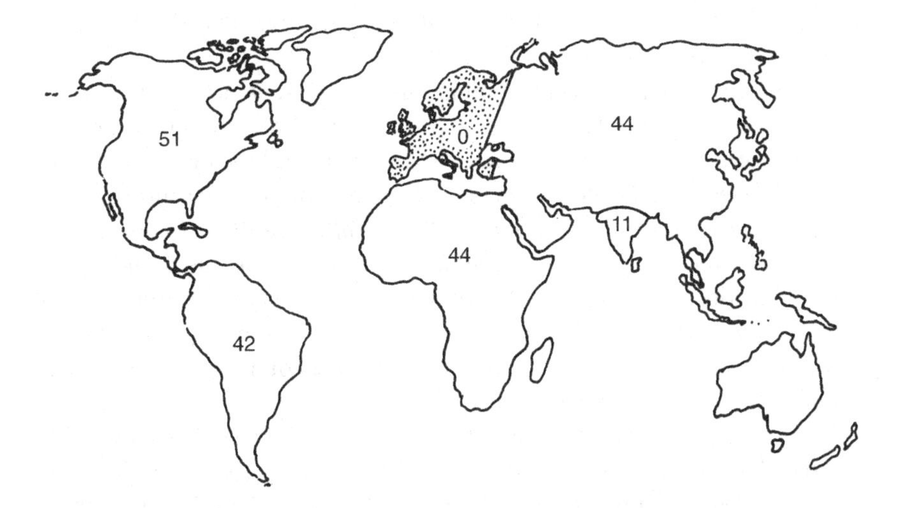

Fig. 3.1. The sylvatic reservoir of bubonic plague: the number of rodent species (excluding commensal species) that have been reported to have been infected with *Yersinia*. Note the absence of suitable species in western Europe. After Twigg (1989).

epizootic plague and in some epizootics dead rodents have been collected by the barrowful; their dead bodies are often the first indications that plague is present in a locality. It is *only when the epizootic spreads to rats living in urban areas* in loose association with humans that the real danger of a major outbreak of bubonic plague amongst a human population occurs. An epizootic begins to wane when the numbers of resistant rodent hosts rises and the numbers of susceptible rodents falls.

One of the interesting things about *Yersinia pestis* is the ease with which it can accommodate to new species of rodent. During the 20th century, following the introduction by ships to the ports, plague has spread to form sylvatic reservoirs in North America, South America and Africa. In addition, it is present in many species in Asia and India but, despite this readiness to spread, it has not become established in any European species in modern times (Twigg, 1989) so that it was not possible to establish an enzootic for bubonic plague during the age of plagues after the Black Death (Fig. 3.1).

3.5 Black and brown rats

In Asia, rodents other than rats are essential for maintaining *Yersinia* during the enzootic sylvatic phase but, when this changes to a full-blown epizootic, it is via the peridomestic rats and their fleas that the *Yersinia* is transmitted to humans.

There are only two species of rat in Europe today, namely the brown rat, *Rattus norvegicus*, and the black rat, *Rattus rattus* (Davis, 1986). The brown rat is not an accomplished climber and inhabits cellars and lower floors. This species spread from Russia in the early part of the 18th century and so *did not arrive in Britain until some 400 years after the Black Death and 100 years after the disappearance of the plague*. It has since spread to most parts of the world; being the more hardy species, it is today the common rat of temperate latitudes and lives both indoors and outdoors. In the tropics today it is confined to ports and towns and in the countryside lives close to human habitation (Twigg, 1984).

The black rat (or roof rat) occurs in three colour phases, one of which is the melanistic form; it is a descendant of a rat that probably originated in India and spread along trading routes to establish itself as the common rat of both town and countryside in the tropics and may have arrived in England some time in the Middle Ages (Matheson, 1939), although a variety of dates have been suggested (see Twigg, 1989). It is widely spread away from seaports in the *tropics* when there are no indigenous rodent competitors. In temperate latitudes, however, it is confined to buildings because, for most of the year, the outdoor temperatures are too low for its liking. Being a good climber it has readily gained access to ships, which have transported it over much of the world. In buildings it climbs to the higher levels and lives in the roofs and ceilings rather than the basement; it rarely, if ever, inhabits burrows, tunnels in the ground or aquatic habitats. Davis (1986) suggested that the black rat in northern France and the British Isles may have persisted in towns and the grain ports in the Middle Ages, but their numbers were small; the population in a particular town disappeared after a few years but may have been re-established as a result of new introductions. *Rattus rattus* may have lived in small numbers in rural areas in the much warmer Mediterranean region of France.

The black rat, therefore, is today a native of the Mediterranean area and generally does not persist in England without recurrent introductions (Davis, 1986): it is found in the ports (but does not spread more than three-quarters of a mile inland; Davis, 1986) and in those inland towns that are connected to them by canals. Modern populations of the black rat in England have depended for their survival upon frequent topping-up by the importation of rats in cargo from the ports and now that canal traffic has ended and containers are replacing loose cargo, not only the inland but also the port populations of *Rattus rattus* have begun to die out (Twigg,

1989). The dynamics of the black rat in the ports of England in the 20th century have been discussed in detail by Matheson (1939), Twigg (1984, 1989) and Davis (1986) and, for example, the black rat was found widely in central London in the 1950s where the multistorey, centrally heated buildings with abundant food sources from the many restaurants provided ideal conditions (Twigg, 1989).

In summary, therefore, the black rat (*Rattus rattus*) was the only rodent species present that could possibly have carried bubonic plague in the Black Death and during the subsequent plagues in the 15th, 16th and 17th centuries. It requires the warmth of human habitations and does not spread far from them. The Black Death acted virtually independently of season and climate in the British Isles (Chapter 4), whereas when bubonic plague appears in warm countries it is circumscribed by climate and breaks out only at certain clearly defined and predictable times (Twigg, 1989). It is inconceivable that the black rat could have transmitted bubonic plague rapidly and widely in winter in 1348 to 1350 in northern Britain and northern Europe, when the Black Death was raging, or that the disease could have been propagated over mountain passes in the Alps. This pandemic even reached the polar regions and there was said to be an epidemic in Greenland (Kohn, 1995).

3.6 The role of the flea

Bubonic plague can potentially circulate between a flea, a rat, *Yersinia* and humans. We have seen how, in reality, for plague foci to be maintained and for epizootic–enzootic cycling to continue, the detailed dynamics are more complex and other indigenous, *resistant* rodents have to be involved. In this section we introduce the fourth component of bubonic plague, namely the rat flea. At least 30 species of flea have been proved to be vectors and, since more than 200 species of rodent can carry plague, the host-vector permutations in the Asian subcontinent are formidable and the population dynamics complex (Christie, 1969). But in England in the 14th, 15th, 16th and 17th centuries there were probably only one species of flea (*Xenopsylla cheopis*, the oriental rat flea) and only the black rat that we have to consider.

Adult rodent fleas live on warm-blooded animals; they are small, wingless, laterally flattened and adapted for clinging by means of hooks on their legs, which are adapted for impressive jumping, the means by which they transfer to their different hosts.

The mouthparts of the flea include a central stylet, the epipharynx, which is enclosed by two blade-like piercing organs, the maxillary laciniae. These three components comprise the fascicle and this pierces the skin and enters a venule; the flea sucks blood into its oesophagus by the action of its cibarial and pharyngeal pumps. This behaviour is important in two respects: firstly, pathogens are more likely to be present in an infected host's bloodstream than in the subcutaneous tissues and, secondly, if an infected flea injects pathogens into a healthy host, these go straight into its bloodstream and thus are very likely to establish an infection (Christie, 1969). A flea takes in a large number of bacteria in a blood meal from an infected rat and these pass into the stomach of the flea. The bacteria become established in the stomach in only about 12% of fleas where, by dividing rapidly, they form a solid mass. The proventriculus of the flea is a bulbous structure provided with seven rows of spines, which interlock and act as a valve shutting off the stomach when the encircling muscles contract; when they relax, the valve opens and the ingested blood enters the stomach. The resulting gelatinous culture of *Y. pestis* eventually glues the spines together and blocks the valve. It is then known as a 'blocked' flea (Bacot & Martin 1914; Twigg, 1984); it continues to feed still more voraciously and becomes dehydrated and hungry. But the blood it sucks cannot get into the stomach and simply distends the oesophagus, and when the pharyngeal pump stops, the distended wall of the oesophagus recoils and drives blood down into the wound, taking with it plague bacilli which go straight into the blood of the bitten host (Christie, 1969).

3.7 Flea survival

An understanding of flea biology and ecology is an important key to understanding the dynamics of *Yersinia* infection and the etiology of bubonic plague outbreaks. It takes time for the bacilli that the flea has ingested along with its blood meal to multiply sufficiently to reach an infective concentration and this varies in different flea species. In *X. cheopis* the average time between feeding and infectivity is 21 days (range = 5 to 31 days), whereas it is 53 days in *Diamanus montanus*, the flea of the Californian ground-squirrel. This time varies with external temperature and humidity. Fleas survive for different times after becoming infective: in one investigation *X. cheopis* survived a mean of 17 days (maximum 44 days) and *D. montanus* 47 days (maximum 85 days); the longer a flea survives, the more often it can feed and pass on the infection. Not all infected fleas become infective: the yersiniae may not multiply to block the proventriculus, and such fleas may live a very long time (Christie, 1969).

The long survival times of uninfected fleas account for their persistence in the wild in spite of such hazards as fluctuating temperature and relative humidity. At 50 °C and high humidity fleas survive a long time, but they die quickly when the temperature is over 80 °C or under 50 °C, especially if the atmosphere is dry. Experimentally determined survival periods, vary from country to country: infected fleas survived for 47 days in India and for 130 days in the USA. However, although fleas can enter diapause and survive for long periods in the microclimate of deserted burrows, the external conditions of warm temperature and high humidity are essential if the flea is to play its part in the development of the epizootic and its escalation into an epidemic of bubonic plague in humans.

3.8 Flea reproduction

The survival of the flea species in the wild also depends on their ability to produce and raise more fleas, and reproduction is strongly dependent on environmental and other factors. *Xenopsylla cheopis* can lay about 300–400 eggs in its life; the temperature and humidity of the environment greatly affect both egg-laying and the development of larvae. A temperature of between 18 °C and 27 °C and a relative humidity of 70% are ideal for oviposition by *X. cheopis*, whereas temperatures below 18 °C inhibit it. The egg hatches in 2–14 days, but it takes much longer if the temperature and humidity are unsuitable. The life cycle through egg, larva, pupa and imago is completed in 2–3 weeks if the conditions are right: pupation lasts 8 days at 18 °C, 6 days at 22 °C but it is completed in 4 days at 29 to 35 °C. However, if conditions are not satisfactory, the flea can remain in the pupal stage for long periods, possibly for at least a year, and emerge only when the conditions change and the selected host becomes active in its neighbourhood. The flea must feed within 1–3 days after eclosion and it begins egg-laying 1–4 days after the first feed, so that conditions must be exactly right if the life cycle is to continue. The importance of climate in flea biology has been emphasised by Twigg (1989) because the part of the life cycle that is passed away from its homoiothermic host is unprotected against fluctuations of temperature. Temperatures below 7 °C are deleterious to all stages except the adult.

Twigg (1989) has summarised the calculated climatological fluctuations in central England from 900 to 1900 (Fig. 3.2) and it is evident that at no time, and certainly not during the age of plagues, was the average July to August temperature above 18.5 °C and suitable for flea hatching. A *sustained* high temperature is necessary to yield the high flea numbers that are

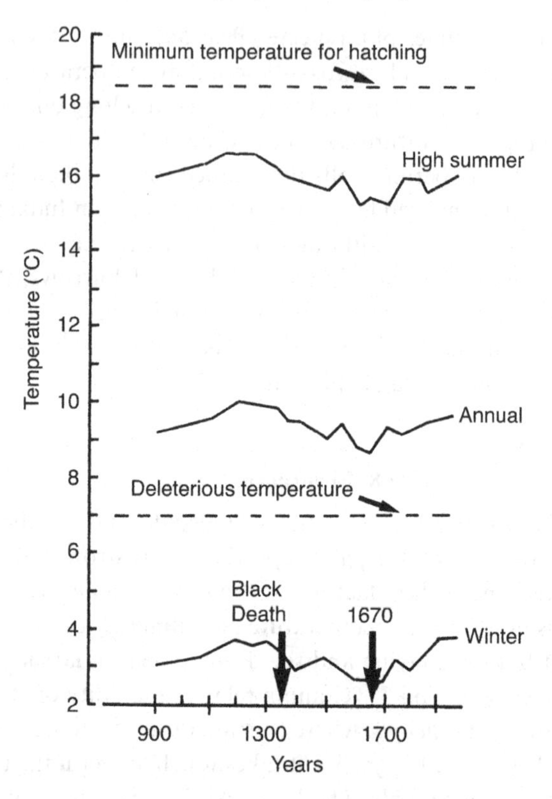

Fig. 3.2. Estimated temperatures in central England, AD 900 to 1900. The age of plagues is indicated by bold arrows. The minimum temperature for the hatching of fleas and the temperature below which conditions are deleterious to all pre-adult stages of the flea are indicated. From Twigg (1989).

essential to promote a rat epizootic. Figure 3.3 shows the mean July temperatures over western Europe in the 20th century and this confirms that the British Isles does not have a climate capable of sustaining regular seasonal outbreaks of flea-borne bubonic plague in the summer months and certainly not in winter (Twigg, 1989). Indeed, in Europe, it is only in the southwest area and in the Mediterranean coastal region together with the Iberian peninsula that possibly suitable conditions are to be found (Fig. 3.3); the hinterland of Marseilles experienced an outbreak of plague that was probably bubonic in 1720–22 (Chapter 12) and there were a number of outbreaks at Barcelona through the age of plagues that may also have been bubonic (section 11.4.2).

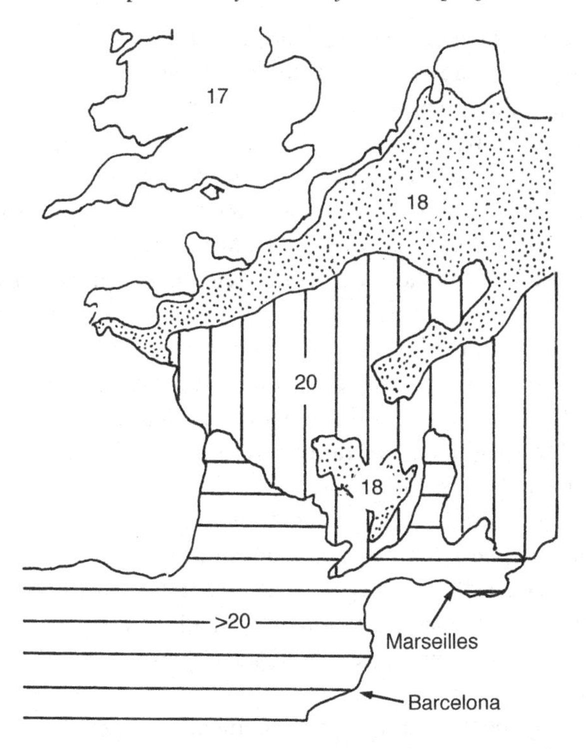

Fig. 3.3. Mean temperature (°C) in July in western Europe in the 20th century. Note that it is only in Spain and the Mediterranean coast of France that summer temperatures are suitable for sustaining seasonal outbreaks of flea-borne bubonic plague. After Twigg (1989).

3.9 Population dynamics of bubonic plague

The numbers of rodents that are present in a locality are an important factor in determining the dynamics of bubonic plague. A species may be highly susceptible, but if it is rare it is not important in the spread of plague. The rodent breeding system is in the spring and summer in colder climates and if there are outbreaks of bubonic plague they are likely to occur then. Rodents breed throughout the year in warm climates and may have 5 to 9 litters and a total of up to 35 or 40 young and, in this way, the population density of rodents builds up quickly and outbreaks can then occur at any time. Thus the turnover of a rodent population can be rapid because as fast as they breed so infections (including *Yersinia*) and a variety of predators act to reduce population numbers, so that regular population cycles are generated (Scott & Duncan, 1998) that prevent the build-up of a stable, plague-resistant population. This helps to maintain the *Yersinia*–flea–

rodent cycle, yet some resistance on the part of the hosts is essential for the maintenance of a plague focus. A totally susceptible population would soon die out, as did the prairie dogs when *Yersinia* arrived in Colorado. It is essential, therefore, if the enzootic is to be maintained, that there is a balance between susceptible and resistant hosts: voles and gerbils are susceptible to plague but mostly do not die from it; ground-squirrels are highly susceptible, but voles and field mice in the same area are more resistant and they, not the ground-squirrels, maintain the enzootic. The infection can persist even when conditions become unfavourable for the rodent or the flea, or both; the rodent goes into hibernation and at low body temperature becomes more resistant, the adult flea can survive without food or host for a year and the pupal stage can be prolonged until a new host enters the burrow and vibrations from its body stimulate emergence from the pupa (Christie, 1969).

Bubonic plague in the wild is, therefore, a disease of arid plains, rocky escarpments, steppes, prairies and semi-deserts where *Y. pestis* passes quickly from generation to generation, causing little inconvenience to its semi-resistant mammalian hosts and none at all to *Homo sapiens* unless he or she leaves an urban or village dwelling and goes out into the focus in the countryside.

A focus may be temporary or permanent. In the former, plague is imported accidentally from another area, but conditions are not suitable and the outbreak is short lived; the flea and the host are perhaps not adapted to each other; a flea may be able to survive by feeding on the blood of a rodent but be unable to reproduce on it. The flea is essentially a nest-dweller and if the available host is not a burrowing animal then the infection does not become established in the area. However, an area can become established as a permanent focus of plague if the following factors obtain and there is nothing unfavourable in the environment: (i) there are many rodents in the area and some (such as gerbils) keep large, permanent burrows; (ii) some of the rodents are partially immune but nevertheless support a prolonged bacteraemia with *Y. pestis*; and (iii) fleas are also present in large numbers and they are a long-lived variety and adapted to the host. Activity in such a focus will wax and wane and there will be changes related to season or to the breeding pattern of rodents and fleas, and weather changes such as floods or drought will affect it (Christie, 1969). It is evident that these factors have never been present in Europe and that persistent enzootic bubonic plague there is an impossibility.

An enzootic of a plague focus, therefore, is not static but oscillates and, superimposed on these dynamics, are the periodic surges of the epizootics with the potential of spreading bubonic plague to man. As Christie (1969) said,

Man harvests a field and the rodent inhabitants are driven to seek shelter and food elsewhere, probably nearer to man. A storehouse is emptied of its grain or rice, and again the rodents, with their fleas and their bacilli, are turned out, though an odd rat or an odd flea may accompany the grain to a distant market or a distant port. In all these clear circumstances there is a danger that first man's peridomestic and then his commensal rodents and then man himself will get the plague. But enzootic plague can swell into the epizootic for no obvious reason, though there must be a bio-ecological one; and an epizootic, if conditions are right, can slip over into an epidemic which may be world-wide in its impact.

3.10 Evolution of bubonic plague virulence

A genetic basis for the supposed evolution of plague virulence has been presented by Rosqvist *et al.* (1988) and their interesting hypothesis has been summarised by Lenski (1988) as follows: it is suggested that less virulent strains of *Y. pestis* were harboured by rats and fleas during endemic phases. Single point mutations could have given rise to hypervirulent strains, which spread to cause the plague epidemics. They have elucidated the genetic determination of virulence of *Y. pseudotuberculosis*, which is closely related to *Y. pestis*; the two are essentially indistinguishable from DNA hybridization data (Bercovier *et al.*, 1980) and *Y. pseudotuberculosis* infections in rats provoke immunity to *Y. pestis*. Previous work has implicated two outer-membrane proteins in mediating the invasion of mammalian cell cultures by *Y. pseudotuberculosis*: invasin, which is encoded chromosomally (Isberg & Falkow, 1985), and Yop1, which is encoded by a plasmid (Bolin *et al.*, 1982). Rosqvist *et al.* (1988) demonstrated that mutations in one or other of the genes encoding these proteins have little effect on the virulence of *Y. pseudotuberculosis* in mice, but when mice were administered bacteria containing mutations in both genes, the LD_{50} (dose lethal to 50% of the sample) went down dramatically, indicating a heightened degree of virulence. *Yersinia pestis* apparently does not express either invasin or the Yop1 protein, consistent with its much greater virulence relative to *Y. pseudotuberculosis* and with its inability to grow invasively in mammalian cell cultures.

Wolf-Watz and colleagues have previously shown that virulence plasmids pYV019 and pIB1, carried by *Y. pestis* and by *Y. pseudotuberculosis*, respectively, have a high degree of sequence homology (Portnoy *et al.*, 1984). Both carry the *yopA* gene, although only *Y. pseudotuberculosis* expresses the corresponding Yop1 protein (Bolin *et al.*, 1982). Rosqvist *et al.* (1988) have sequenced the *yopA* genes from *Y. pestis* and *Y. pseudotuberculosis* and find only 15 nucleotide differences among 1230 base-pairs; the non-functional *yopA* gene in *Y. pestis* is presumably derived from a functional ancestral state. When Rosqvist *et al.* (1988) introduced the functional gene from *Y. pseudotuberculosis* into *Y. pestis*, they observed a corresponding reduction in the virulence of *Y. pestis*. Thus *Y. pestis* has apparently undergone a mutation in the past that caused the loss of function of the *yopA* gene, with a concomitant increase in its virulence.

Rosqvist *et al.* (1988) believed that this supports the hypothesis that single mutations played an important role in triggering plague epidemics. But mutations alone cannot drive epidemics. The necessary genetic variability in the pathogen must exist and so must the appropriate selective conditions for the spread of hypervirulent mutants. Hence there remains the equally perplexing question concerning the selective pressures that were responsible for the increase in the frequency of hypervirulent strains, once they appeared by mutation. Indeed, for many years, conventional wisdom favoured the view that evolution would select those pathogens that had the least harmful effects on their hosts (May & Anderson, 1983). On the other hand, pathogenicity, or virulence, is often associated with transmission (Anderson & May, 1982a) and mathematical analyses (Levin & Pimentel, 1981; Anderson & May, 1982b; May & Anderson, 1983) have shown that the evolution of pathogens is highly dependent on this coupling between transmissibility and virulence.

Lenski (1988) suggested that the dramatic declines of the human population in Europe during the great plague epidemics of past centuries were presumably accompanied by comparable declines in the population of susceptible rodents although, as we have seen, no such population existed. Rosqvist *et al.* (1988) hypothesised that not only might these epidemics have been triggered by the appearance of hypervirulent strains of *Y. pestis*, but the declining populations of susceptible hosts may, in turn, have favoured less virulent strains.

Hinnebusch *et al.* (1996) found that three genes in *Y. pestis* change the bacillus from a harmless, long-term inhabitant in the flea mid-gut to one that causes blocking in its foregut (section 3.6). Their experiments focused on three hemin storage (*hms*) genes of *Y. pestis*; they gave oriental rat fleas, *Xenopsylla cheopsis* (the normal vectors of bubonic plague), blood meals that contained either normal *Y. pestis* or a mutant form missing the *hms* genes. Only those fleas that were infected with normal bacteria developed blocking, which was accompanied by the usual high rate of mortality.

These results suggest that the *hms* genes are required for *Y. pestis* to cause blocked fleas. The failure of the mutant *Y. pestis* to block fleas could not be attributed to rapid elimination of this form of the bacteria, because the same percentage of fleas infected with either normal or mutant *Y. pestis* strains were heavily infected after 4 weeks.

Xenopsylla cheopis were then infected with either the normal or mutant forms of *Y. pestis*, both of which were tagged with a fluorescent dye so that their passage through the gut of the flea could be followed. After the first week, it was clear that the mutant bacteria remained in the mid-gut whereas the normal bacteria had migrated to the foregut in many fleas which eventually became packed with bacteria. Hinnebusch *et al.* (1996) suggested that the mutant bacteria may fail to colonise the foregut because, being less cohesive, they are disrupted and flushed back into the mid-gut during feeding. These studies are being extended in an attempt to explain the observation that the blockage of the flea foregut breaks down at temperatures above about 27 °C. Do such temperatures suppress the products of *hms* or other genes? Again, these studies are of considerable interest and contribute to our understanding of bubonic plague, a major historical and a minor present-day scourge but, we suggest, are not relevant to the biology of haemorrhagic plague.

3.11 Spread of bubonic plague to humans

Yersinia pestis can spread from a focus to humans in the following ways (Christie, 1969):

(i) It can escape from the focus in the body of a rodent that strays near human habitations and then shares its fleas with peridomestic rodents. *Yersinia* is then readily spread from rat to humans and the result today may be a few cases of plague in a remote Asian village or the escape may be on a larger scale. The important point here is that the basic mechanism of a village outbreak and a major epidemic of bubonic plague is the same: the transfer of fleas from wild rodent to domestic rat, and from domestic rat to humans. The role of the rat is entirely that of liaison. The black rat is not a reservoir of plague and it is for this reason that it acts so effectively; its role is to die and pass on its infection. One or two dead rats may be found in a village compound, or they may be swept up by the barrowful in a South African township (Annotation, 1924) or litter the streets of an eastern metropolis (Liston, 1924).

(ii) When humans invade a natural focus, for example as a hunter, bubonic plague is caught directly from the wild rodent reservoir, usually on a small scale from trapping, skinning or eating a small animal, although 60 000 hunters in total caught plague from marmots which they hunted for their skins between 1910 and 1911 in Manchuria (Wu *et al.*, 1936; Chandler & Read, 1961).

(iii) Occasionally humans get plague from eating a domestic animal (a goat or camel) that has roved over a natural plague focus.

(iv) However, and this is a further important point, it is movement into new, untrodden regions, as in war or mass migration that brings humans most often into direct contact with enzootic bubonic plague (Christie, 1969).

3.12 Clinical manifestations of bubonic plague in humans

The three classical manifestations of bubonic plague are: (i) bubonic, (ii) septicaemic and (iii) pneumonic.

Bubonic and septicaemic plague are not distinct forms but differ only in the intensity of the infection and the speed of its development: in any outbreak of plague in humans today, patients with bubonic and septicaemic plague are found together, in the same house or village, or lying in adjacent beds in the same hospital ward. They have been infected in the same flea-borne way and *they are not normally infectious one to another.* This is in complete contrast, as we shall see, to haemorrhagic plague, which was clearly transmitted by person-to-person infection. Today, probably between 30% and 50% of patients with bubonic plague will die if untreated, whereas nearly all those with septicaemic plague will die. If a person is infected and recovers they usually become immune. Every case of bubonic plague may become septicaemic, but the term is usually kept for patients in whom the disease is overwhelming from the outset and who die with little or no evidence of any bubo. The onset may be deceivingly mild and yet the patient may be dead in 3 days, but mostly the patient is rapidly prostrated and shocked and all the serious signs of bubonic plague are more acute and severe (Christie, 1969).

Pneumonic plague is different in that it is a respiratory infection acquired by direct contact with another pneumonic patient. The plague bacillus always comes initially in a flea from a rat, and the bitten patient develops bubonic or septicaemic plague: in about 5% of such cases, before the patient dies, *Y. pestis* reaches the lungs and, if the patient lives long enough, he or she coughs out the organism in the sputum (which may be bloody) and contacts inhale it and get pneumonic plague. From then on, one patient infects another with pneumonic plague by direct inhalation of *the bacillus without the intervention of a flea*: the onset is abrupt and *severe with rapid prostration*; breathing is shallow, distressed and very rapid; sputum is watery, teeming with yersiniae and soon mixed with blood. The face is dusky, the temperature high, and the patient may bleed. There is always pulmonary oedema and often a pleural effusion, but the symptoms are those of respiratory distress and shock. The patient dies about the third, never later than the sixth, day and without modern medical treatment, pneumonic plague is invariably fatal. This distinction from the mortality in untreated patients suffering from bubonic plague is of importance when we attempt to determine the possible role of *Y. pestis* in the plagues in earlier centuries.

The bubo is the characteristic symptom of bubonic plague; it is a lump formed by a swollen lymph node and is of variable size. It is found most commonly in the groin but its location depends on where the flea bites which, in turn, depends on how the victim is clothed. Christie (1969) summarised this as follows:

> An Indonesian peasant . . . wears only pants and a hat: the flea can bite him anywhere, with least effort on the legs. A Libyan farmer has boots and breeches and flouncing robes, and your flea needs all its wit to get to his skin: the arm or the neck may be easier than the leg. When a patient gets plague from skinning some animal the infection will be through his hands and the bubo in his axilla: if he eats the flesh, *Y. pestis* may settle on his tonsils and the bubo be in his neck.

The bubo appears early in the illness, on the first or second day; in the septicaemic type there may be no bubo at all or it may be so small as not to be noticed and have no time to enlarge before the patient dies. Usually, the bubo is very painful and tender and, in patients who live long enough or survive, it breaks down and discharges pus.

There is wide variation in the onset and course of bubonic plague which may be mild enough to be overlooked (termed pestis minor) or it may be overwhelming. The incubation period is typically 2–6 days after exposure, i.e. very short when compared with haemorraghic plague (see section 5.5). It is important to record accurately the details of the course of the disease so that we may compare these with such accounts as we have of the plagues of earlier centuries. Once again, we can rely on Christie (1969), who wrote from a wealth of personal experience of bubonic plague as follows:

> Typically the onset is sudden with chills and rigors and rise of temperature to 102 °F or 103 °F (38.8 °C to 39.4 °C). The patient has a severe, splitting headache

and often pains in the limbs, the back and the abdomen. He may curl away from the light or, as the painful bubo develops, take up some attitude in bed that relieves the pressure on the painful swelling. He becomes confused, restless, irritable or apathetic, his speech slurred as if drunken, he is unable to sleep, sometimes wild or maniacal. He may vomit. He is usually constipated, but diarrhoea can be an ominous symptom. His eyes are suffused, occasionally blood-shot, and, as the disease advances, he may bleed into his skin, or internally into his stomach or intestine, or from his kidney . . . Rarely he is jaundiced. Within a day or two he is prostrate with all the symptoms of shock. His temperature may come down and he appears better on the third day or so, but this is deceiving: he is worse the next day and dead soon after. Most patients died between the third and sixth day: if they are alive on the seventh day they may struggle through to recovery. In the last stages the patient may have a cough and other signs of respiratory embarrassment, but mostly they die without obvious signs or symptoms of pneumonia.

Most importantly for our purposes, he continued,

The picture is non-specific: *it might be any severe septicaemic illness, or typhus, typhoid, malaria and the like* [our italics]. The only distinguishing feature is the bubo.

3.13 The significance of pneumonic plague

An understanding of the etiology and epidemiology of pneumonic plague is of particular importance when we try to determine the identity of the causative agent in the Black Death and subsequent plagues in Europe. It is evident from the foregoing that bubonic plague could not suddenly jump over 100 miles in a cold climate in a vast metapopulation where there were no resistant species of rodents but only the sedentary black rat and, consequently, many workers have resorted to the person-to-person transmission of pneumonic plague (Morris, 1977; Gottfried, 1978) as an explanation of its spread. They attempt to counter Shrewsbury's (1970) perfectly reasonable statement that pneumonic plague cannot occur in the absence of the bubonic form and that it cannot persist as an independent form of plague.

Their evidence is poor. Gottfried (1978) simply says 'The seasonal patterns of the epidemics of 1433–1435, 1438–1439 and 1479–1480 all hint at the presence of pneumonic plague. Further, William of Worcestre tells us his nephew died two days after contracting the plague in January, 1479–1480, a characteristic sign of pneumonic plague. Although pneumonic plague was not nearly as common as bubonic plague, there is no evidence, either medical or historical, to deny its existence altogether'. Morris (1977) also claimed that there was a high percentage of pneumonic cases in the Great Pestilence of 1348–50 and that, in many places, the plague first appeared in its pneumonic form. He added that the Manchurian epidemics of 1910–11 and 1920–21 were exclusively pneumonic. These references to plague in Manchuria in the 20th century come from the work of Wu (1926; Wu *et al.*, 1936), who affirmed that 'an intimate relationship exists between rodent plague and human bubonic affections are [*sic*] due, in an overwhelming majority of instances, to transmission of the virus [*sic*] from the rodents through their fleas'. Pneumonic plagues almost always originate in human cases with secondary lung involvement: 'evidence tends to confirm experiences in areas like South-East Russia, Transbaikalia and Manchuria that usually primary pneumonic plague is traceable to bubonic cases with well marked secondary lung involvement' (Wu *et al.*, 1936).

Pneumonic plague was largely absent from the southern Chinese provinces and was usually less than 4% of the total cases. It was more conspicuous in the northern provinces (12%), although Wu believes that this is principally because of seasonal influences, with low temperatures at the time of the outbreaks bringing the patients into close contact and thereby increasing the chances of respiratory infection from any secondary lung infection. Nevertheless, in addition to the Manchurian pneumonic epidemic of 1910–11, repeated plague outbreaks, often pneumonic in character, were reported from the Narinsk district to the southwest of Issyk-kul Lake and the Prjevalsk (Karakol) district to the east of the same lake but Wu *et al.* (1936) say 'These are evidently also due to *epizootics among tarabagan-like marmots*' [our italics]. Finally, a pneumonic outbreak took place in 1929–30 in the Alma-Atinsk (Verni) district situated in the northeast of Issyk-kul Lake, 'said to have been due to *an epizootic among the local hares*' [our italics].

According to Wu, many authors take for granted that bubonic cases with secondary lung involvement are the source of pneumonic outbreaks that occur if meteorological and social factors are favourable, especially cold weather, which creates unhygienic conditions in tightly shut and overcrowded houses. Most infections in the Manchurian epidemics occurred indoors, especially at night-time, when the workers returned to their comparatively warm but crowded shelters to rest and sleep. But Wu pointed out that pneumonic plague epidemics also occur in summer and in countries with a warm climate; for example, the high incidence of lung pestilence in Upper Egypt during the hot and dry plague season.

Where an outbreak of bubonic plague has settled into the pneumonic form, transmission of *Yersinia* will be largely person-to-person and the epidemic will probably follow Reed and Frost dynamics (section 2.5). There is general agreement that the time from infection to death is short, probably about 5 days, so that the infectious period is even shorter. Consequently, the Reed and Frost equations predict that the outbreak will be short-lived (see Fig. 2.4) unless it is restarted from the rats and their fleas. This is quite unlike the dynamics of haemorrhagic plague, which has a very long incubation period and the epidemics are consequently of extended duration. We conclude that pneumonic infection probably markedly exacerbated the mortality of many outbreaks of bubonic plague, as in Marseilles in 1720–22 (Chapter 12), but its main effect was probably within the household and family and such neighbours that came to visit. It is impossible that a mortally sick person who was rapidly prostrated when infected with the pneumonic form and who was only 3 days away from death could have spread the disease over long distances either by land or sea as occurred during the Black Death and in many of the plague outbreaks thereafter.

3.14 Pathology

When *Y. pestis* is injected into humans by a flea most of the bacteria are phagocytosed and killed by the polymorphonuclear leucocytes, which enter the infection site in large numbers. However, a few bacilli are taken up by tissue macrophages that are unable to kill them but provide a protected environment for the organisms to resynthesize their capsular and other virulence antigens. The re-encapsulated organisms kill the macrophage and are released into the extracellular environment, where they resist phagocytosis by the polymorphs. The resulting infection spreads quickly to the draining lymph nodes, which become hot, swollen and tender, with haemorrhagic necrosis, and there is usually a gelatinous oedema in the surrounding tissue giving rise to the bubo. The infection spreads through the lymphatic vessels and invades the blood stream to cause lesions in the spleen, liver, kidney and other organs of the body. In pneumonic plague, the lymphatics of the lung are rapidly invaded by the baccilli and the condition at autopsy is one of haemorrhagic pneumonia. Culture of almost any organ will be positive for *Y. pestis.*

An autopsy of a seaman who died of authentic bubonic plague in 1900 was reported by Savage & Fitzgerald (1900) and the important points are given below. This autopsy report can be directly compared with the few accounts available of the examination of cadavers who died from haemorrhagic plague (see Chapter 8).

> Hypostatic congestion was well marked. All traces of rash had disappeared. There was diffuse ecchymosis over the face, neck, and shoulders. A sanious froth oozed freely from the mouth. The glands in the right groin were visibly enlarged and it was difficult to separate the individual glands owing to the great infiltration around them; in some places this was haemorrhagic. Sections of one of the enlarged right inguinal glands showed great vascular engorgement with numerous blood extravasation, both in the gland and especially in the periglandular tissues. Throughout the gland, but especially in patches, there was well-marked infiltration with small cells, apparently leucocytes. Cultures from the largest glands showed a mixed growth of *Yersinia pestis* and other organisms, but bacilli were absent from the left inguinal and femoral glands.
>
> No fluid in the pericardium; heart collapsed, flabby, and empty; heart muscle pale and very much softened; sections showed degenerative changes in the muscle fibres of the wall. Both lungs were greatly congested, emphysematous in patches which, on section, were nearly black and dripped blood.
>
> The spleen was a dull purple colour and very considerably enlarged, its under surface hyperaemic and bloodstained, its veins engorged and substance softened. A mixed growth of *Y. pestis* and other bacilli was found in cultures of spleen and kidney. The liver was not markedly enlarged; the entire surface was studded with a number of whitish, irregularly shaped bodies, varying in size from a pea to a bean, which on section showed a soft, friable, caseous pinkish-white material. They showed on microscopic examination a varying amount of degenerated liver substance and a number of small, more deeply staining masses which consisted almost entirely of *Y. pestis*.

The important point to note here is that there are only limited signs of necrosis in this autopsy report, in stark contrast with the post-mortem examinations of those who died from haemorrhagic plague.

The major defence against *Y. pestis* infection is the development of specific anti-envelope (F1) antibodies that serve as opsonins for the virulent organisms, allowing their rapid phagocytosis and destruction while still within the initial infectious locus (Fig. 3.4). Although the V and W antigens are associated with virulence, a number of avirulent strains may also possess them, and some individuals possessing high anti-VW antibody titres will nevertheless undergo a second attack of this disease. Thus the

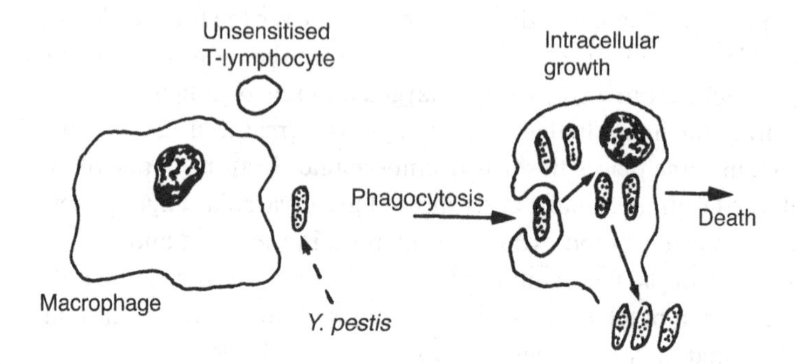

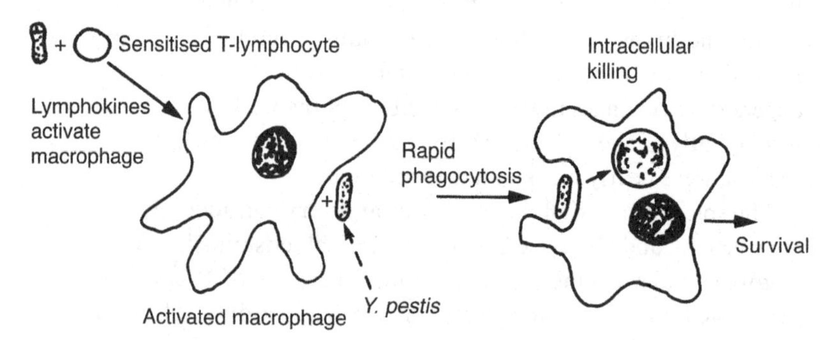

Fig. 3.4. Cell-mediated protection against *Yersinia pestis* in unsensitised (*top*) and sensitised (*bottom*) T-lymphocytes.

immune mechanism against this disease is complex and involves a combination of humoral and cellular factors. The convalescent host is solidly immune (at least for a time) to virulent rechallenge, the inoculum being eliminated as though the organisms were completely avirulent. Killed *Y. pestis* vaccines, especially when given with a suitable adjuvant, induce some measure of host protection, although this will be less effective than that afforded by the live infection.

3.15 Case studies of the dynamics and epidemiology of bubonic plague in India in the 20th century

3.15.1 Mixed epizootics in Bombay City, 1905–6

We are all indebted to the painstaking and pioneer work of the Plague Commission in India at the end of the 19th and the start of the 20th

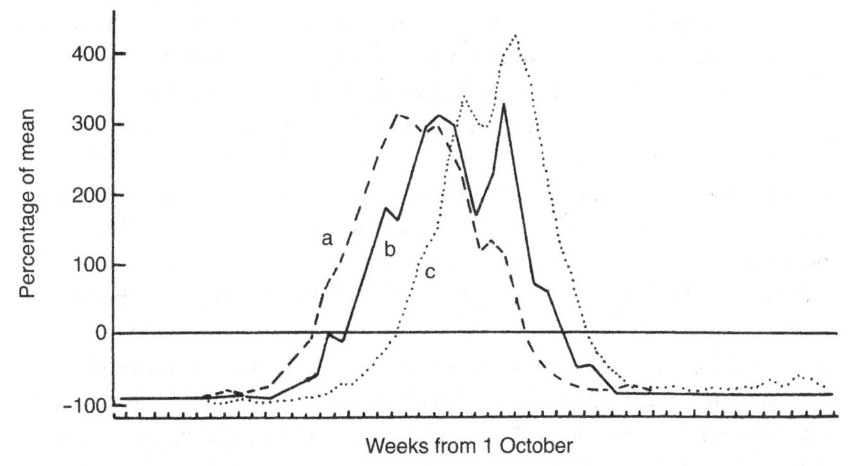

Fig. 3.5. An epidemic of bubonic plague at Bombay, India, from October 1905 to September 1906. (a) Plague-infected *Rattus decumanus* (epizootic). (b) Plague-infected *Rattus rattus* (peridomestic). (c) Human plague deaths. All results expressed as a percentage above and below the mean. The sequence of the infections can be seen clearly. From the Plague Commission in India (1907a).

centuries; the elucidation of the epidemiology of bubonic plague stemmed directly from their work. In Volume 7 of their reports (1907a,b) they describe the interrelated rat epizootics, *Rattus decumanus* and *Rattus rattus*, that existed in Bombay City at that time. They found that there was little difference in the liability to infection in males and females, that the greatest incidence of infection was in persons between 11 and 20 years of age and that Hindus and Muslims suffer most severely from the disease.

Figure 3.5 shows the relative seasonal numbers of plague-infected *R. decumanus* and *R. rattus* and human plague deaths. It is obvious that the *R. decumanus* epizootic curve precedes that of *R. rattus* by a mean interval of 10 days, which, in turn, was followed some 14 days later by the development of the human plague. The plague season in Bombay lasted from the end of December to the end of May, although plague-infected rats of both species and deaths of humans from plague were recorded in every month of the year. During the 'off-plague' period, infected *R. decumanus* were, on average, some five times more common than infected *R. rattus*. Overall, the incidence of plague was twice as great in *R. decumanus* as in *R. rattus*, although the two rat species are equally susceptible to infection. It was noted that *R. decumanus* harboured twice as many fleas as *R. rattus*.

Rattus rattus is essentially a house rat that is common in native houses and has a wide distribution in Bombay Island. *Rattus decumanus* is typically a wandering rat, but does occur in the lower floors of inhabited buildings and is confined to Bombay City; it does not occur in the outlying villages of the island because of the absence of gullies and drains there.

The report concluded that the persistence of the plague was associated mainly with the *R. decumanus* epizootic and that this species was directly responsible for causing the epizootic in *R. rattus*. The epidemic of bubonic plague in humans, in turn, was directly attributable to the *R. rattus* epizootic.

We conclude that this study illustrates well the dynamics of a bubonic plague epidemic in humans. As always, it was completely dependent on a pre-established epizootic in local rodents, and the maintenance of the enzootic during non-epidemic periods was critically dependent on a balance between susceptible and resistant rats and their alternative rodent vectors.

3.15.2 Epidemiology of bubonic plague in India in the 20th century

The first authenticated plague epidemics in modern times in India occurred in 1895–96 and this pandemic reached its peak in 1907. Since then there was a continuous overall decline in human mortality, on which have been superimposed clear epidemics, as shown in Fig. 3.6 (Seal, 1969). Note the continuing persistence of endemic bubonic plague over a period of more than 40 years. Sharif (1951) in his study of the endemicity of plague in India suggested that the infection was entrenched in three groups of foci in northern, central and southern India (Fig. 3.7):

(i) The northern foci consist of three endemic centres at the foot of the Himalayas, perhaps forming part of a big sub-Himalayan focus. These centres were considered to be responsible for plague outbreaks in East Punjab, Uttar Pradesh and districts of Bihar north of the River Ganges.
(ii) The focus in central India (Madhya Pradesh) comprises the watersheds of the Vindhya, Bhanrer and Maikal ranges and the Mahadeo hills.
(iii) The three southern foci are situated in: (a) the watersheds of the Western Ghats of Bombay and Mysore States, (b) the watersheds located in the districts of Salem, Coimbatore, Nilgiri and Madura in Madras State, and (c) the hilly regions of Hyderabad State.

The endemic centres in southern India may have been established after Bombay became infected in 1896 (see section 3.15.1). On the other hand, it

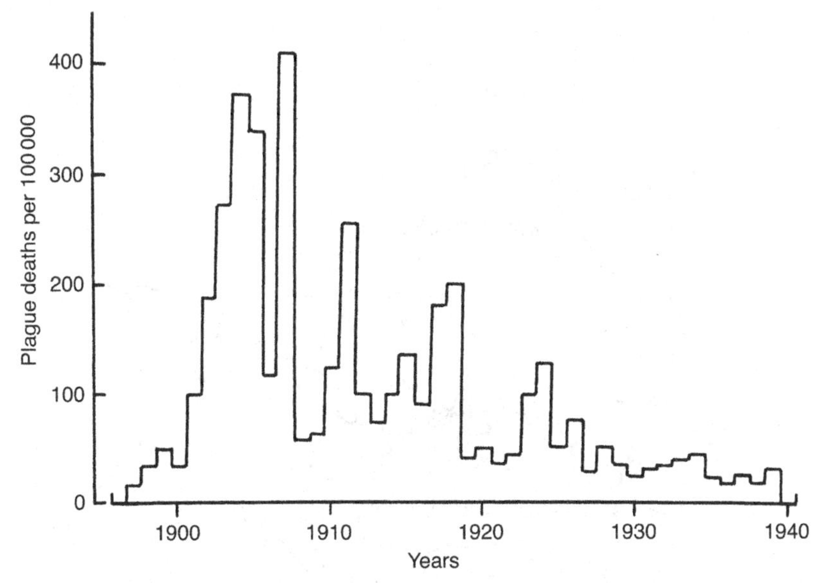

Fig. 3.6. Bubonic plague mortality in India per 100 000 population, 1896–1939. From Seal (1969).

is possible that the endemic foci in the Himalayas were of long standing, plague infection being known in the Kumaon and Gharwal districts since 1823. The infection might have persisted there as a relic of a great pestilence of the 17th century and might have been responsible for occasional plague outbreaks until 1877. Thereafter, it remained latent for some time, to become active again early in the 20th century. In central India and Madhya Pradesh, plague started after the infection of Bombay in 1896.

Plague is essentially bubonic in India; true septicaemic plague is rare, although some bubonic types have septicaemic manifestations. Primary pneumonic plague is also rare, and this is an important point; it generally occurs after lung involvement in a bubonic-septicaemic case leading to plague pneumonia, and subsequent contacts of such cases may develop primary pneumonic plague. Such outbreaks have been known to occur in India (Seal, 1969) but they have generally remained confined to a single family or a few families only. The incidence of pneumonic plague in India is generally below 1% and has never exceeded 3% in any year since 1895.

Plague is both urban and rural in India, the latter predominating. It appears that plague has failed to gain a foothold in many of the towns in India, perhaps because of unsuitable climatic conditions and the lack of any efficient flea vector (as in Madras and Assam). Regular heavy annual floods may also be responsible for keeping certain states (e.g. Bengal) free

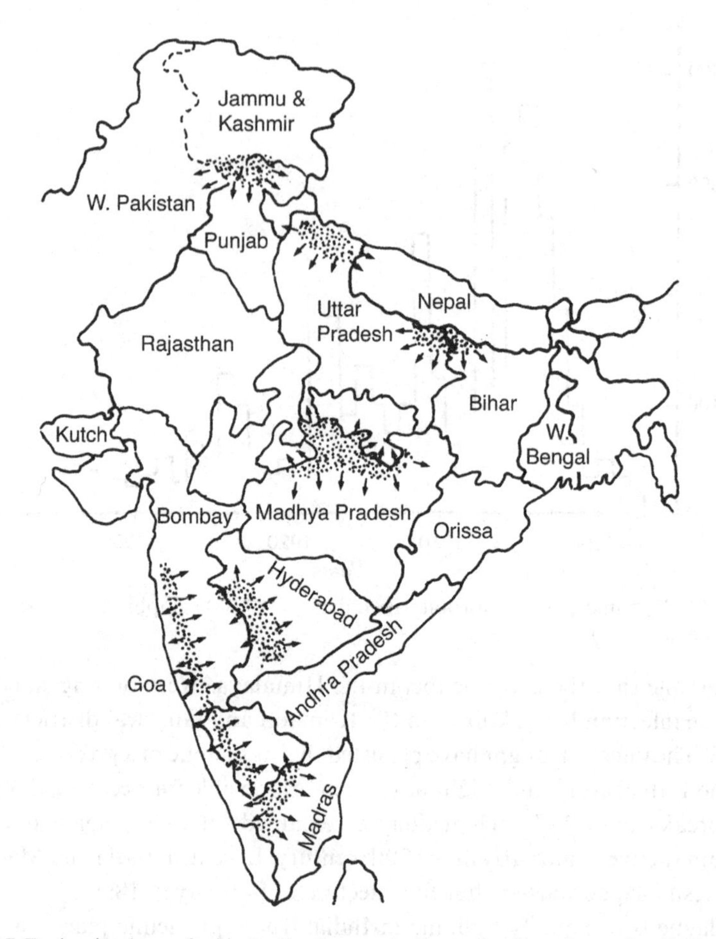

Fig. 3.7. Endemic plague foci in India. Arrows indicate the directions of radiation of the plague. After Sharif (1951).

from plague. Another factor which may play an important part is the distribution and proportion of various types of rodent. Given a suitable flea vector, a large proportion of *R. rattus* will make for easier and quicker spread of plague among humans than a similar proportion of other rodents. On the other hand, replacement of one rodent by others, as in Bombay, may disturb the balance and plague may recrudesce or be imported with consequent severe outbreaks. However, given a suitable climate, a sufficiently large rodent population, effective vectors and the plague bacillus, *Y. pestis*, the infection may often become firmly entrenched among the rats of towns and persist there for many years (Seal, 1969).

The optimal conditions for bubonic plague in India are 20–25 °C and a

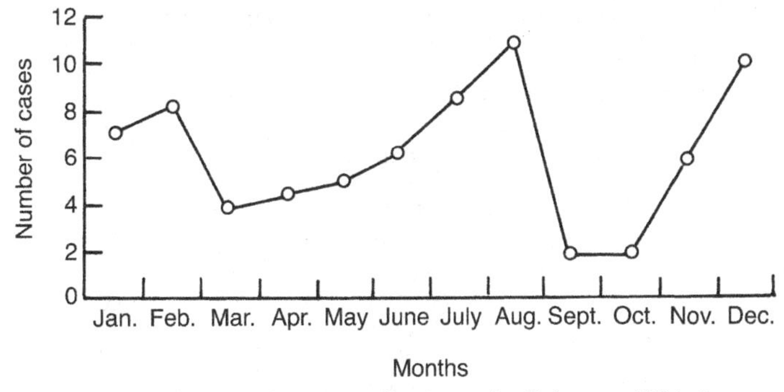

Fig. 3.8. Seasonal incidence of pneumonic plague in Calcutta, 1904–7.

relative humidity above 60%. Seasonal conditions in India influence the numerical importance and longevity of rat fleas and the multiplication of *Y. pestis* in both rats and fleas. These optimal conditions impose severe constraints on the dynamics of bubonic plague in India, resulting in marked differences in the epidemiology of the disease in different areas of this vast subcontinental metapopulation.

At high latitudes in northern India, the atmospheric temperature attains the critical level only during the late summer and early autumn, so that it is at this season of the year that a plague epidemic is liable to occur; a decrease in latitude is therefore associated with earlier occurrence. In the subtropical regions, on the other hand, where either the temperature or humidity are unfavourable during the summer, the plague epidemics have a vernal periodicity, as in northern India.

Pneumonic cases (only 3% of the total) also have a cyclic, seasonal incidence (Fig. 3.8) and are probably also driven by climatic factors; in turn, they can contribute to the determination of the seasonal pattern of the epidemics because of their influence on the transmission of infection (Seal, 1969).

We see that the complex plague dynamics in India are driven by the enzootic/epizootic and endemic/epidemic balances and are strongly dependent both on the biology and resistance of the peridomestic rats and the reservoir population of other rodents and on other external factors, such as temperature and humidity. These multiple factors are not equally operative in all places.

3.15.3 Effects of population size

Bubonic plague in India was primarily a disease of the smaller towns: the intensity of an outbreak in 1897–98 was inversely proportional to the size

Table 3.1. *Population-dependent death rates during outbreaks of bubonic plague in India, 1897–98*

Place	Population	Death rate per 1000
Bombay	806 144	20.1
Poona	161 696	31.2
Karachi	97 009	24.1
Sholapur	61 564	35.0
Kale	4431	104.9
Supne	2068	102.5
Ibrampur	1692	360.5

Sources: Hankin (1905) and Twigg (1984).

of the community, with the maximum mortality in the villages rather than towns (Hankin, 1905); this is clearly shown in Table 3.1. Furthermore, although the plague spread readily from village to village, Hankin observed that it did not appear to be carried great distances in the epidemic form. This is in complete contrast to the behaviour of haemorrhagic plague in Europe (see section 13.9).

3.16 Conclusions: key points about the biology of bubonic plague

The detailed studies of the great bubonic plague in India that began at the end of the 19th century give a very clear picture of the biology of this disease. *Yersinia pestis* is a disease of rodents that sometimes spreads accidentally to humans. Endemic bubonic plague grumbled on for at least 60 years in Asia during which time there was a dynamic balance between its susceptible and non-susceptible rodent hosts. On occasions, it spread to specific locations, usually villages, causing an epidemic among the peridomestic, susceptible rodents, causing their deaths and the infection of humans via the rat flea. An outbreak of human bubonic plague was often presaged by the appearance of dead rats and this is an important point because, as we shall show, there are no records of rats dying in Europe in the epidemics between 1347 and 1670. The continuation of the epidemic in the peridomestic rodents is dependent on the dynamic balance between the resistant, partially resistant and non-resistant strains of the susceptible rodents. Thus, in Asia, bubonic plague in humans is confined largely to the villages where the inhabitants and their peridomestic rodents are in close contact with the rural, non-susceptible rodents, although, as in Bombay in

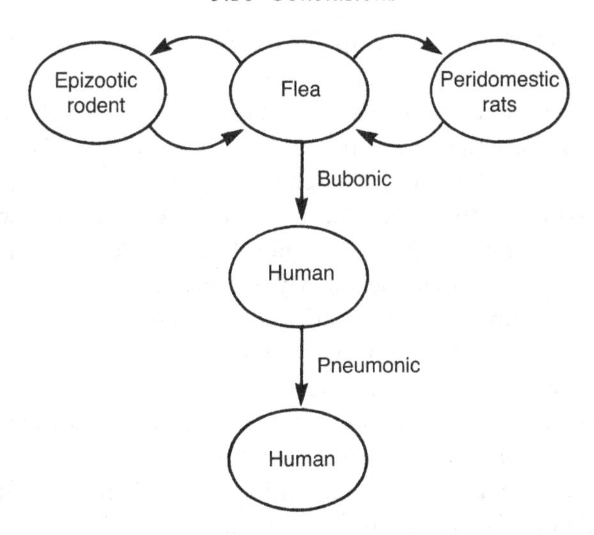

Fig. 3.9. The biology and transmission of *Yersinia pestis*, illustrating the central role of the flea in maintaining the epizootic–enzootic dynamics.

1905–6, when all the conditions are suitable, a major epidemic in humans can erupt in a city.

The mortality from bubonic plague in India is impressive, with 12 million people dying, but it must be remembered that this occurred over 60 years over a vast area of the subcontinent with an enormous indigenous population.

The biology of *Y. pestis* is summarised in Fig. 3.9; it illustrates the central role of the flea in maintaining the epizootic–enzootic dynamics by transmitting the bacterium between the different rodent hosts and to humans and it emphasises the point that the disease does not spread to humans unless it is already pre-established in the local rodent population. This is another critical point: *Y. pestis* could not have been the causative agent during the Black Death, which spread rapidly from the shores of the Mediterranean almost to the Arctic, unless there was an indigenous European population of rodents in which bubonic plague was already well established. Figure 3.9 may be compared with the simple dynamics of a 'normal' infectious disease shown in Fig. 2.3A in which the epidemics are short lasting and can be described by the Reed and Frost equations (see section 2.5). Bubonic plague in humans, with its more complex biology, does not follow Reed and Frost dynamics.

Endemic bubonic plague did not spread rapidly in the 19th century, but the development of much more rapid sea travel via steamships allowed the dispersion of infected rats to the warmer parts of the globe and, where conditions were suitable, *Y. pestis* infected the local rodents. Dispersal was assisted by the development of the railways but, even so, the spread of bubonic plague was slow and, although it was eventually quite widespread and grumbled on in the local rodents, there were relatively few human deaths in South Africa and America. Again, bubonic plague did not follow Reed and Frost dynamics in these newly emerged infestations.

The spread of bubonic plague is critically dependent on climatic conditions, particularly on temperature, which is the major environmental factor governing the biology of both the rodents and the flea. It is for this reason that endemic bubonic plague does not spread away from warmer regions as, for example, the Mediterranean coast and southern Europe (see Fig. 3.3). Outbreaks in ports in more northern latitudes did not spread far and had to be reinforced by fresh introductions of rats and fleas.

During the 300-year period from the Black Death to the Great Plague of London, the only species of rat in Britain that could have carried bubonic plague was the black rat, an animal of warm buildings that was confined largely to ports. Steamships in the 20th century have brought bubonic plague to England via infected seamen and rats on several occasions and the disease has sometimes been transmitted to the black rats of the ports but plague has never spread and there have been few fatalities. It is impossible that bubonic plague could have been established as an epizootic and have spread rapidly and widely inland, sometimes in winter months, via the agency of black rats and their fleas.

Many writers, aware that the facts concerning the rapid spread of plagues in historical times were clearly at variance with the etiology of bubonic plague, have fallen back on invoking interhuman transmission via pneumonic plague (see section 3.13). Christie (1969) showed clearly that this was not so: bubonic plague always begins with an infection in rodents and only a small percentage of the cases in humans develop the pneumonic form in the terminal stages of the disease. Although this would invariably have been lethal and have exacerbated the spread of the infection and the death toll within the household, such grievously ill patients would have been unlikely to be able to move any distance to spread plague to other villages.

4
The Great Pestilence

The pandemic that struck Europe in 1347 was probably the most serious outbreak of a lethal infectious disease in history in terms of the percentage of the population that died and the speed with which it spread. This outburst involved the whole of continental Europe, the Channel Islands, the British Isles, Iceland and Greenland (Shrewsbury, 1970; Kohn, 1995).

This pandemic was called the Great Mortality or the Great Pestilence by contemporary writers. It was named the Black Death in English historical writings in 1823 and was introduced into medical literature as such in 1833 because of the black blotches, caused by subcutaneous haemorrhages, that appeared on the skin of the diseased humans near the time of death (Kohn, 1995). Shrewsbury (1970) stated that the latter name was inappropriate because, although the cadaver of a victim of the plague may exhibit a purplish discoloration, the corpse did not turn black. However, since the pseudonym 'The Black Death' is so firmly entrenched in historical writing we shall continue to use it.

Although contemporaneous written records are scattered and scarce, the Black Death has been discussed exhaustively (see Creighton, 1894; Nohl, 1926; Hirshleifer, 1966; Deaux, 1969; Ziegler, 1969; Shrewsbury, 1970; Gottfried, 1983; Twigg, 1984; Carmichael, 1986, 1997; Horrox, 1994; Kohn, 1995; Ormrod & Lindley, 1996), covering, in particular its origins, the pattern of spread, the number of deaths from the disease and the role of the rat as a carrier. These authors almost all assume without questioning that the Black Death was an epidemic of bubonic plague. Many bizarre side-issues are covered in these writings: (i) the religious response and the attitude of the Church, where the plague was considered as Divine punishment (Nohl, 1926; Horrox, 1994); (ii) pseudo-scientific contemporaneous explanations of the plague, such as astrological causes and the dangers of corrupted air and earthquakes (Horrox, 1994); (iii) the persecution of the

Jews, who were thought either to have deliberately spread the plague or to have polluted society and brought on God's vengeance (Nohl, 1926; Horrox, 1994); (iv) the erotic (Nohl, 1926) and diabolic elements in the plague, including devil worship following the moral collapse of the Church (Nohl, 1926); (v) the effect of the Black Death on the Church and religion (Bolton, 1996); (vi) the politics of pestilence – government in England after the Black Death (Ormrod, 1996); (viii) the Black Death and English art (Lindley, 1996).

4.1 Arrival of the Black Death in Europe

Ziegler (1969) concluded that the Great Pestilence began in central Asia and Twigg (1984) recorded that the Russian archaeologist Chwolson had shown unusually high death rates in 1338 and 1339 near Lake Issyk-Koul, Semiriechinsk in central Asia. He continued

> Stewart (1928) in his account of the spread of Nestorian Christianity has drawn attention to two old cemeteries, fifty-five kilometres [34 miles] apart, which contained tombstones indicating they belonged to Nestorian Christians in Semiriechinsk. One of these graveyards contained 611 stones and for the years A.D. 1338–9 three inscriptions stated that the persons buried had died of plague. He further pointed out that during those two years the number of inscriptions was exceptionally large. He also was of the opinion that plague originated in eastern Asia at about that time and that it spread rapidly to Asia Minor, North Africa and Europe and reached the Crimea in A.D. 1346. Pollitzer (1954) has also referred to this large number of burials in the Nestorian graveyards and says quite categorically that it is therefore certain that plague was in evidence in central Asia a few years before the infection of the Crimean ports in 1346 and that from there it was carried by ship to Europe.

Vernadsky (1953) stated that 85 000 people died in the Crimea and the plague was probably carried thence to Constantinople, where the epidemic raged in 1347. The Black Death was then taken to the sea-coasts and the eastern Mediterranean, including Greece and Egypt, became infected. Thus, although the introduction of the epidemic in Europe is said to be from the Crimea, via the well-established sea trade routes, Twigg (1984) suggested that there may have been parallel introductions from Syria, since ships plied from there to the Italian and French ports. All the accounts describe the plague spreading rapidly almost as soon as docking had taken place.

4.2 The plague in Sicily

Nohl (1926) provided a translation of the account given by Michael of Piazza, a Franciscan friar, of the events that followed the arrival of the plague in Sicily. He wrote his history some 10 years later (for *notes* see p. 85).

At the beginning of October . . . 1347, twelve Genoese galleys . . . entered the harbour of Messina. In their bones they bore so virulent a disease that anyone who only spoke to them was seized by a mortal illness and in no manner could evade death. The infection spread to everyone who had any intercourse with the diseased [*note 1*]. Those infected felt themselves penetrated by a pain throughout their whole bodies and, so to say, undermined. Then there developed on their thighs or on their upper arms a boil about the size of a lentil which the people called 'burn boil' (*antrachi*) [*note 3*]. This infected the whole body, and penetrated it so that the patient violently vomited blood [*note 5*]. This vomiting of blood continued without intermission for three days, there being no means of healing it, and then the patient expired [*note 6*]. But not only all those who had intercourse with them died, but also those who had touched or used any of their things [*note 1*]. When the inhabitants of Messina discovered that this sudden death emanated from the Genoese ships they hurriedly expulsed them from their harbour and town. But the evil remained with them and caused a fearful outbreak of death. Soon men hated each other so much that, if a son was attacked by the disease, his father would not tend him. If, in spite of all, he dared to approach him, he was immediately infected and could by no means escape death, but was bound to expire within three days [*note 6*]. Nor was this all: all those belonging to him, dwelling in the same house with him, even the cats and other domestic animals [*note 2*], followed him in death. As the number of deaths increased in Messina many desired to confess their sins to the priests and to draw up their last will and testament. But ecclesiastics, lawyers and attorneys refused to enter the houses of the diseased. But if one or the other had set foot in such a house to draw up a will or for any other purpose, he was hopelessly abandoned to sudden death [*notes 1 and 6*]. Minor friars and Dominicans and members of other orders who heard the confessions of the dying were themselves immediately overcome by death, so that some even remained in the rooms of the dying [*note 6*]. Soon the corpses were lying forsaken in the houses. No ecclesiastic, no son, no father and no relation dared to enter, but they paid hired servants with high wages to bury the dead. But the houses of the deceased remained open with all their valuables, with gold and jewels; anyone who chose to enter met with no impediment, for the plague raged with such vehemence that soon there was a shortage of servants and finally none at all. When the catastrophe had reached its climax the Messinians resolved to emigrate. One portion of them settled in the vineyards and fields, but a larger portion sought refuge in the town of Catania . . . But the plague raged with greater vehemence than before. Flight was no longer of avail . . . Many of the fleeing fell down by the roadside and dragged themselves into the fields and bushes to expire. Those who reached Catania breathed their last in the hospitals there. The terrified citizens demanded from the Patriarch prohibition on pain of ecclesiastical ban, of burying fugitives from Messina within the town, and so they were all thrown into deep trenches outside the walls. The population of Catania was so godless and timid that no one among them would have intercourse or speak to the fugitives, but each hastily fled on their approach . . . Thus the people of Messina dispersed over the whole island of Sicily and came also to Syracuse and with them the disease, so that in Syracuse innumerable people died. Sciacca, Trapani, Girgenit and Messane, thus, were also infected by the plague, but particularly Trapani, which was completely depopulated [*note 1*]. The town of Catania lost all its inhabitants [*note 1*] so that it ultimately sank into complete oblivion. Here not only the 'burn blisters' [*note 3*] appeared, but there developed in different parts of the body gland boils in some on the sexual organs, in others on the thighs, in others on the arms, and in others on the neck. At first there were of the size of a hazel-nut [*note 3*] and developed accompanied by violent shivering fits, which soon rendered those attacked so weak that they could no longer stand upright, but were forced to lie in their beds consumed by violent fever [*note 4*] and overcome by great tribulation. Soon the boils grew to the size of a walnut, then to that of a hen's egg or a goose's egg, and they were exceedingly painful, and irritated the body [*note 3*], causing it to vomit blood by vitiating the juices [*note 5*]. The blood rose from the affected lungs to the throat, producing on the whole body a putrefying and ultimately decomposing effect [*note 5*]. The sickness lasted three days, and on the fourth, at the latest, the patient succumbed [*note 6*]. As soon as anyone in Catania was seized with headache and shivering [*note 4*], he knew that he was bound to pass away within the specific time [*note 1*] . . . But the pestilence raged from October 1347 to April 1348 [*note 7*].

This account provides a clear picture both of the characteristics of the Black Death and of the behaviour of the inhabitants of this stricken population in Sicily. It can be supplemented by the account given by the Florentine humanist Giovanni Boccaccio when the plague swept through Florence which corroborates many of the details given by Michael of Piazza (see Schevill, 1928):

Unlike what had been seen in the east, where bleeding from the nose is the fatal prognostic [*note 5*], here there appeared certain tumours in the groin or under the arm-pits, some as big as a small apple, others as an egg; and afterwards purple spots in most parts of the body; in some cases large and but few in number, in others smaller and more numerous – both sorts the usual messengers of death [*note 3*]. To the cure of this malady neither medical knowledge nor the power of drugs was of any effect; . . . whichever the reason, few escaped [*note 1*]; but nearly all died the third day from the first appearance of the symptoms [*note 6*]; some sooner, some later, without any fever [*note 4*] or other accessory symptoms. What gave the more virulence to this plague, was that, by being communicated from the sick to the hale [*note 8*], it spread daily . . . Nor was it caught only by conversing with or coming near the sick [*note 8*], but even by touching their clothes [*note 9*], or anything that they had before touched . . . Such, I say, was the quality of the pestilential matter, as to pass not only from man to man [*note 8*], but, what is more strange, it has been often known, that anything belonging to the infected, if touched by any other creature, would certainly infect and even kill that creature in a short space of time [*note 9*]. One instance of this kind I took particular notice of: the rags of a poor man just dead had been thrown into the street. Two hogs came up, and after rooting amongst the rags and shaking them about in their mouths, in less than an hour they both turned round and died on the spot [*note 2*].

A number of important points concerning the epidemiology of the Black Death emerge from these accounts and are noted in the texts above:

(1) The disease was characterised by a high infectivity and high mortality (perhaps close to 100% in Catania and Trapani).
(2) Domestic animals, it is said, also rapidly died of the disease.
(3) The disease was characterised by the early appearance of a boil on the thighs or upper arms, and in Catania these were also found on the sexual organs and neck. They were initially the size of a hazel nut, which increased to that of a walnut and, finally, to that of a hen's or goose's egg. The disease was also characterised by the burn blisters and by the purple spots.
(4) The patient suffered from a violent fever, although this is not verified by Boccaccio.
(5) The disease was characterised by bleeding and the vomiting of blood and (most importantly) by a generalised necrosis.
(6) These writers inferred that the incubation period was short, but this is not necessarily correct, although it is clear that the progress of the disease was remarkably rapid, with death occurring 3–4 days after the appearance of the symptoms.
(7) The epidemic burnt out (apparently completely) in 7 months, corresponding to the duration of later plague epidemics and indicating from the Reed and Frost model that the infection had a long serial generation time.
(8) The disease was spread by the movement and dispersal of infected people, apparently by person-to-person contact.
(9) Boccaccio averred categorically that the clothing of diseased persons could carry the infection.
(10) These accounts of the infectivity and devastating lethality of the disease suggest that the pestilence came to a virgin population in continental Europe and its offshore islands in which the majority of the inhabitants had no naturally acquired resistance.

The foregoing near-contemporaneous accounts are of value when assessing the nature of the causative organism in the pandemic of 1347–50. Twigg (1984) raised an additional and interesting point concerning the transfer of the infection from the Crimea to Sicily, if it really was via the Genoese galleys. The voyage from the Crimea to Genoa in northern Italy via Sicily was difficult and he considered that it could not have been completed in less than 4 or 5 weeks and would have taken, at worst, almost 3 months. All the accounts of the arrival of the plague on the European mainland or Sicily describe the disease spreading rapidly almost as soon as the galleys docked. Creighton (1894) says that the Italian traders escaped the plague, yet 'the infection appeared in Genoa in its most deadly form a day or two after the arrival of the ship, *although none of those on board were suffering from the plague*' (our italics). He also says 'for we know that there were no cases of plague on board ships, although the very atmosphere or smell of the new arrival seemed sufficient to taint the whole air of Genoa and to carry death to every part of the city within a couple of days'. It is difficult to equate the details of these accounts, particularly the long voyage of symptomless and apparently healthy carriers followed by the immediate spread of the disease when they landed, with *any* infectious disease and certainly not with bubonic plague.

4.3 Spread of the Black Death in continental Europe: a metapopulation pandemic

The Black Death arrived in Sicily in October 1347, brought, according to Michael of Piazza, by 12 Genoese galleys to the port of Messina and apparently within a few days the plague was firmly established in the city. The citizens drove the sailors back to sea and, in so doing, it is believed that the disease was spread around the Mediterranean. But this action was too late and, with the *slightest contact with the sick seeming to guarantee rapid infection*, the population panicked and fled into the vineyards of southern Sicily carrying the plague with them (Ziegler, 1969). Michael of Piazza recounts (above) that once the inhabitants of neighbouring Catania realised the enormity of the disaster, they introduced strict control over immigration, but to no avail; the same pattern of behaviour was to be repeated throughout Europe, although cutting themselves off from their neighbours was rarely efficacious. The plague spread quickly over Sicily, ravaging with particular violence the towns and villages at the western end. It is the rapidity of spread that characterised the Black Death. From Sicily it spread to North Africa via Tunis, Corsica, Sardinia and to Barcelona on the Iberian peninsula, and also to southern Italy, following the main trade routes (Ziegler, 1969).

In addition to Sicily, the other centres for the entry of the plague into southern Europe were Genoa and Venice, where it arrived almost simultaneously in January 1348 (i.e. some 3 months after the galleys docked in Messina) and the disease reached Marseilles on the southern coast of France 1 month later (February 1348) where 56 000 are said to have died. From Marseilles the plague spread westwards to Montpellier, Narbonne and Carcassone, which it reached between February and May 1348. Thence it progressed to Toulouse and Montauban and arrived in Bourdeaux, a seaport on the Atlantic coast of southwest France, in either June or August 1348. Twigg (1984) estimates the *average* rate of spread to be:

Marseilles to Carcassone = 5 miles per day
Carcassone to Bordeaux = between $1\frac{1}{2}$ and 3 miles per day

The plague also moved northwards from Marseilles and was in Avignon in March to April; it arrived in Lyons in early summer, in Paris in June and Burgundy in July and August. Twigg (1984) gives the average rate of movement of the plague as follows:

Marseilles to Avignon = $1\frac{1}{2}$ miles per day
Avignon to Lyons = 2 miles per day
Marseilles to Paris = $2\frac{1}{2}$ miles per day

The relentless spread of the Great Pestilence is illustrated in Fig. 4.1; it was most rapid in the early stages from December 1347 to June 1348, by which time it had spread through Italy and much of France, Spain and the Balkans. It continued its movement northwards, eventually spreading into Norway, Sweden and the Baltic by December 1350.

Certain places, such as Milan, Liège and Nuremberg, escaped the disease. A small area east of Calais and a very large area north of Vienna were also plague free and Twigg (1984) makes the interesting point that the climate in the latter is continental and one would expect the summer temperatures to be high enough to encourage breeding of fleas. On the other hand, the pandemic spread freely with great mortality in Norway (Kohn, 1995) and also across the Alps and Pyrenees.

The mortality throughout Europe during the Great Pestilence was truly terrible. Historical estimates of the mortality from the plague vary from 25% to 75% of the population of Europe, where at least 25 000 000 are estimated to have died between 1347 and 1351. Eighty per cent of the population of Marseilles died; the Pope at Avignon, where half of the population died, consecrated the Rhône to permit corpses to be thrown into it for Christian burial; more than a third of the population of Italy are said to have perished (Kohn, 1995). In Florence, with a population of some 100 000, 100–200 deaths were reported every day through the late spring but this toll suddenly rose to a daily mortality of 400–1000 (Carmichael, 1997).

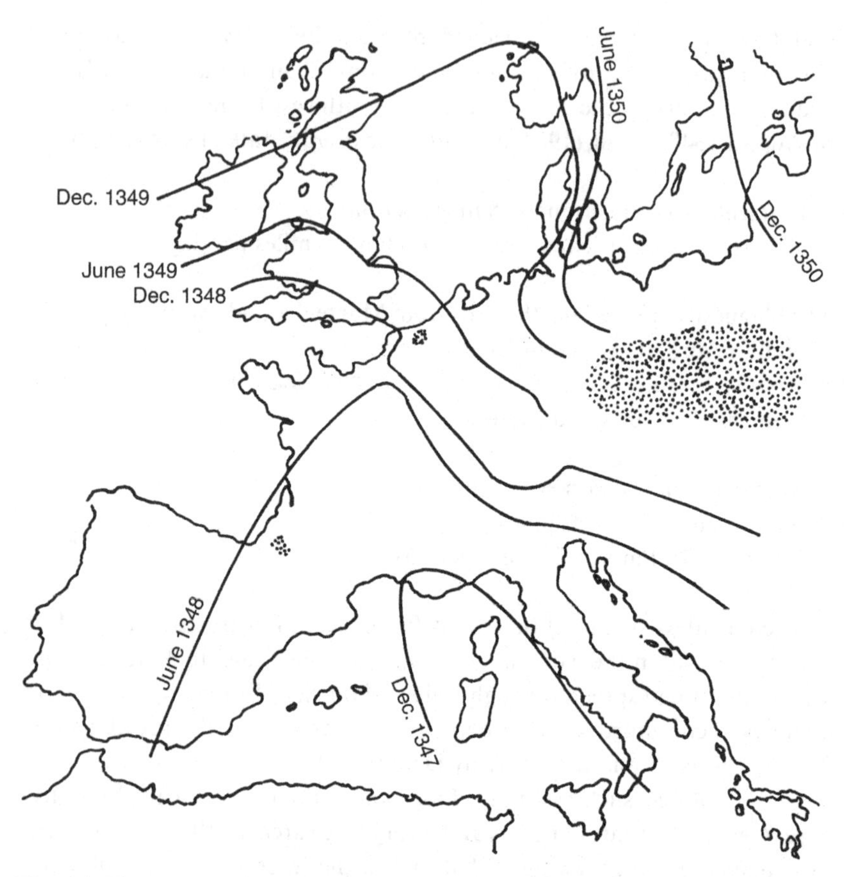

Fig. 4.1. The spread of the Black Death across Europe. Areas denoted by dots escaped the plague. After Twigg (1984), Zeigler (1969) and McNeill (1977).

4.4 The pestilence arrives in England

There seems to be general agreement that the plague entered Britain at Melcombe Regis (now called Weymouth), an important town and port in Dorset on the south coast at that time, although Bristol and Southampton have also been suggested as entry points. Ziegler (1969) quoted from a 14th century chronicle from the Grey Friars at Lynn as follows:

In this year 1348, in Melcombe, in the county of Dorset, a little before the Feast of St John the Baptist, two ships, one of them from Bristol, came alongside. One of the sailors had brought with him from Gascony the seeds of the terrible pestilence and, through him, the men of that town of Melcombe were the first in England to be infected.

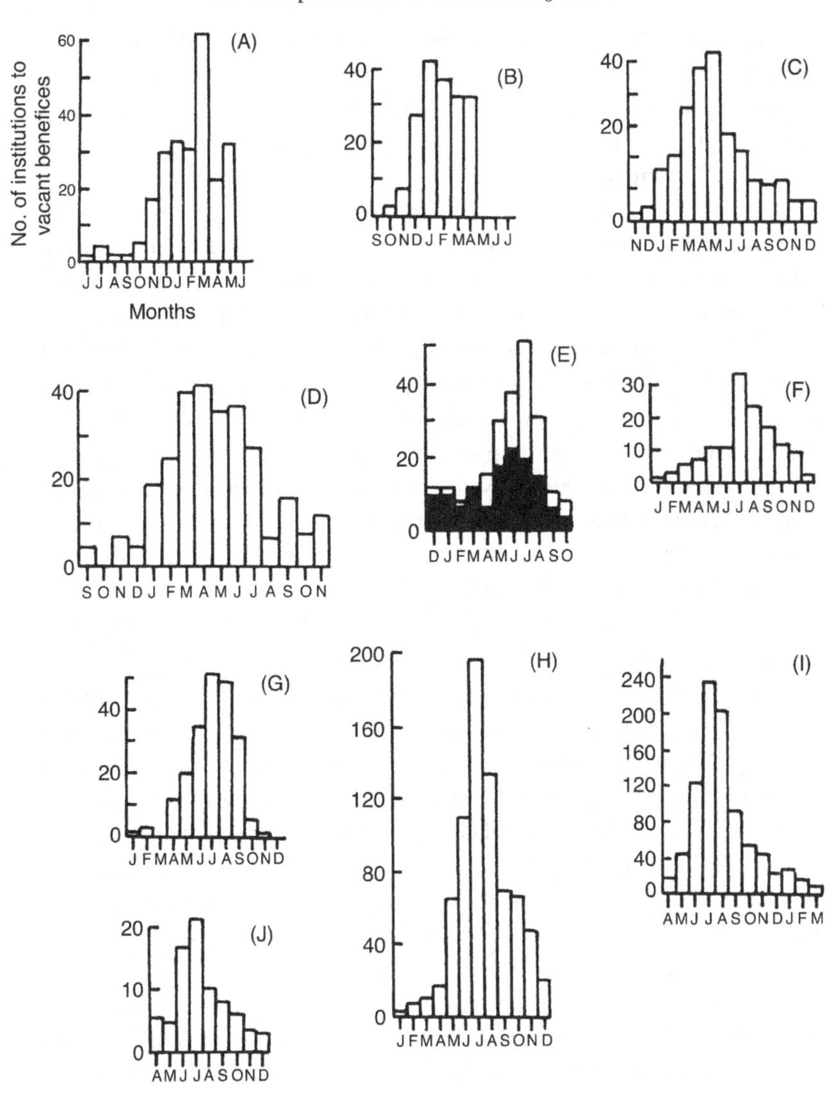

Fig. 4.2. Institutions to vacant benefices in 11 dioceses during the Black Death, illustrating that most epidemics were long-lasting and followed Reed and Frost dynamics. (A) Salisbury. (B) Bath & Wells. (C) Winchester. (D) Exeter. (E) Gloucester (filled area) and Worcester (open area). (F) Hereford. (G) Lichfield. (H) Norwich. (I) Lincoln. (J) Ely. Data from Shrewsbury (1970).

Shrewsbury (1970) suggested that the plague probably came to Melcombe Regis either from Calais, which was then an English possession, or from the Channel Islands, which were in constant communication with

England and were severely visited by the plague in 1348 prior to its appearance in England. We do not know whether this was the only introduction of the pestilence into England or whether there were additional arrivals of infectives at the ports (perhaps on the east coast) during the next 12 months.

Its arrival in 1348 is variously dated as June, July or early August; it spread from Dorset to Bristol and thence, by way of Oxford, to London, which it reached at the end of October or the beginning of November 1348. It continued through the winter but the main force of the epidemic did not begin until the spring of 1349 and Fig. 4.2 shows that in each diocese the epidemic followed Reed and Frost dynamics (section 2.5) and persisted for some 7 months. There is wide variation in the estimates that have been made of the total number dying from the pestilence in London, some even suggesting a total of 100 000 (Kohn, 1995). Ziegler (1969) concluded that a figure between 20 000 and 30 000 deaths from a population of 60 000 to 70 000 would be a good estimate, a percentage mortality that was probably in line with that suffered in other cities.

Ziegler (1969) wrote that, although there were few impressionistic reporters and dispassionate medical records of the Black Death in England, a number of archives contribute to the provision of the fuller picture of the progress of the Black Death through the metapopulation than in any other country. The most complete source available is the ecclesiastical records, and the epidemiology of the Black Death has been traced by studying the vacancies among the beneficed clergy at that time, a practice that Shrewsbury (1970) did not think was completely satisfactory, mainly because his findings were not consistent with the spread of bubonic plague. Among the lay documents, Ziegler (1969) has also studied the manorial Court Rolls and the Account Rolls; taken together they provide a picture of life on the medieval manor and show the incidence of the pestilence in each. However, many fewer of them remain than is the case with ecclesiastical documents.

It was reported that the summer of 1348 was exceptionally wet, which the populace believed to be the cause of the pestilence, although Shrewsbury pointed out that both *Rattus rattus* and *Xenopsylla cheopis* dislike damp conditions. It is said that the Black Death always struck hardest at seaports and coastal districts, and that stretches of marshland and fen acted as barriers against the spread of the epidemic. Sparsely populated hilly districts were usually only slightly affected, although villages close to communication routes were often comparatively severely hit, all indicative of person-to-person transmission of an infectious disease spread by travellers. According to a contemporary observation, the common people bore

Table 4.1. *Maximum percentage mortality rates of the beneficed clergy by county during the Great Pestilence*

County	%	County	%
Bedfordshire	37	Huntingdonshire	34
Buckinghamshire	37	Leicestershire	36
Cambridgeshire	52	Lincolnshire	51
Cheshire	33	Northamptonshire	37
Cornwall	56	Nottinghamshire	36
Derbyshire	58	Oxfordshire	34
Devonshire	51	Shropshire	43
Gloucestershire	47	Somerset	47
Hampshire	49	Staffordshire	34
Herefordshire	48	Surrey	56
Hertfordshire	35	Warwickshire	36
		Worcestershire	48

Data source: Shrewsbury (1970).

the main toll and among them it fell most heavily on the young and vigorous (Shrewsbury, 1970).

4.5 The Great Pestilence moves through the Midlands to the north of England

The Great Pestilence struck the central part of England haphazardly, moving at average rates that have been variously calculated as between 1 and 10 miles per day. The maximum percentage mortality rates of the beneficed clergy by county are given in Table 4.1; those cited by Shrewsbury (1970) varied from 33% (Cheshire) to 58% (Derbyshire). Shrewsbury (1970) quoted from an unpublished thesis by Lunn (1937), who commented that the registers of York, Lincoln and Lichfield show that the mortality of the Black Death was remarkably uniform when its effects were spread over large areas; each of these three registers reveal a 40% death rate among the clergy. Again, Shrewsbury challenged this conclusion because it conflicts with the epidemiology of bubonic plague, where a much lower mortality would be expected.

There is an interesting account (quoted by Ziegler, 1969), of sheep dying from the plague in Leicestershire, that was given by a canon of Leicester Abbey who was an eye witness of the events: 'In this same year a great number of sheep died throughout the whole country, so much so that in one field alone more than five thousand sheep were slain. Their bodies were so corrupted by the plague that neither beast nor bird would touch them.'

Within the Diocese of Lichfield, Lunn (1937) concluded that the epidemic began in April 1349 and was virtually over by October 1349, a duration of 7 months. It moved from south to north and was at its peak in Coventry in May and June, in Derby in July and August, in Shropshire in August and reached Cheshire during September. The Great Pestilence is believed to have begun in Derbyshire in May 1349 and Cox (1910), following an inspection of the episcopal registers, stated that the county was pre-eminent in desolation and lost two-thirds of its beneficed clergy within 12 months. He declared that in Derbyshire no class of the community seemed to have been spared, and no place was too remote or healthily situated to escape desolation. Lunn (1937) affirmed that the average plague mortality rate of this county's beneficed clergy was almost twice as high as that of Staffordshire.

In Cheshire, Middlewich was 'a chapelry of Sandbach, and six miles distant from the mother church. Encumbered by the necessity of burying their dead at Sandbach, the parishioners of Goostrey experienced serious inconvenience during the excessive mortality of 1349. Corpses of plague victims were left to rot at the roadside . . . so great were the perils and hazards of the way . . .'.

But elsewhere the great woodlands, for which Cheshire in the Middle Ages was famous, impeded the progress of the plague. There are several instances of reduced rents and of lands lying idle through lack of tenants at Middlewich, Russheton, Chelmondeston, and Kingsley in 1350, all attributed to 'The Great Pestilence' in the manorial records (Shrewsbury, 1970).

In 1939, R. Sharpe France presented a manuscript on the history of the plague in Lancashire, which he had collated from many sources. The following account is based on his paper but there is one caveat that should be borne in mind: Sharpe France was convinced that the pandemic of 1348–50 was caused by bubonic plague and it is not always clear which of his statements are based on authenticated records from Lancashire and which have been adapted from accounts elsewhere. The pestilence in Lancashire lasted from 8 September 1349, until 11 January 1350 and the mortality was 'almost unbelievably high'. The disease was swift in its action, 'one day people were in high health, and the next day dead and buried'. Sometimes death occurred within 12 hours of a person being infected, and usually within 3 days at the most. Apparently few of the upper class died, but of the common people and the clergy 'a multitude known to God only'. In the affected classes it was generally the youthful and healthy who were carried off, the old and ailing usually being spared. In Preston, 3000 people died; in Kirkham, 3000 people fell victim to the pestilence; 800

parishioners of Poulton-le-Fylde died; in Garstang 2000 died; deaths in Cockerham numbered 1000.

Of the inhabitants of Lancaster 3000 are said to have died, although White (1993) pointed out that this figure was too large: Law (1908) described how the Archdeacon of Richmond gave the total plague mortality figure of 13 180 for his deanery in his claim for probate dues, but she warned that his figures are certainly exaggerated and that the contemporaneous manuscript that recorded these figures must be read with caution as it contains several glaring errors. As the jurors allowed the archdeacon only £30 3s 4d of his claim for £113 10s, it seems evident that they knew that his figures were exaggerated.

In contrast to this story of widespread and severe mortality, Lunn (1937) concluded that Lancashire suffered little from the ravages of the Black Death. The county in the Middle Ages was sparsely populated and churches were few and far between. In many parts, easy travel met with insurmountable obstacles and the pestilence made little impression upon those who were cut off from the main stream of civilised life.

Barnes (1891) suggested that there are good grounds for believing that the epidemic came to Cumbria; he believed that the Scottish army on emerging from the forest of Selkirk would more probably have entered England by the western route rather than via Berwick in the northeast:

> They had appointed a rendezvous in the forest of Selkirk, to avail themselves of the mortality which was then desolating England. Scarcely had they passed the borders, when they were seized by the pestilence. Five thousand of them dropped down dead, and many were cut off by the enemy who had found means to draw a considerable body to the field . . . The few Scots who returned from the invasion communicated the pestilence to their countrymen (one-third of whom, according to Fordun, perished). The patient's flesh swelled excessively and he died in two days illness, but the same author tells, that the mortality chiefly affected the middle and lower ranks of the people.

There is also evidence that Carlisle was involved in the pestilence because a royal grant of various liberties was made to its citizens in February 1352 in recognition of its importance as a frontier fortress against the Scots and because it was then wasted and more than usually depressed by the mortal pestilence lately prevalent in those parts as well as by the frequent attacks. The conclusion that the Black Death struck severely at Carlisle is supported by the accounts of Richard de Denton, which he presented in 1354, claiming that, because of the mortality from the pestilence lately raging in those parts, the greater parts of the manor lands attached to the King's Castle at Carlisle were still lying uncultivated. For 18 months after the end of the plague the entire estate had been allowed to go to waste for lack of labourers and tenants. Mills, fishing, pastures and meadow lands could not be let during that time for want of tenants willing to take the farms of those who had died in the plague. The jury found that Richard de Denton had proved his facts and accepted the greatly reduced value of the estate.

4.6 Spread of the epidemic in northeast England

Archbishop Zouche of York sent out a warning order on 28 July 1348 about the Black Death, which was rampant in mainland Europe but some 8 months before the first cases were recorded in Yorkshire. Miller (1961) affirmed that the pestilence appeared in the county in March 1349; it was at its peak in York from the end of May to August and persisted there for nearly a year (Ziegler, 1969). The plague had probably spread to Yorkshire from Nottinghamshire, but Shrewsbury (1970), who was persistent in attempting to shape events in accordance with the biology of bubonic plague and its association with ports, suggested that

> The Great Pestilence may have entered the county through its port of Kingston-upon-Hull as a maritime importation by coastal shipping from East Anglian ports. Hull was certainly attacked by plague, and it would seem severely, for in March 1353, Edward III made a grant to its townsfolk because a 'great part of the men' had 'died in the late pestilence and the survivors are so desolate and poor that without succour they cannot pay . . . charges on their town'.

The mortality in Yorkshire was not as severe as in neighbouring Lincolnshire; nevertheless 223 benefices of the 535 parishes in the diocese of York were vacated by death and it is believed that between 42% and 45% of the parish clergy died. The archdeaconry of York covers a large area and shows the usual wide variation in mortality. The deanery of Doncaster lost 59% of its clergy, whereas virtually no benefices were vacated between Doncaster and the Humber. In Pontefract the figure was 40%, but with few casualties in the eastern flats. The city of York, surprisingly, with 32%, was relatively lightly affected. Sellers (1913) quoted a report by Clarkson that Richmond was so grievously ravaged by a plague at this time that about 2000 of its inhabitants died. However, this report is not supported by any evidence and it is almost certainly exaggerated because it is improbable that Richmond had a population of more than 1000 persons of all ages at any time in the 14th century (Shrewsbury, 1970).

Thompson (1914) concluded that, in the archdeaconries of Nottingham and York, the two extremes, mountainous districts and marshland, were comparatively immune from pestilence, while normal agricultural country and the lower highlands suffered most heavily. This is borne out by the figures of the percentage mortality of the beneficed clergy for the three deaneries of Cleveland:

Bulmer	Arable and pasture	51%
Ryedale	Hilly	28%
Cleveland	Moorland	21%

Events at the Abbey of Meaux, 6 miles north of Hull, in August 1349 give valuable details of the infectivity of the Black Death. 'God's providence ordained at that time', wrote the chronicler of the monastery, 'that in many places the chaplains were kept alive to the very end of the pestilence in order to bury the dead; but after this burial of the lay folk the chaplains themselves were devoured by the plague, as the others had been before them'. The abbot and five monks died in a single day, on 12 August 1349; the Prior, the Cellarer, the Bursar and 17 other monks died also and, out of 50 monks and lay brethren, the chronicler claims that only 10 survived. The infection clearly spread rapidly and the six synchronous deaths on 12 August shows that they were multiple infections from an original source. The 80% mortality testifies to the infectivity and lethality of the pestilence. However, 10 persons in this closed community survived; most of them must surely have been exposed to the infection and hence either recovered or were resistant.

Bradshaw (1907) suggested that the Great Pestilence moved north from Norfolk during the spring and early summer of 1349, with the result that the peasants living along the Salters' Track in County Durham were the first people in the diocese to experience it. He suggested that it may have reached Sunderland before the middle of July, because by that time 'four of the bishop's tenants at Wearmouth were dead'. He noted that the business at the summer halmote (a court) at Chester-le-Street on 14 July proceeded normally; but at the halmote at Houghton-le-Spring on 15 July the peasants were in a state of panic, and a similar panic was manifest at Easington on the following day. There was no alarm at Middleham halmote on 17 July, however, possibly because this village lay off the main road, and the halmotes of Stockton on the 17th, Sadberge and Darlington on the 20th, Wolsingham on the 21st or 22nd, and Lanchester on the 23 July, all seem to have been conducted normally. Shrewsbury (1970) quoted Bradshaw as follows: 'He deduces from the contemporary records that plague erupted as an epidemic disease at first in the south-eastern part of the county, and he believes that it raged with especial virulence at Billingham, where 48 tenants of the Prior died of it. Although Billingham was not a large village he is convinced that at least half of its population perished'.

The Black Death subsided in the winter but it left behind a ruined and dispirited people. 'No tenants came from West Thickly because they are all dead' is one entry on the bishop's rolls, and in another place we are told that only one tenant was left at Rowley whilst across the river at Bishopwearmouth 'a very large number of houses were fallen in ruins for want of tenants'. Comford (1907) stated that in the priory of St Cuthbert in Durham city so many of the clergy died of plague that there were not enough priests left to administer the sacraments, and she affirmed that the tenants of the Hospital of St Giles Kepier also suffered severely. Their visitation by the disease was accompanied by a failure of their crops and a murrain among their domestic animals.

Shrewsbury (1970) gave three references from the royal archives of the visitation of Newcastle-upon-Tyne by the pandemic:

20 November 1350	A pardon to the burgesses and other men in consideration of their losses, as well as on account of the deadly pestilence.
1 December 1350	Certain rents shall be borne even though several merchants and other rich men who used to pay the greater parts of the tenths and fifteenths and of other charges incumbent on the town have perished in the mortal pestilence raging.
12 November 1352	Another pardon for the burgesses and townfolk on consideration of their damages and losses sustained on account of the pestilence lately affecting the town.

4.7 The consequences of the Black Death in England

The enormous mortality of the pandemic had many far-reaching social and economic consequences that have been explored by Langer (1964), Ziegler (1969) and Bolton (1996).

The economy in England had been declining well before 1348 and for at least 25 years agricultural production, exports and the area of cultivated land had been shrinking. Furthermore, the working population had expanded far beyond the work available by 1348, which led to chronic underemployment; there were far too many villeins available to do the work of the landlord. Ziegler (1969) suggested, therefore, that it was questionable whether the Black Death did indeed bring about as fundamental a revolution in land tenure and social organisation as has been suggested.

However, it is generally accepted that wages more or less doubled, whereas the prices of agricultural products fell steeply during and directly after the Black Death because of the lack of demand. The landlord was partially sheltered against these difficulties by the extra income that accrued in 1349 and 1350 from the entry fines levied on the estates of the dead before the heir was allowed to take them over. There was undoubtedly greater mobility of labour during and directly after the Black Death and any landlord unready to make concessions to his tenants might well have found that they had vanished to seek a better master (Zeigler, 1969).

If a third of the peasants of a given area disappeared during the Black Death there would have been inevitably serious dislocation in the short term, but Zeigler (1969) made the important point that one of the most striking features was the speed of recovery shown by the medieval community. In some areas the process of adjustment would have been relatively simple; in others, where the Black Death did its worst damage, it would have been painful and protracted. Ziegler (1969) described these events as follows:

> On the estates of Crowland Abbey, where eighty-eight holdings were left empty, all but nine of these were quickly taken up; not by peasants from other villages who might have deserted land elsewhere and so left another gap to fill but by people with names already known on the manor who, one must presume, were landless residents before the plague. The estate of the Abbey, in fact, had sufficient surplus of man-power to fill even the huge vacuum left by the plague. At Cuxham, nine out of thirteen half-virgates were still vacant by March, 1352 and in this case recourse was had to importing tenants from outside the manor. Within another three years all the vacancies were filled.

He concluded that the Black Death was a stimulus towards the greater mobility of labour (and hence towards the disintegration of the manorial system) in spite of legislation designed to prevent it and that, in general, the countryside was soon back to near normal conditions, although some of the poorer villages were unable to resist the effects of the pestilence and the temptations offered by richer neighbours. Some villages never recovered fully, but such cases were rare in the Midlands, where a boom-town like Leicester, which was at the centre of a prosperous agricultural area with rapidly growing trades and industry, could quickly make up its strength by recruiting not only peasants but free men.

The foregoing describes events and social changes that followed the Black Death in southern and central England; no such comparable information exists for northern England but it is interesting to compare these events with the social upheaval that occurred at Penrith in Cumbria after the major mortality of the plague that devastated the town 350 years later, as described in section 13.17. Here again, the ecological niches were quickly filled and the community returned to pre-crisis population dynamics.

4.8 The Black Death: conclusions

There seems little doubt that the Black Death in England was the same pestilence that swept northwards through continental Europe in 1347–50. This infection also penetrated to Iceland, Greenland and the Channel Islands and the accounts of witnesses on the Continent and the ecclesiastical records in England allow us to chart its progression and hence the speed of its spread.

4.8.1 Reservations expressed by Shrewsbury

Shrewsbury (1970), in his detailed *A History of Bubonic Plague in the British Isles*, raised doubts as to whether all the subsequent epidemics were the result of bubonic plague, although he repeatedly averred that it was the causative agent in the Black Death: 'There is not the slightest doubt that during the late spring, summer and early autumn months, "The Great Pestilence" was an epidemic of bubonic plague, engendered and sustained by an unrecognised epizootic of rat-plague'. When he studied the epidemiology of the pestilence, particularly in northern England, he found that the evidence of the regional distribution, biology and the magnitude of the outbreak were completely at variance with the known biology of bubonic plague (see Chapter 3) and he therefore discounted the evidence and his own analysis and reluctantly concluded that the pestilence could not have spread on such a scale to certain regions. We cite the following examples of this confusion.

The five least densely populated counties in England in 1348 were Cumberland, Northumberland, Durham, Westmorland and Cheshire, with densities ranging from 20 to 25 persons per square mile. Shrewsbury stated that

> The densities are so low that it would have been biologically impossible for bubonic plague to have spread over any of these counties in the fourteenth century, though it might by chance have been introduced into one or more of their towns and villages . . . The epidemiology of bubonic plague renders it improbable that *Y. pestis* could have been distributed by rat-contacts as epizootic plague in any English county in 1348–9 having an average density of population of less than 60 persons in the square mile.

We have presented evidence above that shows how the Great Pestilence did indeed spread through these northern counties. Shrewsbury continued:

In the fourteenth century England was a much wilder country than it is now. There were large areas of uninhabited forest, waste, and fenland, and the moorlands of Yorkshire, Somerset, Devon, and Wales were devoid of human and rat life, except for a sprinkling of hamlets on their margins. Everywhere the population was more thinly scattered than it is today, though it is important to remember that nearly all the habitations were grouped in villages [and] there were next to no outlying farms. This fact is essential for a correct understanding of the epidemiology of 'The Great Pestilence' in England. Outlying farms, with their relatively large colonies of house-rats, would have served as links in the train of transmission of rat-plague from one village concentration of house-rats to another. In the absence of such links many – probably very many – villages must have completely escaped the disease. From what is now known about the etiology of bubonic plague it was therefore a biological impossibility for the whole or even a major part of England to have been ravaged by 'The Great Pestilence' of 1348–50.

He then proceeded to detail the devastation and remorseless spread of the Black Death and recorded Lunn's conclusion:

The united testimony of the three registers of York, Lincoln and Lichfield, show the mortality of the Black Death to have been remarkably uniform, when its effects are spread over large areas. This rate of 40 per cent which the registers of Lichfield, York and Lincoln have produced is surely no mere coincidence. Four out of every ten of the parish clergy in England died in the Black Death . . . Whether this death-rate of 40 per cent among the clergy exceeds that of the general average can never be determined.

Shrewsbury challenged this conclusion on two grounds.

The first is that it conflicts with the epidemiology of bubonic plague, the incidence of which over large areas must inevitably have been most erratic. The second is that it is based on the assumption that all the deaths recorded in the registers were plague-deaths and takes no account of the mortality among the clergy in the absence of that disease.

Surely, the lack of this correction factor for non-plague deaths would have only a small effect on Lunn's statistics and Shrewsbury was attempting here to modify the results of his own and other writers' analyses to fit with his understanding of bubonic plague.

With reference to the plague in the Diocese of Lincoln, Shrewsbury said:

Thompson concludes his study with the statement that 'the percentages of vacant benefices in each rural deanery work out with such unanimity that the mortality of the beneficed clergy . . . may be taken as a fair guide to the general death-rate'. Unfortunately the biological nature of bubonic plague contradicts his conclusion, because if there is one characteristic of the disease that is absolutely constant it is the erratic manner of its spread and consequently its haphazard attack in every community. It is in fact the unanimity of these clerical statistics which precludes their use as a guide to the national death-rate from bubonic plague; but their unanimity might represent the activity in England of another lethal disease during the normally cold months from December to March when the rat-fleas, *X. cheopis* and *C. fasciatus*, hibernate in temperate climates. The characteristics of this other manifestation of 'The Great Pestilence' – for such it would have been to contemporary recorders – must have been the similarity of its clinical picture to that of bubonic plague: its winter incidence; its high infectivity; its transmission by direct human contacts; its preference for the well-fed and, especially with regard to the death-rate among the clergy, its high mortality in the middle-aged and elderly. Classical epidemic typhus fever completely fulfils these requirements.

Once again, he was dismissing evidence because it does not fit with the epidemiology of bubonic plague. He was swayed by the erratic spread of this disease and appeared to be presenting evidence for two separate causative agents operating simultaneously in the Great Pestilence, surely an unlikely scenario.

When discussing the plague epidemics in Hull in the 1370s Shrewsbury quoted Sellers as writing: 'But there is one marked difference between these attacks of plague of the second half of the century and the Black Death. They were localised in towns, escape by flight was possible to the wealthy, but in 1349 death was in the soil, town and country suffered alike, the fortunes of the wealthy and poor were equalised'. Shrewsbury continued 'If such a "marked difference" had existed, then the "Black Death" was certainly not an eruption of plague; but the statement unfortunately betrays only a sad ignorance of the immutable etiology of that disease, which can no more change its nature than "the Ethiopian his skin, or the leopard his spots"'.

In Yorkshire, Shrewsbury quoted Sellers' view that careful statistics

prove that more than two-thirds of the parish priests in the West Riding, and 35 of 95 in the East Riding, died during the pestilence, the inescapable inference from her statements is that – at any rate for the West Riding – more than two-thirds of the population died of bubonic plague, which, in view of the nature of that disease and the wide dispersion of the county's population, is an impossibility.

He also stated:

If, as seems extremely probable, the bulk of the scanty population of the county of Durham in the fourteenth century occupied the valleys of the Tees and the Wear, the rest of the county must have been virtually desolate, with no conceivable possibility of the spread in it of bubonic plague. *In all probability the only microbial diseases that could have spread outside those valleys under the prevailing conditions were smallpox, typhus fever, and epidemic influenza* [our italics]. Nevertheless Gee declares that the county was not spared by 'The Great Pestilence' and he asserts that there is documentary evidence that its death-roll was heavy. For example, he says that in the Cursitor Roll special provision for a land title was made in the event of the death of the assigns during the pestilence which was then raging. Moreover, he adds, 'the papal registers for the next forty years give in their concessions to monastic houses conclusive proof of the virulence of the outbreak in the north' . . . Wilson finds little contemporary information about the effect of 'The Great Pestilence' upon the beneficed clergy in this county; but he notes an ominous gap in the diocesan register of Carlisle for the six years 1347 to 1352, which he assumes signifies a severe plague-mortality among the clergy. It is improbable that such an assumption is justifiable because the geography of this county, the sparseness of its population, the unease of its living conditions, and the etiology of bubonic plague, conjoin to make it unlikely that *P.* [= *Yersinia*] *pestis* penetrated farther into the county than Carlisle at this time . . . The references to the presumptive activity of bubonic plague in Wales have most been supplied by Rees, who affirms that Snowdonia, especially Anglesey, was overrun by 'The Great Pestilence'. He may be correct for Anglesey, which appears to have been populous in the fourteenth century, but it is extremely improbable that Snowdonia was sufficiently colonised by the house-rat to support an epizootic of rat-plague. In any case he nullifies his affirmation by his succeeding statement that while certain hamlets had a heavy mortality others were little affected 'and even entire commotes [Welsh manors] escaped lightly'.

We concur with Shrewsbury's observations, detailed above (see also in the following section) that there is evidence that the Black Death was not an outbreak of bubonic plague. But there are many additional pieces of information and much more cogent reasons, which we have advanced in Chapter 3 and present in the following chapters (see summary in section 13.3), which demonstrate that it is a biological impossibility that bubonic plague had any role in the Great Pestilence. Nor is there any evidence to suggest that more than one lethal disease was raging during the period of the Black Death.

4.8.2 The death toll of the Black Death in England

It is impossible to determine accurately the mortality from the disease in England. The exaggerated claims of the Archdeacon of Richmond of the plague mortality (section 4.5) are a reminder of the misleading statistics of the time. This practice of inflating the numbers dying was continued when the total dying in the plague at Penrith in 1597–98 was given on a plaque in the church; this value greatly exceeds the size of the population (section 5.2). Clapham (1949) suggested that a third of the population of England may have died during the Black Death, in agreement with many other historians, but felt that 20% to 25% may be a more likely figure. He says:

Possibly the traditional figure of a third of the population dead may be correct: many scholars have accepted it. But modern experience of [? bubonic] plagues suggests a fair number of spots which would remain immune and affect the total. Perhaps 20 or 25 per cent may be nearer the mark than 33.3 . . . We hear of depopulated but not deserted villages. When villages can be studied in groups, what surprises us is the continuity of their life. On the vast estate of the Bishop of Winchester in Wessex there is 'no sign of chaos or complete depopulation', and following the Pestilence there is 'no revolution either in agriculture or in tenure'. Across the country, on the East Midland and Fenland manors of Crowland, it is just the same: the estate accounts run on, and what is most important, there seems always to be someone ready to take over a vacant holding.

Trevelyan (1945) believed that a third or possibly a half of the population died of plague in less than 2 years. Shrewsbury (1970) could not accept that from 20% to 50% of the total population died in the Great Pestilence and believed, like Clapham, that the estimates of mortality were greatly exaggerated because there were no deserted villages, no interruption in agriculture or land tenure and no appreciable change in the manner or customs of English life. He continued:

Bubonic plague is principally an urban disease – but there were relatively few urban aggregates in England in the fourteenth century. There was not even a profusion of villages, thickly sprinkled over the land and linked by outlying farms, to enable *Y. pestis* to be spread by haphazard rat contacts, erratically and unpredictably, among the colonies of house-rats which could only maintain themselves in close association with human aggregates. Yet without that fundamental epizootic spread of rat-plague no epidemic of the human disease could possibly have occurred.

With an average population density in 14th century England of one individual to every 8 acres, Shrewsbury considered that, under the existing social conditions, bubonic plague was biologically incapable of destroying as much as 20% (let alone higher estimates) of the population. Consequently, he considered that

in all probability the national death-roll from 'The Great Pestilence' did not exceed one-twentieth of that population. If 'The Great Pestilence' had been in its entirety a disease with a malignity approximately equal to that of bubonic plague – such as haemorrhagic smallpox – that was spread by human contacts, a death roll of one-third of the nation might conceivably have been achieved. But bubonic plague was not so spread; had it been 'The Great Pestilence' would not have excited anything like the terror it aroused.

Shrewsbury was arguing for a 5% mortality from the plague, which is a very much lower value than that given by other historians, on the grounds that (i) the population could not have recovered so quickly and so well from a higher level of mortality and (ii) bubonic plague could not have spread so extensively in such a low-density population. We suspect that this is another example of special pleading on the part of Shrewsbury in his efforts to convince us that bubonic plague was responsible for the Black Death.

However, equipped with the detailed records contained in the parish registers, we are able to make a good estimate of the mortality in the plague at Penrith that occurred 250 years later in 1597–98. Coupled with the famine that immediately preceded it, over 50% of the population died in a 3-year period (section 5.2) but, although this mortality crisis had subtle and long-lasting effects on the population dynamics (section 13.17), the community rapidly returned to steady-state conditions with the same average number of annual births and deaths. Human populations, to the outside observer, can adapt remarkably quickly and effectively to major mortality crises. Epidemics in the cities of northern Italy frequently caused a 40% to 50% mortality (section 11.3). We conclude that the mortality in the more populous parts of England during the Great Pestilence may well have been at least 30%.

4.8.3 Epidemiological notes on the Black Death

Some additional important points concerning the epidemiology of the Black Death can be identified:

(i) Some areas were untouched by the pandemic, such as Milan and Bohemia (Fig. 4.1).

(ii) Small villages and towns were equally affected and in this respect the plague of 1348–50 differed from later outbreaks: 'But plague henceforth is seldom universal; it becomes more and more a disease of the towns, and when it does occur in the country, it is for the most part at some few limited spots' (Creighton, 1894). In contrast, in the outbreak of genuine bubonic plague in India in the early years of this century, the intensity was in inverse proportion to the size of the community, the maximum mortality being in villages rather than towns (Hankin, 1905).

(iii) There is no doubting the horrific infectivity and lethality of the disease during its remorseless spread, but we have no evidence to answer the question of why some people who were obviously exposed to infectives survived. Were they resistant, immune or did they recover?

Shrewsbury (1970) also drew attention 'to the fact that certain individuals in a population, of which all the members are equally exposed to the pestilential atmosphere, escape the disease'.

(iv) There are scattered references to the death of domestic animals associated with the Great Pestilence, as in the account of Michael of Piazza (section 4.2, *note 2*).

(v) It is difficult to accept unreservedly the account of the arrival of the Great Pestilence in Sicily in the Genoese galleys (section 4.2) which must have been at sea for a long time and which had no cases of plague on board and yet the epidemic had spread to every part of Genoa within 2 days. Twigg (1984) has made a careful study of shipping and voyages between Genoa and the Crimean ports at the time of the Black Death and concluded that the evidence is not consistent with the causative agent being bubonic plague. Indeed, from what we have learnt about the rapid spread, the short time between infection and death and the high infectivity it is difficult to conceive of any disease that could have remained extant for the 4–5 weeks of the voyage (Twigg, 1984) with an apparently healthy crew who on arrival caused widespread infection within 2 days of coming ashore. Presumably, if the story of the Genoese galleys were true, and it is widely accepted, the crews would have had several carriers of the disease among their number, raising again the question of immunity to the disease.

The understanding and epidemiology of many diseases is complicated by chronic infection in immunodeficient individuals, who become 'silent' or inapparent carriers of the virus or bacterium, sometimes essentially for life. Anderson & May (1991) gave the example of hepatitis B: the virus infects people everywhere in the world, with the highest rates of infection in sub-Saharan Africa and east Asia. Although the majority of cases are not serious in developed countries, a minority lead to acute hepatitis with jaundice, causing some deaths. There are estimated to be about 200 000 new hepatitis B infections in the USA each year, of which about 25% experience acute hepatitis, leading to around 10 000 hospitalisations and 250 deaths. More generally, the outcome of the infection depends on the immune response of the individual: an adequate immune response leads to the production of antibodies that clear the infection and produce lifelong immunity whereas an inadequate immune response allows continued viral replication and, if this is maintained for 6 months or longer, such viral production usually persists indefinitely and the infected individual becomes an asymptomatic carrier of infection.

We agree entirely with Twigg that the Genoese galleys did not bring bubonic plague and, on balance, we do not believe that their crews were all 'silent' carriers of the disease. Even if the original story is only partly true, in all probability the arrival of the galleys was coincidental with the outbreak of the Black Death and we discuss this point further in section 13.8.

4.8.4 Effects of malnutrition

There are a number of reports in which an outbreak of haemorrhagic plague follows a severe famine; for example, the plague at Penrith in 1597–98 was preceded by the famine of 1596. There are also a number of epidemics recorded in France that were preceded by famine, sometimes in conjunction with bad weather conditions. Such observations led to the suggestion that the extreme mortality was because of a compromised resistance or immune response in the malnourished community. The regular epidemics of smallpox in England in the 17th and 18th centuries were driven by cycles of malnutrition associated with oscillations in the price of grain (Duncan *et al.*, 1993a,b; 1994a,b; Scott *et al.*, 1998a,b) and the mortality from scarlet fever in the 19th century in England was determined, in part, by nutritive levels linked to the price of wheat (Scott & Duncan, 1998).

The nutrition of the bulk of the population of England was not completely satisfactory in the 14th century, even in good periods, and Drummond & Wilbraham (1991) have estimated that the average daily peasant diet in England in the 15th century consisted of 1 pint of milk, 1 pint of whey, 2 oz of cheese, 1 oz of bacon, 2 lb of muslin bread, 2 oz of pease. This diet was probably adequate in terms of calorific intake, with sufficient protein and fat, but was deficient in vitamin C and probably also in other vitamins and trace elements. The bulk of the population may, therefore, have had an impaired resistance to a new disease that might have accounted for the tremendous mortality.

4.9 Seasonality of the outbreaks of the Great Pestilence in different localities in England

The time of year when outbreaks of true bubonic plague occur is thought to be important because of the close relationship between flea breeding and the local climate and hence considerable interest has focused on the months in which the different dioceses experienced the major mortality during the Great Pestilence. We are indebted to Russell (1948) for his

Table 4.2. *Summary of the monthly deaths in dioceses in England in 1348–49*

Diocese	Date of start of plague	Peak mortality
Salisbury	Nov. 1348	Mar. 1349
Bath & Wells	Dec. 1348	Jan. 1349
Winchester	Jan. 1349	Apr.–May 1349
Exeter	Jan. 1349	Mar.–June 1349
Gloucester	Mar. 1349	May–Aug. 1349
Worcester	Apr. 1349	July 1349
Hereford	Apr. 1349	July 1349
Lichfield	Apr. 1349	July–Aug. 1349
Norwich	Apr.–May 1349	June–Aug. 1349
Lincoln	May 1349	July–Aug. 1349
Ely	June 1349	July 1349
York	July 1349	Sept 1349

Deaths taken from institutions to vacant benefices.
Sources: Shrewsbury (1970) and Twigg (1984).

analyses of the Inquisitions *Post Mortem*, to Shrewsbury (1970) for his extensive studies of the institutions to vacant benefices and to Twigg (1984) for his overview of the available information and for drawing attention to the shortcomings of the data: Russell (1948) had only small samples and the vacant benefices did not necessarily reflect truly the mortality of the bulk of the population. However, these are the only statistics available to us.

Before the pandemic, the monthly mortality records for England showed two peaks, one in January–February and the other in October–November, with a minimum in June and July. Shrewsbury suggested that the winter peak was because of deaths from smallpox, typhus fever and infections of the respiratory tract and he attributed the autumn peak to deaths from diphtheria, measles, erysipelas and infections of the intestinal tract. The institutions to vacant benefices in 11 dioceses are given in Fig. 4.2 and it is immediately apparent that the previous bimodal distribution of monthly deaths was replaced by a single peak during the Black Death. Of course, not all these deaths were from plague, some would have been from other causes and, without further evidence, it is impossible to identify exactly the beginning and end of the outbreak in each locality. Although there are differences in the patterns of these outbreaks, we can see that the pestilence lasted, on average, about 7–8 months in each diocese with a single major peak.

The results from these histograms are summarised in Table 4.2 and the disparate starting dates between November 1348 and July 1349 simply reflect the spread of the plague northwards and eastwards. Thus the pestilence occurred in every month of the year and it is noteworthy that the peak mortality in Bath and Wells was recorded in January; Shrewsbury (1970) commented that this runs counter to the usual epidemiological behaviour of bubonic plague in a temperate climate. It was not confined to the summer months and in some counties the peak of the epidemic occurred in winter. It is evident that the Black Death simply struck in a diocese when the infection arrived from the adjacent county and the epidemic then exploded with peak mortality occurring some 2–3 months later and the pestilence effectively burnt itself out after some 6–9 months, indicative from Reed and Frost modelling of a long serial generation time.

4.10 Was the Black Death an outbreak of bubonic plague?

Although we lack reliable accounts of the etiology of the Great Pestilence and the analysis of the epidemiology has been derived from secondary sources, we may compare its biology with that of authentic bubonic plague in the 20th century, which has been described in Chapter 3 and summarised in section 3.16. It is obvious that there was a different causative agent in the two plagues. The Black Death was a highly infectious, lethal disease, although we cannot say whether any of the population were resistant or survived an infection. Did the survivors (perhaps two-thirds of the population of England) simply escape an adequate exposure? It was probably viral in nature, as shown in section 13.1 (*Yersinia* is a bacterium), and the pandemic struck a naive population in Europe that would have had no built-in immune response. Transmission was direct, person-to-person; the evidence of infection via contagion with clothing is possibly suggestive but not conclusive.

The epidemics followed the typical Reed and Frost dynamics characteristic of a 'typical' infection (section 2.5; Fig. 2.4), rising to a single mortality peak before decaying (see Fig. 4.2). The average duration of an outbreak in a diocese was about 8 months, although it may have been somewhat shorter in individual localities. This is a relatively long epidemic period, which implies that the disease had a long incubation period (see Figs. 2.4 and 2.5) and is completely consistent with the epidemiology of subsequent plagues in England. This is an important point and explains the spread of the Black Death: all accounts agree that the time between the appearance of symptoms and death was very short indeed, perhaps between 3 and 5 days. These dying people would not, in general, have been fit enough to have carried the infection over long distances, although most people in the household would have been exposed. However, with a long infectious period, apparently healthy, subclinical, symptomless individuals would have been moving across the countryside on foot or on horseback and progressively spreading the disease through a metapopulation, although it would probably have been about a month before individuals with symptoms appeared in the next focus. Thus, once infected, apparently healthy individuals would soon be transmitting the disease within their family, within their village and also further on to the next village. Of course, the infection would also have been spread by susceptibles returning from a visit to a nearby village that had symptomless infectives. Local fairs and the movement of goods would have exacerbated the spread. Any infective traveller, setting out on a journey, could have spread the disease widely before showing any of the characteristic signs of the pestilence. The long incubation period also readily explains how the pestilence spread by sea to England, Iceland, Ireland and Greenland. Nobody showing the extreme, terminal symptoms of the disease would have embarked on such a voyage and they would have died before arrival.

The Black Death struck indiscriminately at London, other cities, rural towns, villages and hamlets; those on the main trading routes would have been most at risk from receiving a symptomless visitor. It struck somewhere in every season, even in mid-winter, the establishment of an epidemic being simply dependent on the arrival of an infective from a near-by focus. However, inspection of Fig. 4.2 suggests that the disease may have been more active or virulent in the summer (perhaps because people's behaviour, such as attending fairs, led to a higher R_0); this was also a feature of subsequent outbreaks of haemorrhagic plagues in England.

It is impossible that *Yersinia pestis* could have been the causative agent of the Black Death. The epidemiology of the Great Pestilence, summarised above, is completely different from the biology of bubonic plague, which is critically dependent on a pre-established enzootic/epizootic in rodent populations (both susceptible and non-susceptible species in dynamic balance) throughout mainland Europe and Britain, Iceland, the Channel Islands, Ireland and Greenland. Only the black rat was available and it would have been confined to warmer climates or to close proximity to humans at seaports. It is inconceivable that the Black Death could have spread via rodents over such vast distances so rapidly, particularly across the sea.

As explained in Chapter 3, the biology of bubonic plague is complex and so the disease does not exhibit Reed and Frost dynamics; indeed it is a disease of rodents and it moves very slowly and erratically in the absence of modern transport and usually spreads to humans living in villages close to the rural habitat of the rodents. The lethality and the death toll of the Black Death, even on conservative estimates, was far greater than would be expected for bubonic plague (as acknowledged by Shrewsbury). *Yersinia* is dependent on the biology of its hosts, the rodents and their fleas, and consequently the disease is confined to warmer climates. An introduction of infective rats and fleas by steamship into the ports of northern latitudes in the 20th century has never spread, very few people have died and the outbreak has rapidly disappeared. It is inconceivable that bubonic plague could have spread rapidly in winter, over alpine passes and through sub-Arctic regions including Iceland, Norway and Greenland.

As we have detailed in this chapter, although Shrewsbury believed that *Yersinia* was the causative agent in the Black Death, he was repeatedly aware that the details of the epidemiology did not correspond to the biology of bubonic plague and he had to resort to a variety of devices (including the invention of a second, lethal disease co-existent with the pestilence) to explain his results.

We can now abandon the idea, so firmly promulgated in the literature, that the Black Death was an outbreak of bubonic plague. There is no evidence for such a view. Indeed, apart from the swellings shown by the victims of both diseases, everything suggests that, of all the known infectious diseases, bubonic plague, with its complex biology, is the most unlikely candidate for the relatively simple epidemiology of the Black Death.

4.11 Plagues in England following the Great Pestilence: the 14th century

The Black Death was followed in England by a series of epidemics in the second half of the 14th century, although these were progressively less violent, but Ziegler (1969) suggested that the second epidemic of 1361, by any standards other than those of the Black Death, was catastrophic in its dimensions. Many writers have assumed that these were fresh outbreaks of bubonic plague, but Shrewsbury (1970) believed (and he listed his evidence) that this assumption was not warranted.

There is no doubt that smallpox, measles, typhus fever, and dysentery were repeatedly epidemic in fourteenth-century England; pneumonia undoubtedly occurred in epidemic form in the winter months, and whooping-cough, the enteric fevers, and influenza in all probability were also epidemic at times. It is certain that tuberculosis, 'the white scourge', was responsible for many deaths annually . . . It is probable that diphtheria, erysipelas and poliomyelitis were locally epidemic from time to time and that . . . some of the great medieval epidemics of 'colic' that afflicted the British Isles may have been maritime importations of cholera.

Creighton (1894) also concluded that 'it would be unsafe, therefore, to conclude that all outbreaks of *pestis* in England subsequent to the Black Death, were of bubo-plague itself'. We agree; there is no evidence of bubonic plague in England at that time.

England was free from plague for 10 years after the Great Pestilence before the major eruption occurred in 1360–61. This has been termed the second pestilence, and subsequent outbreaks were referred to as the third, fourth and fifth pestilences. Many of these epidemics were confined largely to the north of England.

The second pestilence was evidently widespread and caused great alarm in England. In some places on the Continent, for example in Florence, it appears to have been as destructive as the Black Death, and Creighton (1894) said that it was marked by the same buboes and carbuncles. The disease was epidemic in London in May 1361 and Shrewsbury (1970) provided the quotation that 'great multitudes of people are suddenly smitten with the deadly plague now newly prevailing as well in the city of London as in neighbouring parts, and the plague is daily increasing'. He pointed out that its well-nigh universal incidence in the community that was attacked and the unique suddenness of its onset were features that were characteristic of epidemic influenza, i.e. it was an infectious disease spread person-to-person and was not bubonic plague. The pestilence of 1360–61 spread northwards to Leicestershire, Warwickshire and Lincolnshire (which it reached in 1362), all of which suffered badly. It reached York and Culcheth in south Lancashire and Liverpool was severely hit.

The epidemic of 1368–69 (probably the third pestilence), which Creighton (1894) described as 'a great pestilence of men and the larger animals', was considered by him probably to be an outbreak of famine sickness, a violent inflammatory fever and very fatal, and has been identified as influenza by some authorities.

In 1379–80 there was a plague that probably began in either 1375 or 1376 and was apparently confined largely to the counties of northern England. The Scots continued to make raids over the border, notwithstanding the danger that they ran from the pestilence and of which they were fully aware. A petition to Parliament was sent by the northern counties that the king would 'consider the very great hurt and damage which they have suffered, and are still suffering, both by pestilence and by continual devastations of the Scots enemy'. The Scots raided Penrith at the time of a fair and returned with immense booty. They passed by Carlisle and laid waste the forest of Inglewood, where they are said to have seized 4000 cattle. However, the Scots suffered severely in consequence because they introduced the plague that they had contracted into their own country. There is no local record of the ravages of this epidemic in Penrith (Barnes, 1891) but most local histories that mention the outbreak refer to the severity with which it overtook the invaders; it was said that the pestilence killed one-third of the people of Scotland 'wherever it came'.

A major epidemic of an unknown disease broke out in Cambridge in 1389 and it appears to have persisted until 1391. It was so serious that it has been compared with the Black Death and is classified by some as the fifth pestilence. Creighton (1894) said that the death-roll in Norfolk and many other counties was comparable to the epidemic of 1348–50. He considered that the outbreak was a mixture of famine-pestilence with bubo-plague: it was active in the great heat of June to September 1390 and attacked the young more than the old. The mortality peak was in 1391, concomitant with a general dearth of food, in consequence of which 'many poor people died of dysentery'. Shrewsbury (1970) did not think that bubonic plague was implicated in this pestilence because that disease did not have a selective preference for 'youths and children'. The pestilence was severe in the north of England, with York suffering badly, although the figure of 11 000 deaths there is certainly exaggerated. There are no specific records to show that this epidemic spread to Lancashire, Cumberland or Westmorland, but in 1399 the people of the north sent a petition to Henry IV praying that he would send assistance because there was a great pestilence sweeping the northern counties with the result that there were not enough able-bodied men left alive to defend the Scottish marches (Sharpe France, 1939).

4.12 Age-specific mortality of four epidemics in the 14th century

Russell (1948), working with a small sample, calculated the age-specific mortality from the Inquisitions *Post Mortem* for four epidemic periods, including the Black Death, in the 14th century and the data are shown in Table 4.3. In the Great Pestilence, the 1–5 year age group has an unusually high death rate from plague whereas the 6–10 year class has the lowest (Twigg, 1984). The death rate from plague in 1375 is lower, particularly in children, but this was not a bad year for the pestilence, except in northern England. The age-specific mortality of epidemics of known bubonic plague in Bombay city in the early part of the 20th century are also given in Table

Table 4.3. *Comparison of the age-specific mortality of four plague epidemics in England in the 14th century and with that of bubonic plague in Bombay City, India, in the early 20th century*

	Plague epidemics in England				India
Age group	1348–50	1360–61	1369	1375	1907
0–5	33	20	—	0	0.3
6–10	7	10	20	0	1.3
11–20	17	14	15	8	1.9
21–40	25	20	9	13	1.1
41–60	33	29	13	9	0.7
Over 60	39	36	23	30	0.4

Figures quoted are the calculated percentage of people dying of plague or bubonic plague in each age group.
Data sources: Russell (1948), taken from the Inquisitions *Post Mortem*; Twigg (1984), taken from the reports of the Plague Research Commission (1907a).

4.3 (data from the Plague Research Commission, 1907a; Twigg, 1984): the results are very different, not only is the overall percentage mortality of bubonic plague very much lower, but it killed predominantly those in the 6–40 year age group.

4.13 Plagues in the 15th century

We have seen that after the Black Death there was a series of outbreaks of epidemic diseases in England and these continued through the 15th century. Shrewsbury (1970) said that the chronology of the pestilence at this time

> is so confused that it is difficult to distinguish individual epidemics and define their extent. Their identification is largely conjectural and a matter of personal opinion, based on one's knowledge of epidemiology, in the absence of explicit descriptions of their clinical characters. I consider that *most of them were not outbreaks of bubonic plague* [our italics]. The governing factor in the persistence of this disease in England was the length of time that an imported strain of *P.* [*Yersinia*] *pestis* retained its virulence among the house-rats, and that factor is unknown. Sooner or later, however, each strain died out spontaneously and then bubonic plague disappeared from the country until a fresh, virulent strain was imported. In the intervals between successive importations local epidemics affecting individual towns and villages occurred; but the importation of a fresh strain was required for the production of a major, *national* outburst of bubonic plague, and after 1350 the next outbreak of this magnitude was in my judgement that of 1563.

He continued:

> If a large number of people showed similar manifestations within a relatively short period of time, then a 'plague' was recorded; but the actual disease might be any one of a dozen or more communicable or transmissible diseases of man. It is therefore quite unrealistic to accept the record of a 'plague' before the middle of the seventeenth century as synonymous with an outbreak of bubonic plague, and even after Sydenham's day, up to the middle of the nineteenth century, bubonic plague and epidemic typhus fever were clinically indistinguishable.

Again, we agree with Shrewsbury; there were no major outbreaks of bubonic plague in England in the 15th century. We give the following brief overview of the outbreaks of these various pestilences in England between 1400 and 1500:

1400 Adam of Usk described a great plague that prevailed throughout England, especially among the young, that was swift in its attack.

1405–7 A major outbreak with reported high mortality in London and in country villages where many families were said to be almost exterminated. Shrewsbury (1970) commented:

> 'If numerous country villages were synchronously involved . . . it must have been a disease of high infectivity that spread rapidly by human contacts, or indirectly by human contamination of foodstuffs, including drinking water, which in the summer were largely consumed uncooked. The most virulent epidemic of bubonic plague could not destroy much more than one-third of the population of London which, according to Creighton, was estimated at 44,770 in 1377, and which certainly had not reached 90,000 by 1407. Moreover bubonic plague is essentially an urban disease and only those villages in the near vicinity of an infected town would be involved in the fundamental rat-epizootic, although occasionally a village at a distance from the epizootic focus would be infected by urban fugitives harbouring "blocked" fleas. The figure of 30,000 deaths in London seems to have had a fascination for fifteenth-century recorders, for it is repeatedly quoted for different outbreaks of disease in that city.'

He repeated his conclusion

> 'In a few weeks available for its activity, the most malignant epidemic of bubonic plague could not conceivably have killed 30,000 people in London; but a water-borne epidemic of cholera or typhoid fever, or an epidemic of malignant dysentery, could have killed that number along the lower reaches of the river Thames during a summer epidemic. Gregory also records that "men and bestys were grettely infectyd with pockys" this year, and as smallpox would not be suppressed by a great frost, an outbreak of virulent smallpox may have been the great plague.'

1420–21 A pestilence, apparently restricted to the north of England (although Barnes, 1891, makes no reference to it); the men of Cumberland represented to Parliament that all the country within 20 miles of the borders had been so depopulated by war, pestilence and emigration that there was a scarcity of able men. In 1423, pestilence broke out in the north of Lancashire and especially in Lancaster where there was a great mortality (Sharpe France, 1939).

1433–34 A severe pestilence in London.

1438–39 There was a famine in these years, with a great dearth of corn, following three wet harvests, which was reported as being associated with a year of pestilence and many commonfolk men, women and children died throughout the realm, principally at York and in the North Country.

1447–54 Shrewsbury (1970) recorded these years as being pestilential ones with an outbreak in 1447 in Lincoln, in 1448 in Oxford, in Reading in 1452 and London in 1454.

1463 'A great pestilence with a dry summer all over England'. There were further, apparently more localised, outbreaks in 1465 and 1467. In the latter, death was described as sudden.

1466 Sharpe France (1939) reported that the plague came again to northern Lancashire and that the ravages of the disease made it necessary for the assizes to be adjourned to Preston.

1471–72 Said to be a widespread pestilence that moved northwards to Hull in 1472, which suffered again in 1476 and 1478 when more than 1500 are reported to have died. Newcastle-upon-Tyne also suffered from an epidemic in 1478.

1479–80 Great mortality reported in London and in many other parts of the country, which Shrewsbury (1970) suggested may have been a recurrence of cholera.

Plague steadily became established in continental Europe, with regular epidemics during the period 1350 to 1500, and developed into a pseudo-endemic state by the 15th century (section 11.7). We have few medical and epidemiological details concerning the epidemics in England during this time, summarised above, but believe that many of them were the consequences of outbreaks of haemorrhagic plague brought in by infectives sailing from the channel ports and from southwest France.

5

Case study: the plague at Penrith in 1597–98

Plague epidemics became steadily more widespread, with a greater death toll during the 16th century, as we describe in Chapter 6. The practice of keeping parish registers began during the last 50 years of this period and, together with other documentary sources, they have proved to be invaluable in determining the epidemiological characteristics of the plague. In this chapter, we present a case study of the plague at Penrith in 1597–98, which has previously been assumed to be a major outbreak of bubonic plague (Howson, 1961; Shrewsbury, 1970; Appleby, 1973). We describe the pattern of events there and the ways in which the individuals responded to this terrible visitation of the pestilence. Equally important, as we shall show, this detailed inspection of the data enables us to suggest the epidemiological characteristics of the disease. Armed with this information, we can then interpret the movement of the plague through the metapopulation of England in the 16th and 17th centuries in subsequent chapters, as it spread from population to population.

Penrith was a market town in the Eden Valley in Cumberland in northwest England and, it is said, the population had suffered from an earlier outbreak of plague in 1554. The registers began in 1557 and we have carried out a full family reconstitution from this date until 1812 (Scott & Duncan, 1998); we draw on the data derived therefrom in the following reconstruction of events during 1597–98.

Penrith had a market as early as 1123 so that local movements tended to be between the surrounding parishes. Geographically, the parish was situated in a valley that was bounded by the Pennines to the east, the Lake District to the west and the Westmorland fells to the south. These factors must have provided significant natural barriers that confined migration largely to within the Eden Valley, whereas Penrith was on the great northwest road, which formed a corridor that provided a steady movement of travellers through the town, some of whom were infectives of smallpox (Duncan *et al.*, 1993a; Scott & Duncan, 1993; Scott *et al.*, 1998a). This area of northwestern England was backward, with marginal farming conditions and with the inhabitants frequently existing under near-famine conditions and it is noteworthy that the population at Penrith, like other towns in the northwest, suffered from a severe famine and heavy mortality in 1596, immediately preceding the plague.

5.1 Traditional account of events in the plague

The first victim of the plague was said to be Andrew Hogson, whose death, as a stranger, is reported in the parish register on 22 September 1597. Andrew Hogson died in his lodgings in one of the little white-washed, stone-roofed cottages that once stood at the north end of King Street, formerly Nether End, which was part of the great northwest road through the centre of the town (Irving, 1935). He is reported to have 'come from some place at a considerable distance and brought the disease with him, it being believed that the period of contagion extended to ten days' (Furness, 1894). He was evidently living in the town before he died and was not just a passing traveller. After the death of Andrew Hogson there were no more deaths of the plague until 14 October 1597 when a daughter of John Railton died.

The entry in the register of Hogson's death is immediately followed by the words 'Here begonne the plage (God punismet) in Penrth. All those that are noted with P. dyed of the infection, and those noted with F. were buried on the fell'. The site of this emergency burial ground, 700 yards to the north of the town centre, was still marked on the 1923 Ordnance Survey Map; it is not strictly 'on the fell' and is now covered by a housing estate. Six hundred and six entries were marked with a 'P', of which 284 were children and 322 were adults, and an additional 213 entries were followed by 'F'. Apparently, some were also buried in the Grammar School yard and some in their own gardens and presumably were not recorded in the church registers. By the end of March 1598, the scale of the crisis led to the need for alternative locations for internment; some of the entries in the burial register now include the symbol 'Sy' and it is believed that 'y' suggested that the victim was buried in a yard and that 'S' implied that it was their own. There have been suggestions that the number in the register represent only those who were buried in the churchyard (Furness, 1894), but since many are recorded in the burial register with an 'F' this is clearly not correct. Between the 13 June (when there were six burials) and the 24

Table 5.1. *Number of plague deaths per family at Penrith, 1597–98*

No. of deaths/family	1	2	3	4	5	6	7	8
No. of families	104	77	34	16	7	3	0	1

Data from reconstitution study (Scott, 1995).

June 1598 (when there were four burials) there is recorded in the published transcript of the registers 'There is no sign here of a gap, but some entries have probably been lost'. Furthermore, there are wills of people who died in the plague whose burials we have not been able to trace in the registers and we conclude that the totals listed above are probably short by at least 50 deaths.

Of those dying during the epidemic, 485 have been identified as belonging to 242 families; a further 74 families appeared to be unaffected. The previous year (1597) was a time of extreme hardship and deprivation for the northwest and a number of those succumbing to the pestilence of 1598 may have been vagrants in search of food. Table 5.1 shows the number of deaths in each family; three-quarters of the families suffered one or two deaths. Sixty-three families became extinct and 79 families had only one surviving parent; a further eight families were left parentless.

Furness (1894) in his *The History of Penrith from the Earliest Record to the Present Time . . .* gave the following account of events in the plague, which, although it was probably anecdotal and conjectural, was of interest because it was written at a time before the true biology of bubonic plague had been elucidated. However, he knew all about buboes.

During this dreadful time the actual state of Penrith can scarcely be imagined. Not a solitary marriage was registered for the whole summer. Houses supposed to contain the infection were shunned, and their inmates suffered to die unaided. People almost feared each other's looks. Evening, the time when the attacks of this disease generally came on, had peculiar terrors during the visitation of the plague. The first paroxysm or period, which included from the evening to the following night, was frequently fatal. The third and fifth days were considered, on the whole, those of greatest danger, and if they survived over the fifth day and the bubo was fully formed, then the patients were considered almost out of danger. All these circumstances were only too well known where the plague had been for some time raging. The wild and furious look accompanying the disease in its incipient state, which ought to bespeak pity for the suffered, dispelled all feelings of humanity from the breast of those whom circumstances brought in their way. The staggering occasioned by an extreme prostration of strength was a warning to his neighbour to flee. The poor sufferer sought his home – perhaps he was the last of his household – and the indifference to recovery, which was considered a most unfavourable symptom, alone relieved the horrors of his despairing and forlorn condition.

The farmers from outside the town would not enter while the plague raged, all regular markets were suspended and temporary markets were set up on the outskirts. Furness (1894) reported that 'the inhabitants of the "dales", who were doubtless especially timorous, came no nearer than Pooley'. Townsfolk paid for supplies by tossing coins into hollowed-out stones containing some crude, supposedly disinfectant fluid, possibly vinegar (Furness, 1894), the so-called plague stones. One such receptacle is preserved in the grounds of Greengarth old folk's home in Bridge Lane at the southern end of the town. The other plague stone, which was located in a small field adjacent to Milton Street (formerly Grub Street) and served the people coming to market from the north, has now disappeared (Irving, 1935; Watson, 1992). This partial isolation foreshadowed the successful quarantine practised at Eyam in 1666 (see section 10.7) but was more rudimentary than the measures introduced at Carlisle (section 7.7) and York (sections 9.1.1 and 9.4).

5.2 Size of the population at Penrith

There was an inscription at St Andrews Church, Penrith, which suggested the following levels of mortality in the plague of 1597–98: Penrith 2260, Kendal 2500, Richmond 2200 and Carlisle 1196. Although parish registers are liable to understate the extent of mortality during epidemics and to record burials in the churchyard rather than deaths in the parish, it is generally regarded that these figures would be well in excess of the population of the parishes of these market towns at that time. It has been suggested that they are the aggregate of the surrounding parishes or, more likely, that the rural deaneries, and not the towns, were named in the original inscriptions (Walker, 1860), although there is no evidence of a substantial mortality in the outlying districts. It is probably another example of exaggerated and suspect recording of plague mortality.

It is difficult to derive firm estimates of a population at a time before such assessments as the Hearth Tax and the Compton Census, although the inherent inconsistencies and problems associated with these have been well documented. For instance, the Hearth Tax is not an entirely reliable source, since there is no certainty that all households were included and possibly up to 40% were never recorded (Husbands, 1984).

Table 5.2 provides estimates of the size of the population at Penrith from different sources. The population size in 1597 (immediately before the plague) and in 1613 (after the major immigration had taken place) has been estimated from a family reconstitution study in which each event in the

Table 5.2. *Population estimates for the parish of Penrith*

Year	Estimate	Source
1554		First plague at Penrith
1587	1700	200 deaths, probably largely because of typhus
1596	1500	153 deaths of people native to Penrith, probably because of famine
1597	1350	Before start of plague. Estimated from family reconstitution study
1598		Second plague at Penrith
1599	858	485 deaths of the plague of people native to Penrith
1610	1134	Appearance of 65 new families during 1600–10
1613	1150	Estimated from family reconstitution study
1623		Mortality crisis (241 deaths recorded in parish register); probably severe famine
1642	1233	Protestation returns for parish (411 × 3)
1673	1147	Hearth Tax for township (270 × 4.25)
1676	1365	Compton census for parish
1688	1147	Denton's survey for parish

Data for 1642, 1673, 1676 and 1688 taken from Clark *et al.* (1989).

registers is collated for every family (see Wrigley, 1966; Scott, 1995; Scott & Duncan, 1998). The calculated values are 1350 in 1597 and 1150 in 1613. The population estimates for 1596 and 1587 are derived from the value for 1597. These results must be regarded as underestimates. Sixty-five families moved into Penrith in the 12 years following the plague, producing a population estimate of 1134 in 1610, a value that agrees well with the estimate of 1150 in 1613 derived from the family reconstitution study and with the estimates derived from the Hearth Tax and other assessments (Table 5.2). It is concluded that the population at Penrith was some 1700 before the famines and plague that occurred during the period 1587–98 and was some 1200 in the early 17th century when the community was settling into a steady state (see section 13.17). Assuming that the population was a minimum of 1350 at the time of the plague, approximately 45% died during the outbreak.

5.3 The three phases of the epidemic: the serial generation time and contact rate

The mean annual number of burials at Penrith was 49 for the period 1557–85; this number rose to 103 in 1586 and to 195 in 1587, when the population experienced a rise in mortality; there was an average of 57

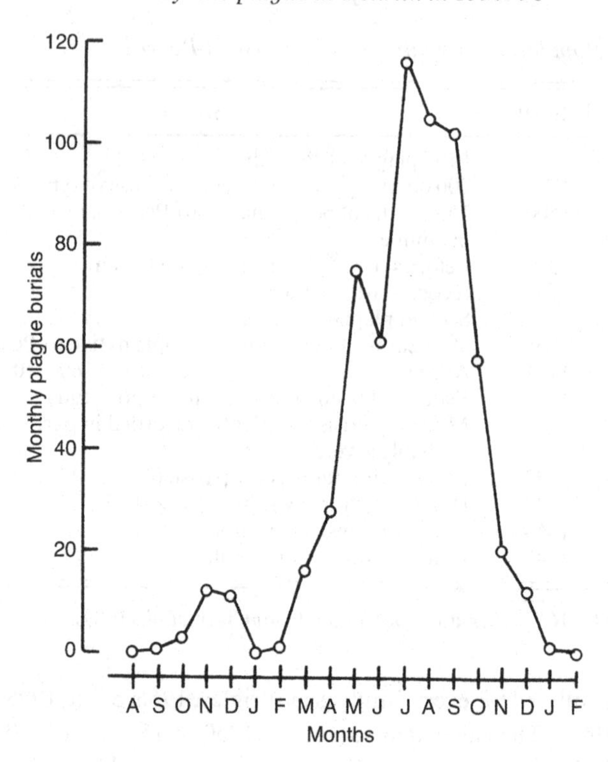

Fig. 5.1. Monthly plague burials at Penrith, August 1597 to February 1599, illustrating the three phases of the epidemic: small autumn peak, almost complete disappearance in winter and the explosion of deaths in the following spring and summer.

burials per annum thereafter until the fateful years of 1596–98. Figure 5.1 shows the monthly plague burials at Penrith. There is a gap in the register of burials between 13 and 24 June, see above, probably because of the pressure of events on the vicar (whose wife and son died in May) and his clerk, and it is almost certain that there were more victims than the 62 recorded. If so, June may have been the month of peak mortality during the epidemic.

The monthly pattern of deaths represents the fingerprints left by the disease upon which the historical epidemiologist may work. It is evident that there were three phases to the pestilence: (i) the epidemic began with a single death in late September 1597 and rose to a small autumnal peak in November–December, (ii) during the winter there were no plague deaths in January and only a single death in February 1598, and (iii) the pestilence appeared again in spring 1598 and rose to a peak in June–July before slowly subsiding in the autumn. After the burial on 6 January 1599, 'HERE ENDETH THE VISITATION' is written in the registers. There is no evidence that it ever returned to Penrith, although plague was widespread through England during the next 70 years.

The third phase clearly follows the typical Reed and Frost dynamics of an infectious disease spread person-to-person (see section 2.5). This phase lasted about 8 months and, in this respect, the outbreak is comparable with the epidemics in the dioceses during the Black Death (see Fig. 4.2) and with plagues in other rural towns, examples of which are shown in Fig. 5.2. The duration of the main part of the epidemic in each of these populations was always long, usually between 7 and 10 months, and this is indicative of a long serial generation time, as shown in Fig. 2.5. This conclusion is supported by the initial events at Penrith, where 22 days elapsed after the death of Andrew Hogson before the next victim was buried, giving a minimum serial generation time of 22 days.

A feature of Fig. 5.1 is the slow build-up of the epidemic in early spring, which is also indicative of a long serial generation time and a low mean contact rate (see Fig. 2.5). Figure 5.3A shows Reed and Frost modelling of a population size $N = 1325$; serial generation time = 25 days; mean contact rate = 3 ($p = 0.0023$) and this may be directly compared with the number of plague burials at Penrith in phase three plotted at the same intervals of serial generation time in Fig. 5.3B. Obviously, a theoretical model which assumes a perfect mixing pattern does not replicate real-life conditions exactly. The ordinate of Fig. 5.3A shows the number of people predicted to catch the disease whereas we have data only for those that died of the disease; some survived infection in other plagues and recovered, although the mortality was probably very high. All Reed and Frost models predict that over 90% of the population would be infected, which was not the case with plague; some families may have fled and some households may have practised successful quarantine measures, i.e. perfect mixing may have been prevented by partial isolation techniques. The theory of Reed and Frost dynamics requires that N is completely susceptible whereas some of the inhabitants may have been immune or resistant. The mean contact rate is dependent on human behaviour and may change during an epidemic; people may have been cautious and shunned their neighbours completely during the later stages of the outbreak.

Nevertheless, the modelling is instructive and it demonstrates how the shape of the mortality curve (see Figs. 5.1 and 5.2) is determined by the long serial generation time and the low effective contact rate. A difference in

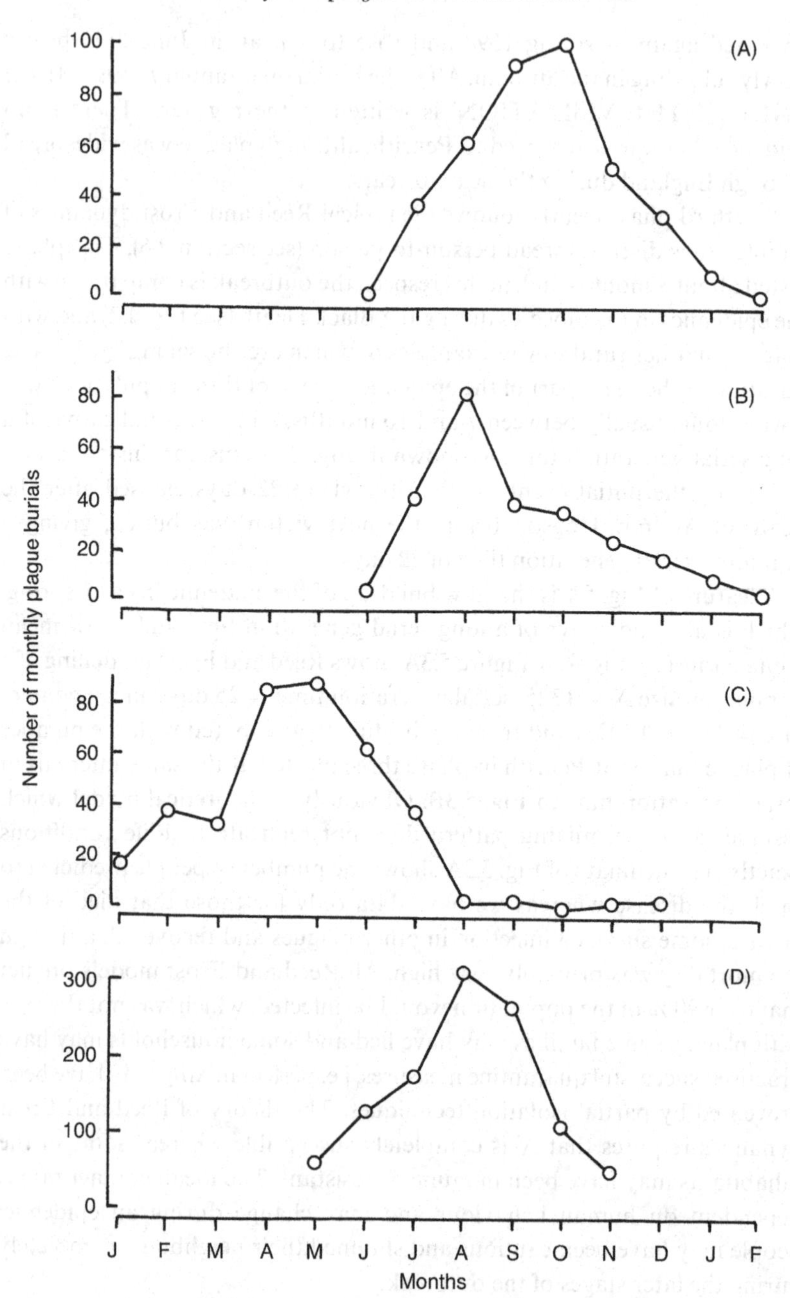

Fig. 5.2. Seasonal pattern of plague mortality in four rural towns in England. (A) Oswestry, Shropshire, 1559. (B) Totnes, Devon, 1570. (C) Ashburton, Devon, 1625. (D) Manchester, Lancashire, 1645. Data from Shrewsbury (1970).

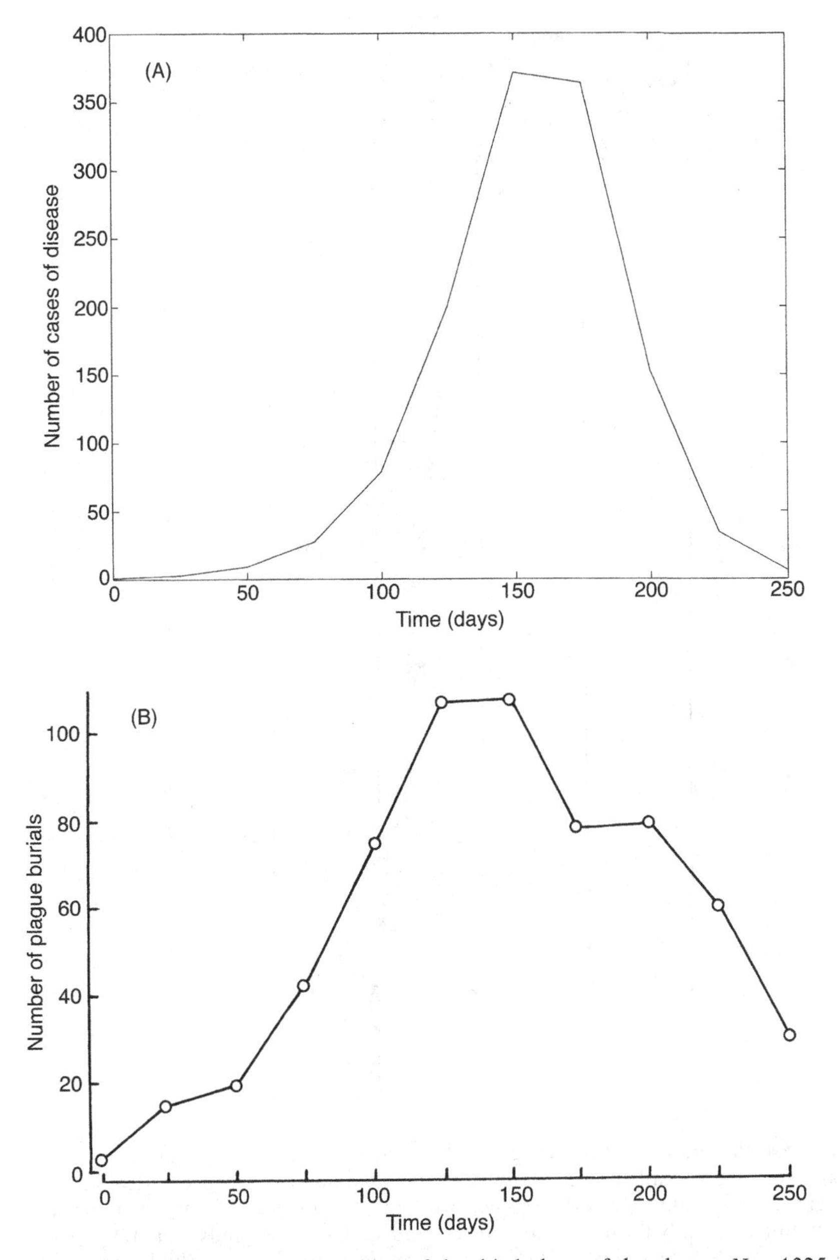

Fig. 5.3. (A) Reed and Frost modelling of the third phase of the plague. $N = 1325$, serial generation time 25 days, mean number of effective contacts = 3. (B) Plague burials at Penrith in phase three plotted at the same intervals of the serial generation time.

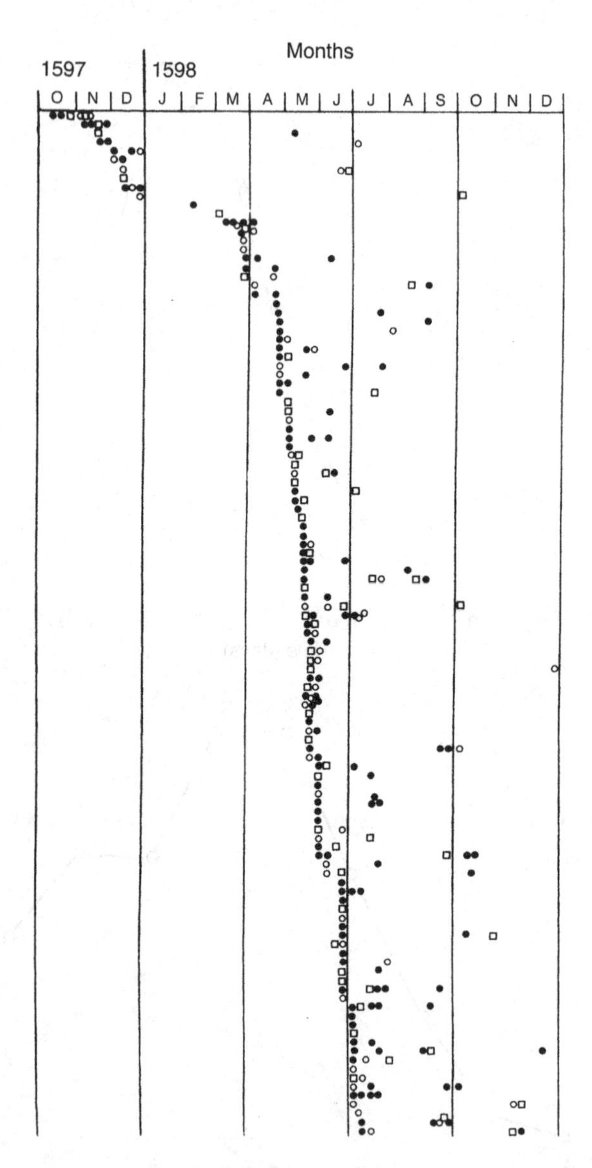

Fig. 5.4. Pattern of deaths during the plague at Penrith 1597–98. The families are arranged sequentially in order of the first recorded death from plague. Reading the figure horizontally shows the spread of the disease through a family; reading the columns vertically illustrates how the disease spread from family to family. Open circles, adult female; open squares, adult male; closed circles, child aged 1–15 years. Recordings begin October 1597, after Hogson's death. Note the high household R_0 in phase one, the almost complete disappearance of the plague in the winter months and the explosion of the epidemic in the spring of 1598 with a high interhousehold R_0.

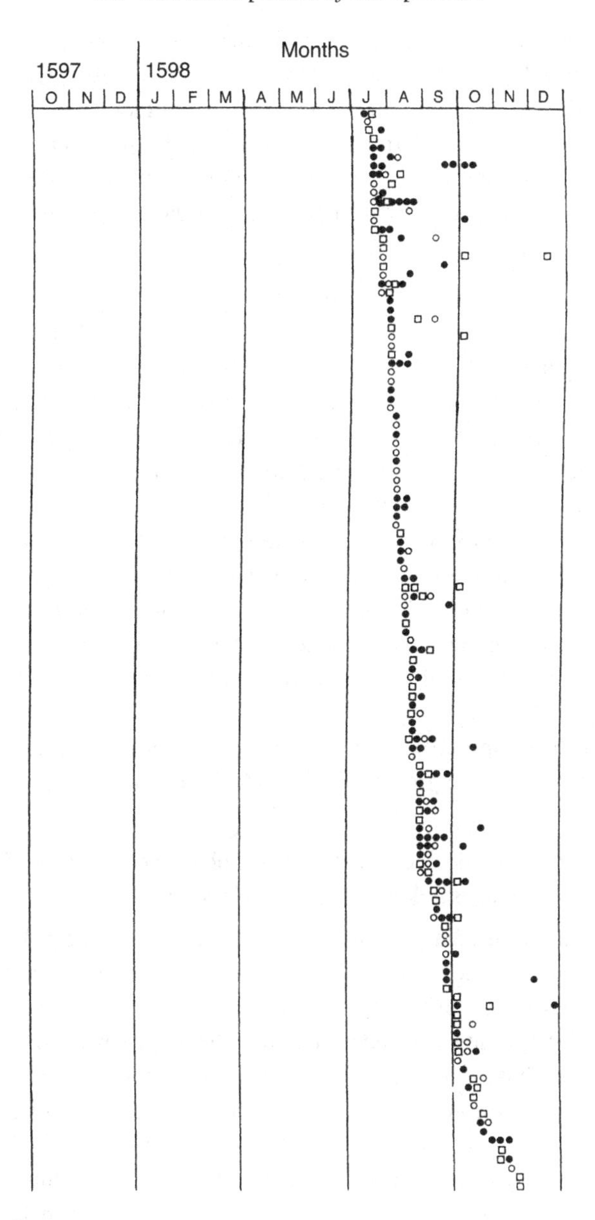

Fig 5.4 (*cont.*)

contact rates will explain minor differences between the dynamics of the epidemics in different populations. But, as we show below, reality is more complicated than simple theory. The effective contact rates within households were much greater than between households and it is upon the latter that the propagation of the epidemic depended. Furthermore, as we shall see, effective contact rates differed between winter and summer.

5.4 Spread of the epidemic at Penrith

The spread of the plague is shown diagrammatically in Fig. 5.4, in which the mortality in each family is shown horizontally (household contact rate). The families are arranged in order of the first recorded death. The data presented are probably not completely accurate because sometimes, for example, there is single record of the death of a member of a household who was listed as a servant and the infection may have occurred within the household of their employer. Nevertheless, Fig. 5.4 shows graphically how the disease spread; reading the columns vertically shows how the disease moved from household to household and the rapid spread *between* households in summer 1666 is clearly evident. The spread of the infection in the first and second stages of the epidemic (September 1597 to May 1598) is illustrated in detail in Fig. 5.5. It has all the characteristics of a lethal infectious disease spread person-to-person and shows a pattern identical with the outbreak of plague at Eyam in 1665–66 (see Chapter 10).

5.5 The epidemic during the first two phases: elucidation of the epidemiological characteristics of plague

Two points are immediately apparent from Figs. 5.4 and 5.5. Firstly, only 10 families were affected in the autumn of 1597 (phase one) and the plague then appeared to die out, with no fatalities in January and only one in February 1598 (phase two). Secondly, once the disease was established in a family during phase one, successive deaths within the household followed (high household contact rate). With respect to the spread of the plague, infections within the household are a dead-end and, for the epidemic to be perpetuated, it must be transmitted to other households. Phases one and two of the outbreak differed markedly from phase three in this respect, transmission *between* families at Penrith was limited during the autumn and winter (Fig. 5.5) whereas the spread was dramatic in summer 1598 (Fig. 5.4).

It is difficult to determine the epidemiological parameters of the plague in the Black Death because there is no detailed evidence of the actual

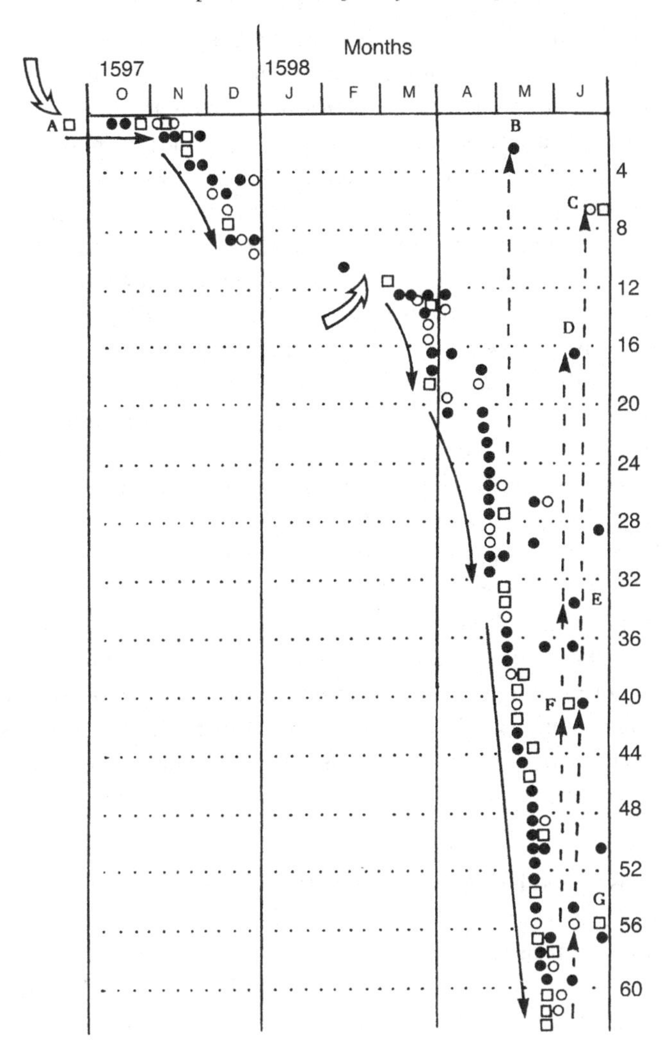

Fig. 5.5. Pattern of deaths during the early stages of the plague at Penrith from the first death on 22 September 1597. The families are numbered sequentially in order of the first appearance of the epidemic (right-hand side). The first death and the outbreak in February 1598 are shown with large, open arrows. The break in January–February 1598 can be clearly seen. The spread of the plague is shown by solid arrows, initially within families and subsequently between families. The steepness of the solid arrows illustrates the rapidity of spread (e.g. see May 1598). Certain individuals are designated by letters: A, Andrew Hogson, stranger; B, C and D, probably not infected from within the family but from individuals in other families lying immediately to the left of the dotted lines; E, F and G, possibly also infected from outside the family. Symbols as for Fig. 5.4.

events; equally, it is difficult to analyse accurately the spread of the infection in the plagues in London because of the size of the population. However, the circumscribed epidemics in towns and villages which we describe below (e.g. Penrith and Eyam) provide admirable material for the critical analysis of the epidemiological characteristics.

Three factors point to a long incubation period for the disease. (i) The duration of the summer epidemic in phase three, as explained above. (ii) In both Penrith and Eyam (Chapter 10) the epidemic began with the death of a stranger or visitor who had taken up lodgings and was living, albeit temporarily, in the town. He must have been infected elsewhere, perhaps some considerable distance away, and brought the disease into the community whilst still not showing any symptoms. Anyone showing any of the typical signs of the plague, which were well known, would know that they had only some 5 days to live and would not be contemplating moving a considerable distance to communities where they were not known. (iii) The death of the first victim, the incoming stranger, was followed by an interval of 15 (Eyam; see Chapter 10) or 22 (Penrith) days.

The early stages of these outbreaks, whilst the epidemic was being established, that we have studied in detail, are much more informative than are the confusing events in mid-summer when the infection spread with devastating rapidity. We have analysed a number of these outbreaks (see following chapters) and have determined from a number of sources the following epidemiological parameters:

Latent period: 10–12 days.
Infectious period before symptoms: 20–22 days.
Period of symptoms: 5 days.
Total infectious period: 25–27 days.
Total time from point of infection to death: 37 days, a very long time and in complete contrast to pneumonic plague (see section 3.13).

Figure 5.6 illustrates the first and second phases of the plague at Penrith; the latent and infectious periods are indicated by appropriate subdivisions of a line against each person dying. The first line is that of Andrew Hogson, the stranger, who died on 22 September; he was infected at a place where the plague was raging on or about 16 August, so that he became infective on about 28 August 1597. Dashed vertical lines are drawn on Fig. 5.6 on 28 August and 22 September, denoting the period during which Hogson, the primary case, was infective and it is evident from this that he infected only the family of John Railton, a cutler; in all other families the initial point of

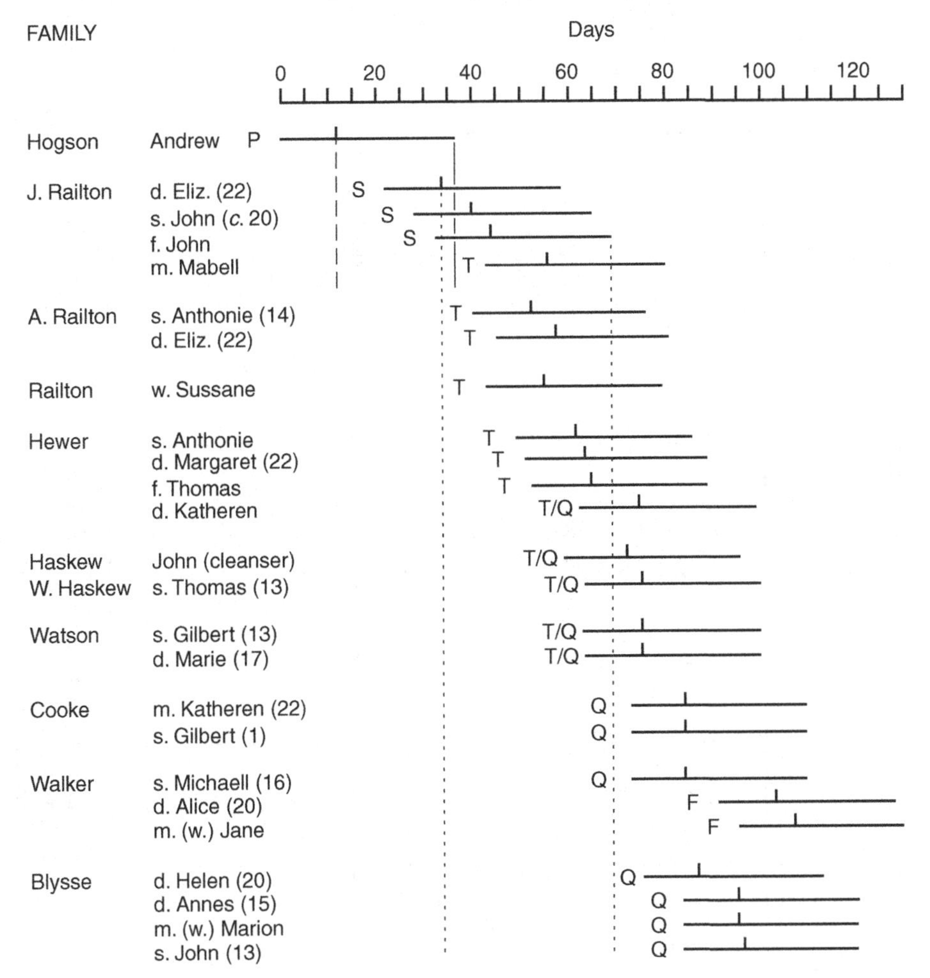

Fig. 5.6. The start of the outbreak of the plague at Penrith and its suggested spread within and between families. In this and subsequent figures the duration of the infection in each victim is shown as a standardised line that is divided into latent and infectious periods. The duration of the infectious period of the primary case (Hogson) is indicated by the vertical dashed lines, during which secondaries can be infected. The vertical dotted lines are drawn between the day on which the first secondary case becomes infectious and the day on which the last suggested secondary dies. Tertiary infections are only possible between these dates. The families or households are listed on the left-hand side of the figure. P, primary; S, secondary; T, tertiary; Q, quaternary; F, fifth generation infection; d., daughter; f., father; m., mother; s., son; w., widow. Age (years) of victim at death given in parentheses where known. Scale: days after 16 August 1597.

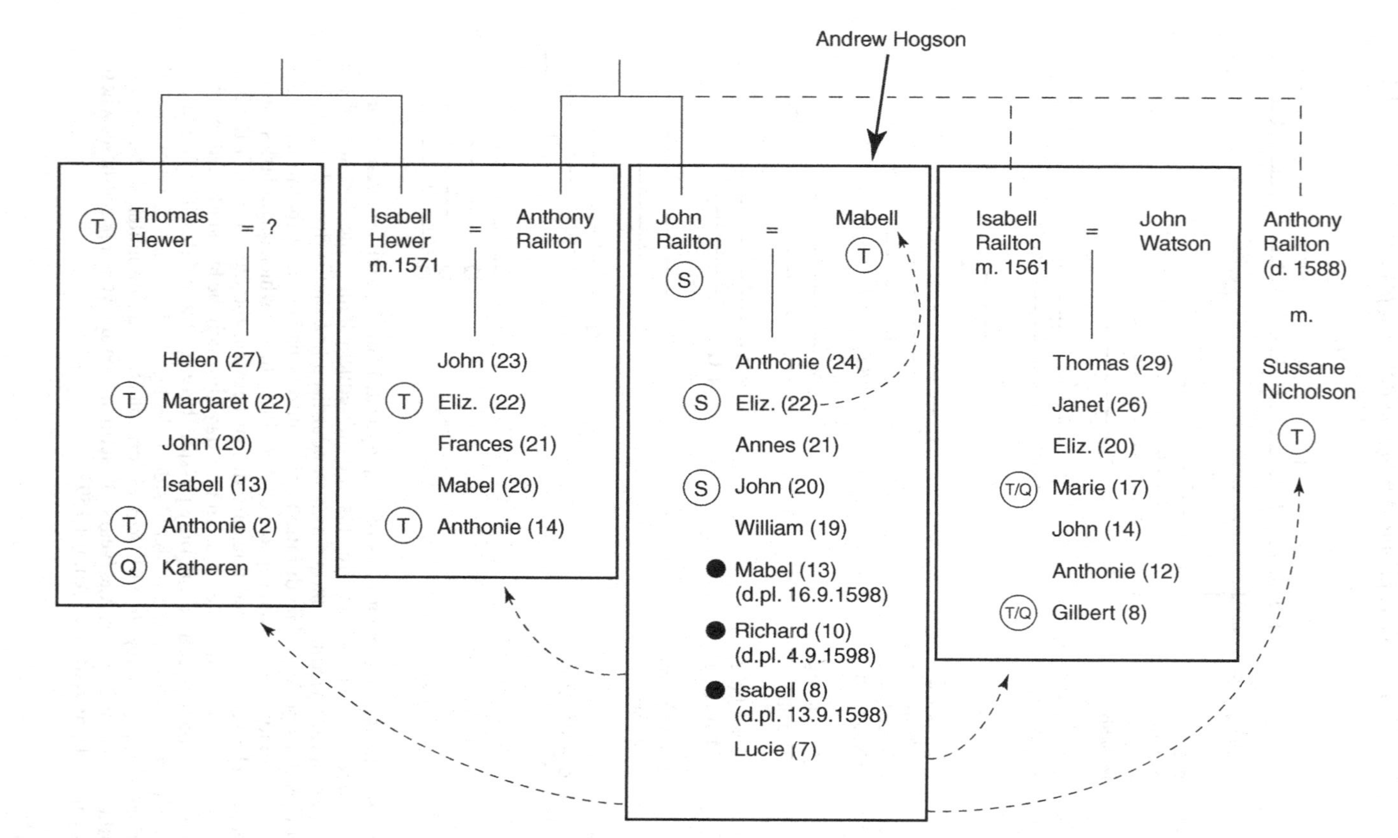

Fig. 5.7. Relationships between the Railton and Hewer families. Ages (years) in 1597 given in parentheses where known. Victims are suggested as secondary (S), tertiary (T) or quaternary (Q) cases. Victims marked with a closed circle died in summer 1598 in phase three of the epidemic. Dotted arrows indicate suggested pattern of spread of the infection between families. Dashed line indicates suggested family relationship. d.pl., died of plague.

infection occurred after Hogson had died (Fig. 5.6). Likewise, vertical dotted lines are drawn between the day on which the first secondary case became infectious and the day on which the last suggested secondary dies; tertiary infections in the community are possible only between these dates.

The three co-secondaries (see section 2.2) in the family of John Railton were, sequentially, daughter Elizabeth (aged 22 years), son John (*c.* 20 years) and the father John, who were infected on 7, 13 and 17 September respectively, all probably before Hogson was showing symptoms. Three co-secondaries infected in rapid succession demonstrate a high household contact rate and it is possible that Hogson was lodging with the Railton family; he probably infected nobody else. Mabell Railton, the mother, was clearly a tertiary infection from one of her two children because she was infected after Hogson died.

The epidemic spread next to the Anthony Railton and Hewer families by tertiary infections; all three families were interrelated, as shown in Fig. 5.7. Only two children of John Railton died in the first phase of the epidemic but three more died 12 months later in rapid succession in September 1598 in the closing stages of phase three. Two of the children of Anthony Railton, Anthonie junior (aged 14 years) and Elizabeth (aged 22 years) were the next tertiaries infected by the secondaries, probably their cousins (Figs. 5.6 and 5.7). They were infected after Hogson had died; the other members of the family survived. The first step in the spread of the epidemic to another household, therefore, was to a closely related family and, as we shall see in the spread of other plagues during the autumn and winter (e.g. Eyam; section 10.3), the transmission was quite probably effected via the youngsters. Sussane Railton, a widow, whose relationship to the Railton families is not clear (Fig. 5.7) was the next tertiary infection.

The next group of tertiary infections were in the Hewer family; Isabell Railton was the sister of Thomas Hewer (Fig. 5.7). Son Anthonie (probably aged 2 years), daughter Margaret (aged 22 years) and the father Thomas Hewer were infected on 4, 6 and 7 October, respectively, and so must have been infected by the three original secondaries in the family of John Railton, as shown in Fig. 5.6. We suggest that the transmission to a new household was probably effected by Elizabeth Railton, who was the same age as Margaret Hewer. Katheren Hewer was probably a quaternary, infected by her brother Anthonie (Fig. 5.6).

After the death of Thomas Hewer, there was a gap of 7 days before the plague struck again with the death of John Haskew, a cleanser, who was infected on 14 October, the same day that Elizabeth Railton (original co-secondary) was buried. He may have been infected at the Railton

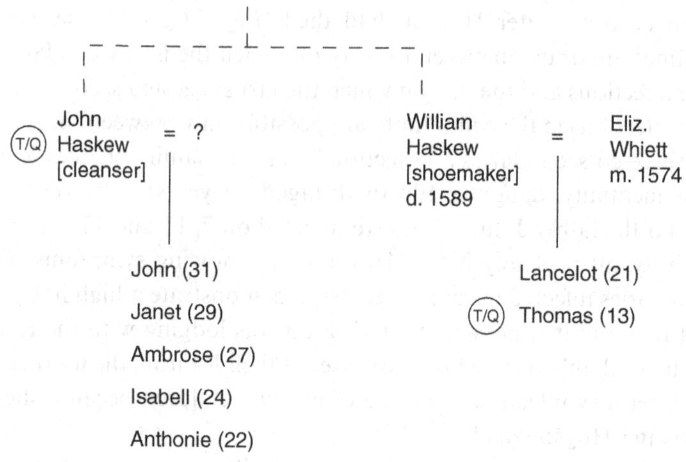

Fig. 5.8. Relationships between the Haskew families. Abbreviations as Fig. 5.7.

household if he attended in his capacity as a cleanser. In Edinburgh in 1585, cleansers were ordered to clean the houses of those infected with plague and were responsible for the slaughter of swine, dogs and cats to prevent the transmission of the disease. The family tree of the Haskews is shown in Fig. 5.8; Thomas Haskew (aged 13), the nephew of John Haskew, was infected 4 days after his uncle on 18 October, possibly when he accompanied him in his work as a cleanser. Probably all the family of John Haskew had left home and he may have been looking after his nephew because his father William Haskew had died in 1589. The other members of the two Haskew families escaped during the first phase of the epidemic, but Anthonie (aged 22 years) died 6 months later on 9 May 1598 during the third phase.

The plague struck next in the Watson family, whose family tree is shown in Fig. 5.7; John Watson had married Isabell Railton in 1561. Two of their children, Gilbert (aged 8 years) and Marie (aged 17 years) were both infected on 18 October 1597; they may have been tertiary infections from John Railton, father and son, when they were showing symptoms in the terminal stages of the disease or, more probably, quaternary infections from their relatives Anthonie Railton or Margaret Hewer (Fig. 5.7). Again, the infection was spread to the youngsters; the others in the Watson family survived.

This transfer of the infection between households of related families can be described in terms of mixing patterns (see Fig. 2.7) and a mixing matrix, as described in section 2.10.

The epidemic spread next to the Cooke, Walker and Blysse families,

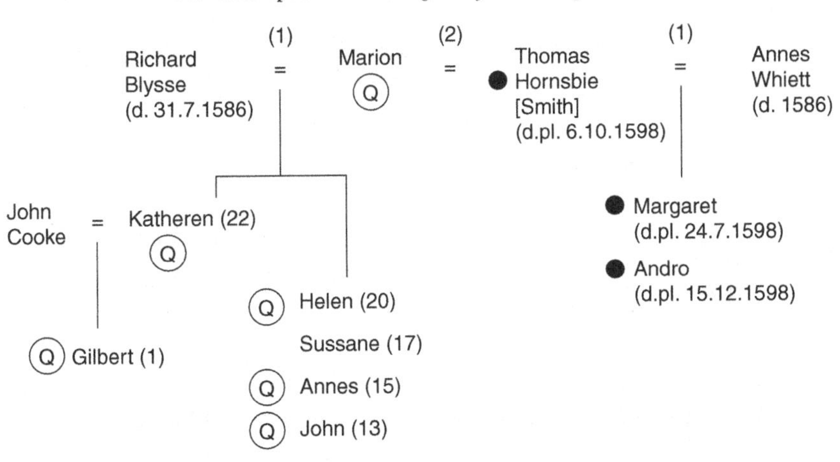

Fig. 5.9. Relationships between the Blysse, Cooke and Hornsbie families. Both Marion Blysse and Thomas Hornsbie married twice. Abbreviations as Fig. 5.7.

where the infection was first introduced between 27 and 30 October (Fig. 5.6). The interrelationships between the Blysse and Cooke families are shown in Fig. 5.9: Katheren Blysse (aged 22 years) the eldest daughter, had married John Cooke in 1595. Katheren and her son Gilbert (aged 1 year) were both infected on 27 October and her sister, Helen Blysse (aged 20 years) on 30 October 1597 (Fig. 5.6). There is no evidence of a relationship between the Blysse family and any of the infectives above, but the most likely candidates would be Margaret Hewer and Elizabeth Railton, both, like Katheren, aged 22 years. One of them may have been visiting the Blysse family, particularly Katheren and Helen, and infected the baby at the same time. The important point is that the infection probably occurred indoors.

Annes Blysse (aged 15 years) and Marion Blysse (her widowed mother) were then infected on 7 November and son John (aged 13 years) on 8 November (Figs. 5.6 and 5.9); they might have contracted the disease from Katheren when she first became infective. Figure 5.9 shows that John Cooke, and Sussane Blysse survived.

There were also two separate plague deaths at this time: Janet, wife of Robert Ladiman and John Lyvocke, a joiner, who were buried on 9 and 12 December, respectively.

Meanwhile the epidemic had also spread to the Walker family (Fig. 5.10); the son Michaell (aged 16 years) was infected on 27 October and he then infected his sister Alice (aged 20 years) and mother Jane (aged at least 47 years) (see Fig. 5.6). Jane Walker was buried on 26 December 1597, the last

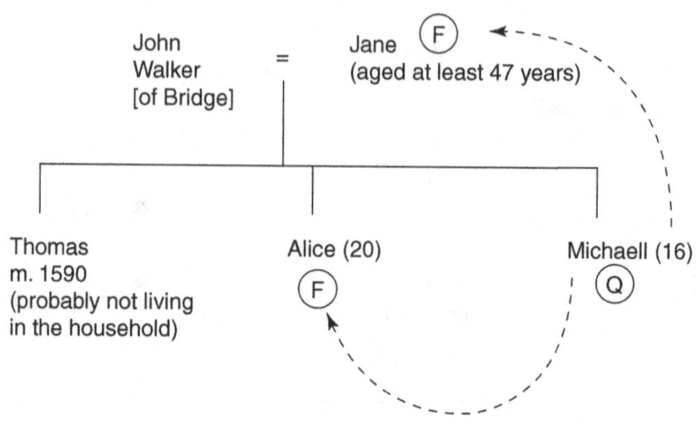

Fig 5.10. Plague victims in the Walker family. Abbreviations as Fig. 5.7.

plague death in phase one of the epidemic (Fig. 5.6); thereafter the outbreak persisted through the winter only by the skin of its teeth and the inhabitants of Penrith must have thought that the epidemic had ceased.

In summary, we have shown that, by combining family reconstitution techniques and epidemiological theory, it is possible to provide a human story set within a scientific framework that underlay the events during the autumn of 1597. The epidemiology summarised in Fig. 5.6 is completely consistent with an infectious disease that was spread person-to-person and it is possible to identify the probable infective in each case. This disease was not bubonic plague; the disease was brought by an infective stranger who lodged in the town; there was no colony of black rats in the cold winter of the Eden Valley, many miles from warmer, southern ports; no fleas could have bred there; the incubation and infectious periods are completely different; the inhabitants of Penrith obviously assumed that it was a 'standard' infectious plague; the epidemic spread within households and between relatives during the first phase in a predictable way, quite unlike bubonic plague; the dynamics of phase three of the epidemic are completely consistent with a Reed and Frost infection with a long incubation period.

The key to understanding the epidemiology of the plague lies in the lengthy incubation period and, because of this, apparently healthy infectives could move around the country on foot or horseback covering considerable distances before they were struck down. Another consequence was that between mid-August, when Hogson was infected somewhere else, and the end of December the epidemic had extended only to quaternary infections (Fig. 5.6).

Only some eight or nine households were infected in the first phase; during the autumn there was a high household contact rate but the epidemic spread with difficulty to other households so that its effective propagation was slow. This transmission was completely different in the summer during the third phase. Everything suggests that people were most readily infected indoors during the autumn and infection probably required more than a casual contact. As at Eyam, the initial infective (Hogson) was lodging in the town for some time and he infected only three members of the Railton family; meeting people in the street seems to have been an ineffective means of transmission in the autumn. However, once the infection was introduced, not everyone in the household died, although they were presumably in close contact; some survivors died later in the third phase of the epidemic.

For a plague epidemic to persist, it was essential that the infectious agent was transmitted to other households and at Penrith during phase one this was usually achieved by visiting close relatives, probably indoors. As at Eyam in 1665, analysis suggests that it was probably the young people, who were free from symptoms and apparently healthy who were carrying the infection round.

There were no plague deaths in January and only one in February 1598, when John, son of John Atkinson de Scill, was buried on the 10th. Thus, in phase two of the epidemic, during the depths of winter, there was a clear break in the transmission of the infection. Jane Walker was buried on 26 December 1597 and it would be predicted that John Atkinson was not infected until about 4 January 1598. There are a number of possible explanations. (i) There may have been someone who was infected and so became infectious during late December and early January but recovered and so did not appear in the burial records. (ii) John Atkinson may have been infected earlier and had a prolonged illness to which he eventually succumbed. Some victims in London *in winter* had infectious periods longer than 42 days (see section 8.1). (iii) One of the people buried in December–January not marked with a 'P' in the register may have died of plague: Elizabeth Smalman was buried on 22 December 1597 and her husband, John Smalman, a 'poor Scottishman', died 22 days later, a typical interval for the plague.

5.6 Explosion of the epidemic in phase three

John Atkinson junior had a critical role in the continuation of the plague through the winter; if he had not been infected, the epidemic at Penrith would have been extinguished. His burial on 10 February was the only recorded plague death between 26 December 1597 and 5 March 1598. It

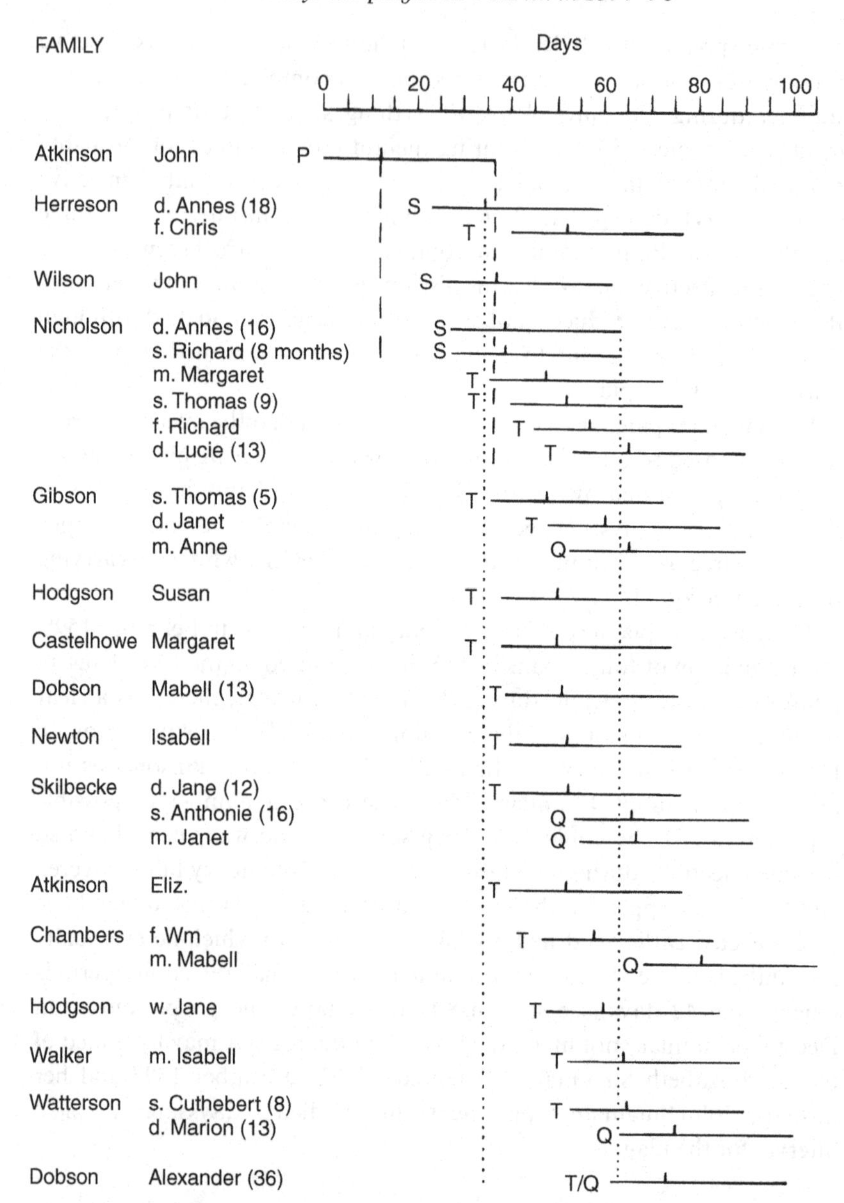

Fig. 5.11. Suggested sequence of infections in the start of phase three of the epidemic at Penrith. The epidemic began again with John Atkinson, who is regarded as a new primary case. Details and abbreviations as Fig. 5.6. Scale: days after 4 January 1598.

is interesting that the gaps between Andrew Hogson (the stranger) and the next recorded plague victim (see Fig. 5.6) and between young John Atkinson and the first victim of phase three of the pestilence are 22 and 23 days, respectively, again indicative of the long incubation period of the disease. John Atkinson junior (considered as a new primary case; probably aged 21 years) infected four co-secondaries in three families (Fig. 5.11); of these Annes (aged 18 years) was the daughter of Xpofer (= Christopher) Herreson, and Annes (aged 6 years) and Richard (aged 8 months) were the children of Richard and Margaret Nicholson. Once more the youngsters had a major role in the transmission of the epidemic to other families; possibly John Atkinson and Annes Herreson were sweethearts.

Phase three of the pestilence gathered momentum in March, with at least 17 tertiary infections from the initial 4 secondaries, including Chris Herreson, Thomas (aged 9 years), Lucie (aged 13 years) and their father, Richard Nicholson. Burials after 1 April 1598 were probably mixed tertiaries and quaternaries and there was a gap between 7 and 14 April after which there was a wave of infections and the epidemic exploded, illustrated by the steepness of the solid arrows in Fig. 5.5.

It is evident from Fig. 5.4 that during phase three of the epidemic the families differed sharply in their household contact rate; in some there was a single burial whereas in others the infection spread through the family. The pattern of the epidemic in phase three was quite different from that in the autumn: in the spring and summer of 1598, the infection passed readily and rapidly between families. For example, 14 people from different families died on 1 August.

5.7 Age- and sex-specific mortality in the plague at Penrith

The plague at Penrith in 1597 occurred only 40 years after the parish registers began, so that there are few records for the older members of the community. However, an age-specific mortality curve for women dying in the plague has been estimated (Fig. 5.12, line a) and this differed from the general cumulative female mortality curve determined from the family reconstitution data base for 1600–49 in the age groups 0–20 years (Fig. 5.12, line b) but agrees closely with the theoretical age structure for Level 4, Women, Model West (Fig. 5.12, line c; Coale & Demeny, 1966), with an expectation of life at birth of 27.5 years. In order to derive an estimate of the age-specific incidence of plague mortality at Penrith that can be compared with other studies, the method described by Schofield (1977) has been used; briefly, the age incidence of mortality during the plague period was

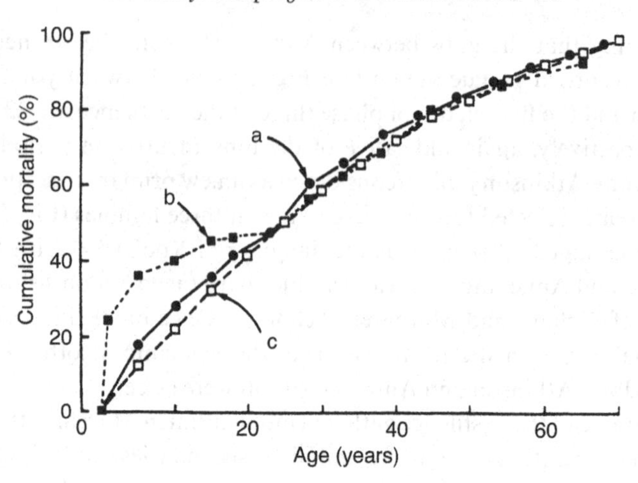

Fig. 5.12. Female age-specific mortality in the plague at Penrith, 1597–98. (a) Estimated percentage age-specific mortality of plague victims (closed circles); (b) cumulative female mortality curve at Penrith, 1600–49 (closed squares); (c) theoretical age structure, Level 4, Women, Model West. Note the close correspondence between lines a and c.

compared with the pattern prevailing before the crisis. The mean annual number of burials before the plague from 1557 to 1597 was 60.2 and they were distributed between the age groups 0–4, 5–14, 15–44 and 45+ in the proportions given in Table 5.3. Subtracting column (1) from (2) gives the observed number of excess burials because of the epidemic for each group; see column (3). This is then compared with the expected excess burials if the age-specific mortality corresponded with the number of people at risk in each age group estimated from the theoretical stable conditions; Model West, Level 4 was used to estimate the percentage of the population in each age group. Those aged over 40 years, for which an age at death cannot be estimated, were divided in the proportions suggested by the model population percentages. For the younger groups (i.e. under 20 years), only 34 burials were unattributed and these were again divided in accordance with the distribution suggested by the model population statistics.

Some authors have suggested that plague mortality was more severe among older children and adolescents rather than an equal incidence among all classes of the population. Hirst (1952) wrote that 'The largest number of cases occur in persons between the ages of 10 and 35 years, the very young and elderly being comparatively little affected' and 'children aged 5–10 years showed the lowest mortality'. In a London parish, Hollingsworth & Hollingsworth (1971) found an increased mortality only for those aged 4–44 years and Pollitzer (1954) stated that adolescents and

Table 5.3. *Excess mortality in plague by age at Penrith, 1597–98*

Age group (yrs)	Mean annual pre-crisis burials 1557–96 (1)	Corrected annual crisis burials (2)	Recorded excess (2) – (1) (3)	Age-structure in model population (4)	Expected excess burials (5)	Excess ratio (3)/(5) (6)
0–4	21.0	84.8	63.8	12	58.2	1.1
5–14	5.2	108.6	103.4	20	97.0	1.1
15–44	22.8	226.4	203.6	46	223.0	0.9
45+	11.2	65.0	53.8	22	106.6	0.5
All	60.2	484.8	424.6	100	484.8	0.9

See text for further details.
Data from family reconstitution study (Scott, 1995).

adults aged up to 45 years displayed the greater susceptibility. It is generally agreed that those aged in their mid-40s and above were less likely to die during a plague outbreak, which might suggest a previous exposure and immunity to the disease. It is noteworthy that the first plague at Penrith was in 1554, 43 years before the great outbreak in 1597. In conclusion, no age group was markedly affected at Penrith, although those aged over 45 years had a lower than expected mortality.

Women were said to be more susceptible to plague than men in parts of Europe, although apparently the reverse was seen in England (Slack, 1985). Hollingsworth & Hollingsworth (1971) and Hirst (1952) have suggested that male mortality was more severe than female, whereas the evidence for Eyam (Bradley, 1977) and Colyton (Schofield, 1977) suggests that the pattern for the sexes was similar. At Colyton, the sex ratio of burials was 93.7 males per 100 females, which was found to be close to the ratio of 92.9 for the 10 years prior to the crisis. Female victims outnumbered males in Barnstaple, Chelmsford and Stratford-upon-Avon (Slack, 1985). Pregnant females, who almost invariably aborted, displayed the highest mortality and, during the 1665 plague at London, a total of 432 deaths due to 'childbed', 'abortive' or 'stillborn' causes appeared during August and September, when mortality was at its highest. Often the cause was more a lack of attention during confinement than any direct consequence of the plague (Leasor, 1962). At Penrith, the sex ratio of plague victims was 137 females to 100 males.

5.8 Wills and testaments of those who died in the plague at Penrith

To read the wills of those who died in the plague allows an insight into the devastating rapidity of the progress of the disease and, more particularly, reveals the tragedy in those stricken families who faced the inevitable with dignity and resignation. They remembered the poor of the parish, often leaving them a quantity of bigg, a poor form of barley, and several of them, as the extracts below show, left money for the building and repair of the bridges in the town. Several (see will (f)) bequeathed belongings that were left to them by their father who had died in the plague but which they had not yet inherited.

The only people who made wills during the plague were those with substantial possessions to leave, and inspection of the extracts below shows that these prosperous members of the town, mostly successful tradesmen, were all interconnected and interrelated with one another and with their witnesses. They were a separate group from the gentry, of whom only three died. Some salient points from the wills are listed below:

(a) Michael Dobson was married on 20 July 1598 and made his will a month later on 27 August 1598. He was buried, aged 20 years, 5 days later, on 1 September.

He leaves to his wife Isabell (who survived the plague) all his 'tenements barnes and yeards with the appurtenance and my tythe estait and tenament right during her wydowehood'.

Witnesses:

Σ John Dobson (probably brother), who, with his wife, survived the plague but four of his children had died of plague before the will was written. He was born before the earlier outbreak of plague in 1554.

Σ Edward Todd, who survived with his wife, but they lost two children after witnessing the will. He may have been born before the earlier plague.

Σ John Burrowe, who survived the plague with his wife and two children.

Σ Thomas Hornsbie, who survived the plague with his wife and two children.

(b) John Steinson made his will on 30 August 1598 and was buried 3 days later on 2 September. He leaves 'all that tenement with appurtances which was my fathers' to his eldest son Thomas (who probably survived the plague) and 'all the rest of goode movable and umovable I do maike and ordaine to my son Richard and Margaret my daughter'. Richard was buried 2 days after his father, whereas it is noteworthy that

Margaret died of the plague some 6 weeks later. A minor bequest was 'to the poore of the parish 6 pecke of ote meale'.

Witnesses:

- Σ Robert Nelson, who survived with one child whereas his wife died of the plague.
- Σ Richard Salkeld, who lost two children in the plague but the rest of the family survived.
- Σ Anthony Steinson (schoolmaster). All the family survived.
- Σ Robert Steinson (shoemaker), who lost one child.
- Σ Thomas Fenton and Michael Graye, both of whom survived with their wives.

(c) Arthur Gibson lost his daughter Elizabeth on 10 July 1598 some 10 weeks before the plague struck his family again; she was buried on the fell. A child was buried on 19 September 1598 and Arthur makes his will 2 days later on 21 September. His wife died on the next day and he and another child (Ann aged 9 years) died on 24 September.

In his will, he wishes

- Σ 'my bodie to be with the bodies of my wyffe and children so manye of us as shall die at this tyme be buried within the parishe church yard of Penreth'.
- Σ 'I give and bequest all my lande tacke and tenements to John my eldest son' [who survived the plague; he may have been living away from home on his own land].
- Σ 'I give to Jane my eldest daughter hawked cowe and also if all my children dye but her I give her twenty sheepe'.

Jane is described as a servant and may have been living away from home. She died in 1606, aged 26 years.

Witnesses:

- Σ Cuthbert Byrd who survived the plague but whose wife and three children died. Cuthbert may have been born before the earlier plague.
- Σ John Castelholme (? alias Castlehow) who survived but whose wife had died of the plague earlier on 22 March 1598.
- Σ John Turner who was born before the earlier plague and survived the plague in 1598.

(d) Robte Holme wrote his will on 25 September 1598, made additions on the 27th and was buried on 3 October, 8 days after beginning his testament. The following are extracts:

- Σ 'I give and bequeath to the buylding of Sandgate Bridge 6s 8d and 4s 3d which Willm Bowman of Kirkoswald oweth me, and to the repairing of Midlegaet Bridge 3s 4d'.

Σ 'I give to Margaret and Frances my two daughters all my lande, houses, tackes and tenements . . .' [They survived the plague but were 23 and 25 years of age and may have left home.]

Σ 'I give to Agnes my wieff two of my best kyne.' [She was his second wife and probably survived the plague.]

Witnesses:

Σ Stephen Stell, who, with his wife and three children, survived.

Σ Willm Emetson, who, with two daughters, survived but whose wife had died earlier of plague on 27 June 1598.

(e) Stephen Jackson made his will on 3 July 1598 but did not die until 1 August. Evidently, he set out his wishes as soon as the plague appeared in his family and not when he first displayed any symptoms: his son aged 13 years was buried on 3 or 5 July and his wife Dorothy was buried on 15 July, 17 days before her husband. Their son John survived. Stephen makes no mention of his wife in the will and he seems to assume that she was doomed. If she was showing the characteristic symptoms on 3 July, her suffering lasted 12 days.

Witnesses:

Σ Lancelet Hind, who survived with his wife although their 2-year-old daughter had died earlier.

Σ John Dobson, who also witnessed will (a).

(f) Isabell Nelson, spinster, made her will on 23 July 1598 and was buried on the following day. She gave 'and bequeath all that my child's portion dewe to me by the last will and testament of Stephen Nelson my lait father'.

(g) Robert Gibbon made his will on 1 October 1598 and was buried 4 days later. He made the following requests: 'Item . . . the tuition and government of my sonne Anthonie [aged 3 years] and my tenement Heath and yard in the head of Penreth with the appurtenances together with his portion to Gilbert Gibbon and William Gibbon my bretheringe during his none age. Item I maike Anthonie my sonne my whole executore of this my last will and testament. And if I and my wyffe dye I gyve all my goods to Willm and Gilbert my bretheringe'.

Robert Gibbon, his wife and their son Anthonie were all buried on the same day, 5 October 1598. There were no survivors.

Witnesses to the will were:

Σ William Gibbon the elder, who, with his wife and two of his children, survived, although they lost one child. William may have been born before the earlier plague.

Σ Gilbert Gibbon, who survived with his two sons, although his wife had already died before he witnessed the will.

Σ Robert Nelson, who was married 2 years previously and survived. His wife died after the will was witnessed but their son, born during the plague, survived.

Σ Edward Todd (brother-in-law), who with his wife survived, although both died in the famine of 1623.

(h) Jefferaye Stephenson made his will on 20 October, the same day that his daughter Isobel was buried, and died 2 days later. He gave 'all his lands meddowes and heridaments . . . all my goods movable and ummovable' to 'Elizabeth my daughter'. He made her 'my whole executrix' and 'I comytt the tuition and government of my daughter Elizabeth to Agnes Stephenson and John Stephenson my brother during her non aige'. He gave further directions 'if my daughter Elizabeth dye' and made other minor bequests as follows:

'Item I give to Sussan Emerson my wief's best cott [coat] and sle . . . es [?shoes]. Item I give to Margaret Todd my wief's best hatt'.

Elizabeth Stephenson was buried on 23 October, the day following her father. He makes no mention of his wife, Janet, in the will and even bequeaths some of her best clothing, although she was not buried until 24 October. Was she already showing symptoms whereas the daughter Elizabeth was not despite dying very quickly?

Witnesses of the will were:

Σ John Stevenson, who survived with all his family.

Σ Edward Todd – see witness to will (g).

(i) Elizabeth Browne, widow, made her will on 29 May 1598 and was evidently suffering from the plague and was being cared for by the Crosbie family because 'and that the said Thomas Crosbie and his weife shall maintain and vel. . . me in my visitation'. She leaves the bulk of her possessions to the Crosbie family but gives to 'Elioner Crosse, one chadgoe and my daughter Jane's clothes' and 'unto Elizabeth Crosbie . . . my lynen web [listed in the inventory at 8s 4d] . . .'

All the Crosbie family survived the plague.

(j) Thomas Sutton, tailor, made his will on an unknown date and was buried on 22 July 1598. Evidently he was aware that his wife might be affected by the pestilence, although, in fact, she was not buried until a month later, on 22 August. He made the following bequests:

'to the poore of the Parish 10s and to the building of the Sandgait bridge whensoever 5s.

to Thomas Sutton my sonne . . . my tenement in Penreth after the death of my wyffe Mabell and I give to her my lease I gave . . . of one other halfe wood land.

to John Sutton my brother John elder sonne my steall cape sword and arrowes I used for service of the Prince.

my other bowes I gyve the one to John of John Turner with 12 shafte and to . . . my quyver and all the arrowes in the same. And to richard Cel. . . my brother in law my good blewe cott and my wedding dublett and to . . . brother my suit of leather apperell.

to Robert Wilson my gymmer bowe with c. . . arrowes and butt . . . shafte. And if God doe call me to his mercie and my wyffe I do gyve to the said Robert one cheste.

to my wyffe the great chest and meale in the same, and if God do call on her, then the chest and meale to be sold to paye my debts.

will that if it please God to call me and my wyffe of this visitation that her . . . friends and myne shall devide the remaine of my goode equallie amongst them.

I appoint Mabell my wyffe to be myne executrice.

I gyve to Henrie Ewrie if we bothe dye a pecke of bigg [barley] and five cartfull of peete.'

Witnesses:

Σ John Turner, who was born before the previous plague. He and four children survived but one child died.

Σ Robert Wilson, a glover, who with his wife and child died in the plague. He was born after the plague of 1554.

Thomas Sutton's widow, Mabell, made her will on 20 August 1598 and was buried on 22 August 1598, 1 month after her husband. She made the following bequests:

'to my sister Jane . . . my best petticoat.

to Robt Wylsons wyffe my next best coat my featherbed and one shyft. [Robt Wilson and his wife Janet were buried 4 September 1598.]

to Anthonye C. . . ell wyffe and her daughter workday coates.

to Reynold Lucas my husbands best fustian doublet and pairr of grene hose and other stockinges and shorte jerkin'.

These wills suggest the rapidity of the course of the disease, with death coming in about 3–5 days from the appearance of symptoms. It is striking that only 2 days before burial (not death) these dying people were still able to think clearly and to make detailed and careful arrangements, including the guardianship of their children. Several members of some families died within a few days of each other (see will (h)) and in one case (will (g)) all three members of the family died on the same day. They were obviously well aware that the remainder of the family would probably die (will (c)).

Some of the older children who were left major bequests and who did not die in the plague may have been living away, perhaps on their own tenements (will (c)). Mabell Sutton (will (j)) was buried 1 month after her husband Thomas; was she infected by him about the time when his symptoms showed or re-infected later? See also wills (b) and (e). Reading the wills suggests the possibility that, once showing symptoms, children may have died more quickly than adults (see will (h)): children were left bequests in the will and were presumably apparently healthy but were buried on the same day as their father.

But not everyone who must have been in contact with an infected person died. Elizabeth Browne, a widow (see will (i)), was cared for in her last days by the Crosbie family, all of whom survived. There are many cases where one or more in the family, often a young child, escaped death and survived the plague. Many of the witnesses survived the plague, in spite of being close to a dying man when they signed his will and in spite of also having plague in their own families. There does not appear to have been any difficulty about finding someone prepared to be exposed to the infection when witnessing the will and some had over six witnesses. We conclude that, within the town, the inhabitants of Penrith did not practice isolation to any great extent.

Were these people who were clearly exposed to infection but survived the plague immune or were they infected but recovered? If the latter, were they subsequently immune? It was believed by the inhabitants of Eyam that it was impossible to catch the pestilence again after recovering from an infection (section 10.8). The parish registers of Penrith do not begin until 1556, 2 years after the plague of 1554 and so the dates of birth of the older members of the community cannot be established. Nevertheless, judging by the dates of marriage and of the births of their first children, some of the witnesses can be identified (see above) as probably having been born before 1554 and so survived the first plague. These men seemed also to have survived the plague of 1597–98 and it is possible that they were immune or acquired immunity from the earlier outbreak of the pestilence.

When the plague in northwest England reached Warcop, the barn where the victim died was burned (see section 7.8) because it was believed that 'miasma' or noxious vapours in the atmosphere were responsible for the disease, and that miasma could be retained in clothes or bedding for long periods and transported, possibly by domestic animals, from house to house (Slack, 1985). In 1666, a Statutory Order was passed that a good quantity of unslaked lime be put into the graves of plague victims, and that the same should not be reopened within the space of a year or more for fear

Table 5.4. *Annual baptisms and burials at Penrith before, during and after the plague*

	1587–96 (mean)	1597	1598	1599	1600	1600–9 (mean)
Baptisms	68.3	46	27	56	67	59.9
Burials	75.3	209	627	19	31	43.6

of infecting others. The inhabitants of Penrith were not worried by such considerations and bequeathed personal clothes and bedding to their friends and relatives and apparently did not think that there was any chance of spreading the infection in this way (see wills (h) and (j)).

5.9 Response of the population at Penrith after the plague

Some 50% of the population at Penrith died between 1596 (a year of famine) and 1598 (at the end of the plague) and yet the mean annual number of baptisms recovered quickly and was at its post-plague level (60 baptisms per year, 1600–1750) by 1600, see Table 5.4 and section 13.17. A population could recover after such a mortality crisis either by new marriages and remarriages and increased fertility, as at Eyam in 1665–66 (Bradley, 1977) or by immigration, as at Norwich after the epidemics of the late 16th and early 17th centuries (Slack, 1985). The birth intervals at Penrith, 1586–95, determined from a family reconstitution study were not significantly different from those after the plague, 1599–1608 and it is concluded that the return to pre-crisis levels was not a consequence of an increase in birth rate.

An estimate of the immigration into Penrith can be derived from the reconstituted family forms. For example, the appearance of a family in the registers where there is no previous record of either partner at Penrith was scored as two immigrants; when one partner at marriage came from another parish, it was scored as one immigrant. The most important immigrations would be the influx of married couples following a mortality crisis to fill available 'spaces' in the community, and at Penrith 65 new families appeared during the first few years of the 17th century and it is concluded that the community was able to return to steady-state conditions so quickly after the plague because of large-scale immigration from the surrounding (unaffected) parishes as soon as the plague had finished.

The large number of families that were eliminated, and the deaths of husbands and wives created openings for marriage, which occurred even

Table 5.5. *The occupations of the head of the households in Penrith affected by plague mortality*

Occupation	No. of families where the head of the household died of the plague	No. of families where at least 1 person died of the plague
Shoemaker/cordwainer	8	10
Smith	5	9
Glover	5	6
Tailor	5	6
Labourer	4	4
Gentleman	3	3
Tanner	2	3
Cutler	1	2
Fiddler	1	2
Wright	2	2
Beadle	2	2
Quarryman	2	2
Maltster	1	1
Saddler	1	1
Locksmith	1	1
Waller	1	1
Gardner	1	1
Joiner	1	1
Cleanser	1	1
Vicar	0	1
Tinker	0	1
Fletcher	0	1
Capper	0	1
Postmaster	0	1
Schoolmaster	0	1
Other classifications:		
Householders (no occupation given)	11	29
Servants: 7 males; 11 females		

while the plague still raged, with 28 in 1598 and 59 marriages in 1599 (the average for the previous 10 years was 19). There were 28 marriages in 1600, after which the average for the next 10 years was 17. Family reconstitution shows that, of the marriages taking place between 1597 and 1600, remarriages accounted for 28% in 1598, 39% in 1599 and 11% in 1600. So, of the 42 husbands and 37 wives widowed by the plague, 25 husbands remarried (60%) but only 12 widows remarried (32%). Five of the widowers married widows.

Table 5.5 shows the gaps left by the deaths of tradesmen in Penrith and the occupations of the head of the families where at least one plague death occurred, and it can be seen that the families of labourers and gentlemen were equally affected; the vicar and schoolmaster also suffered casualties.

5.10 Classification of the epidemics of haemorrhagic plague

The epidemics of the Black Death in each diocese apparently followed Reed and Frost dynamics irrespective of the season of the year, with a single sharp peak in the mortality and the outbreak lasting at least 7 months (see Fig. 4.2), indicative of a long serial generation time. As we shall show in the following chapters, a 'typical' outbreak of haemorrhagic plague during the next 300 years also followed Reed and Frost dynamics, but differed from some of the epidemics of the Black Death in that it began in spring (with the first deaths in March or April), with a slow build-up leading to an explosive epidemic with a peak mortality around August–September. The outbreak lasted at least 7 months and we term these type (i) epidemics.

It is evident that the epidemic at Penrith followed a different pattern and, as we shall see, a proportion of the outbreaks showed a bimodal mortality which we term type (ii) epidemics: plague deaths appeared and peaked in late autumn with the infection confined to a few households during the winter (low R_0); it re-emerged in the spring and then followed type (i) above.

We suggest that these are the two basic types of epidemic of haemorrhagic plague but, of course, there are a number of variants of these basic patterns:

(a) Type (ii) epidemics that did not persist through the winter because R_0 fell to <1, resulted in a very low total mortality. Presumably, there were many minor autumn outbreaks of this type that went unrecorded.

(b) Some type (i) epidemics were persistent and lasted about 15 months, continuing through the winter with a few cases in the following spring.

(c) Some outbreaks *began* in mid-summer (presumably because an apparently healthy infective arrived from a type (i) epidemic) and a major peak in mortality was generated during the autumn. The disease may have persisted through the winter.

(d) An infective arrived in a community in summer and plague spread rapidly through his family (high household R_0) but, because of the low density or small size of the population, few other families were infected (low interhousehold R_0) and so the epidemic did not explode. There were many such minor outbreaks in rural England (see Chapter 9).

6

Pestilence and plague in the 16th century in England

It is bewildering to read Shrewsbury's (1970) account of diseases in England in the 16th century and one gains the impression that plague was rampant somewhere in almost every year of the century. In an attempt to bring some order from confusion, to separate fact from fiction and to distinguish between the possible different causes of mortality, both infectious and otherwise, we have subdivided the outbreaks of disease into the following categories:

(i) The Sweating Sickness that was a scourge for the first half of the century.
(ii) Epidemics apparently confined to London and the southeast corner of England.
(iii) Years in which outbreaks of infectious disease were widespread in England.
(iv) Plagues that were apparently confined to northern England; these are discussed in detail in Chapter 7.

6.1 The Sweating Sickness

The strange disease that became known all over Europe as the English Sweat because of the extreme susceptibility of the English (the disease was absent in Scotland) first appeared (presumably as a result of a mutation in the infectious agent) in the autumn of 1485 and a good account of it is given by Wylie & Collier (1981). Creighton (1894) said that 'the language of historians is that the sweat of 1485 spread over the whole kingdom. We hear of it definitely at Oxford where it lasted but a month or six weeks'. He found reference to it in Bristol and Croyland Abbey but assumed that it spread little outside London and lasted only during the autumn and early winter. The Sweat reappeared in 1508, 1517 and 1528, with the fifth and last outbreak in 1551, after which it disappeared from England forever. Most of the accounts of these outbreaks in England are confined to London except for the fifth Sweat, which began in March–April in Shrewsbury and proceeded to London via Ludlow, Presteign, Westchester, Coventry and Oxford 'with great mortality' (Creighton, 1894).

What was the etiology of the Sweating Sickness? Creighton (1894) considered that

> It is only in the autumn of 1517 that the plague overlaps somewhat on the sweat, and even then it becomes noticeable mostly in the winter following the decline of the sweat. The two poisons had existed in English soil side by side, but had not come out at the same seasons; also the sweat had been mostly a disease of the greater houses, and the plague mostly of the poorer.

He continues:

> Other forms of epidemic fever, in the same pestilential class as the sweat, were coming to the front in England as well as in other parts of Europe. Thus, in 1539, a summer of great heat and drought, 'divers and many honest persons died of the hot agues, and of a great laske through the realm'. The hot agues were febrile influenzas, and the great laske was dysentery. Again, in the autumn of 1557, there died 'many of the wealthiest men all England through by a strange fever', according to one writer, or, according to another, there prevailed 'divers strange and new sicknesses, taking men and women in their heads, as strange agues and fevers, whereof many died'. . . . That epidemic corresponded to a great prevalence of 'influenza' on the continent, which was probably as Protean or composite as the fevers in England. It would not be correct to say that these new fevers or influenzas, with more or less of a sweating type, were the sweat somewhat modified.

These outbreaks were described at the time as 'pestilence' or 'plague'. An attack began without warning, generally in the night or early morning. Chills and tremors were followed by high fever and often with a rash (Kiple & Ornelas, 1997). Shrewsbury (1970) described the Sweats as plagues of 24 hours' duration in which anyone who survived beyond that point recovered; it did not attack infants, small children or the aged. If the victim was to recover, the perspiration diminished, to be replaced with an abundant flow of urine, and recovery was complete within a week, or two at the outside. In grave cases, by contrast, intense headache and convulsions were followed by coma, and death arrived with incredible speed. Many died a few hours after symptoms appeared, although most lingered for 24–48 hours. Surviving the disease seems to have conferred no immunity and some, like Cardinal Wolsey, were said to have endured two or even three attacks in succession (Kiple & Ornelas, 1997).

The disease took at least 6 years to spread from England into Ireland and Shrewsbury (1970) considered that this slow movement showed that the Sweating Sickness was not influenza, the identification generally favoured by medical writers. The rapid onset and haphazard manner of spread led many to suspect that influenza was the cause because it is well known for its ability to mutate from a relatively mild illness to a lethal one. However, there were no reported respiratory symptoms or secondary cases of pneumonia, suggesting that Shrewsbury (1970) was correct.

Commenting on the Sweating Sickness, Slack (1985) said that it impressed contemporaries because it was spectacular, killing in 24 hours: 'It attacked a community suddenly and then was gone. In particular, unlike bubonic plague, it struck the prosperous: aldermen and mayors in London and other towns . . . Yet when we have some precise evidence of its demographic effects . . . it is not impressive. The disease swept through a parish in the space of a very few days, a fortnight at most . . .'. This is an important point: the epidemics were of short duration and were apparently following Reed and Frost dynamics and it is evident that the serial generation time was short, completely different from the epidemiology of the Black Death and the subsequent haemorrhagic plagues.

Slack suggested that the Sweating Sickness may have been an arbovirus infection of some kind; these viruses circulate from one vertebrate host to another via the agency of a blood-sucking arthropod, such as the mosquito that spreads yellow fever. Support for this view has been advanced with the suggestion that the Sweating Sickness epidemics were always preceded by heavy rain and sometimes by flooding, both of which would have encouraged mosquito reproduction. However, arboviruses are usually found in the tropics and the suggestion is not convincing.

The etiology of the English Sweating Sickness has recently been explored in detail. Dyer (1997) has analysed 680 parish registers for details of the 1551 outbreak: it was predominantly a rural disease with a limited demographic impact. He suggested that the disease was initially spread through a zoonosis and later by person-to-person transmission because a national epidemic was capable of very rapid transmission along the lines of communication. This hypothesis has been refined by Taviner *et al.* (1998), who also believed that the preponderance in summer and scattered rural nature suggest a viral infectious agent with a rodent reservoir, although it is not clear what species of rodent was implicated because brown rats had not arrived in England and black rats would have been confined to an urban environment. They suggested that the clinical symptoms, particularly the marked pulmonary component, are characteristic of Hantavirus pulmonary syndrome, which is caused by the acquisition of a virus that normally infects small rodents and consists of a brief and nonspecific prodrome of fever, myalgia, headache and rapidly progressive non-cardiogenic pulmonary oedema. Today, it requires mechanical ventilation in 88% of patients within 24 hours of admission; those that die despite ventilation do so within approximately 72 hours.

6.2 Plagues in London in the 16th century

6.2.1 The first half of the century

The more important epidemics in the first half of the 16th century that were largely confined to London and its environs are summarised below chronologically. This list is based on Shrewsbury's (1970) account and where he thinks that bubonic plague was impossible (usually because of a winter outbreak) he generally falls back (for no apparent reason) on a diagnosis of typhus fever. There were many other epidemics of unknown diseases (excluding Sweating Sickness and typhus) during the 16th century in London, as in 1501, 1504–5, 1506, 1518 and annual outbreaks in 1511–21 (Creighton, 1894).

1513 An epidemic erupted in London in September and was reported to be raging among the sailors of the Fleet and in October it was said to have caused 300–400 deaths a day. On the Continent it was described as an epidemic contagious fever with dysentery and with black spots all over the body, a description that Shrewsbury (1970) suggested was black or haemorrhagic measles. This outbreak continued through December and appears to have the characteristics of haemorrhagic plague.

1514 An outbreak in London in February which Shrewsbury (1970) suggested was typhus fever.

1525 Shrewsbury (1970) considered that the 'great death' in London *in the winter* was an epidemic of typhus fever.

1526 An epidemic that erupted in London in May caused a panic exodus of the citizens and may, according to Shrewsbury (1970), have been bubonic plague, but, since plague was also reported in Guildford and Cambridge at this time, this conclusion can only be conjectural. It was probably haemorrhagic plague.

1529–30 Shrewsbury (1970) averred that bubonic plague appeared in London in June 1529 but apparently subsided rapidly during September, only to reappear in spring 1530 and persist in London until the autumn. This may have been an extended type (ii) epidemic of haemorrhagic plague. Shrewsbury believed that the epidemic was dormant in the winter because of the 'hibernation of the plague fleas' but the conclusion and explanation are not convincing.

1531 An epidemic was violently active in London in autumn 1531, causing a weekly death-roll of 300–400. Henry VIII paid expenses to the poor of Greenwich, who were expelled as a precaution to prevent the spread of infection when he took refuge there. Shrewsbury (1970) believed that this was not bubonic plague, presumably because the little evidence available suggests an outbreak in the autumn and winter. The scale of the mortality and the behaviour of the king may suggest an outbreak of haemorrhagic plague that broke out again after the winter.

1532 There was an outbreak in London in the autumn of this year that continued into November. It caused 99 deaths from plague and 27 from other causes. It apparently began by ravaging Kent in the summer and then spread to London – the reverse direction to be expected from an epidemic of bubonic plague among black rats in the port of London. Contrary to Shrewsbury's (1970) conclusions, there is no evidence that this outbreak was bubonic.

1535 An epidemic erupted in London in midsummer 1535 and had spread all over the city by the end of August, which was warm and wet (regarded as a bad sign); mortality was augmented in September but had disappeared by the end of October.

1536 Shrewsbury (1970) claimed that bubonic plague erupted in London in spring 1536 and became epidemic during the summer, invading Westminster where it was even in the Abbey threatening the coronation of Jane Seymour. Shrewsbury claimed that it died out in October because frost made the fleas hibernate but gave no evidence of this. It reappeared in spring 1537 and spread almost everywhere in London by mid-July; this pattern of a reduced virulence of the plague over the winter months was frequently found and is typical of type (ii) epidemics of haemorrhagic plague. There is a record of 112 plague deaths in 1 week in the city (not excessive numbers for a major outbreak in the metropolis); it spread along the Thames Valley (perhaps by barge on the river) reaching Windsor and Kingston, and also spread southwards to Croydon. Travellers were warned that to escape this epidemic they should ride 26 miles without stopping, i.e. probably about 1 day's ride on horseback to clear the plague area. The epidemic disappeared from London by the end of October 1537. Although there were other epidemics throughout England at this time they do not appear to have come from the outbreak in London.

1538 Shrewsbury (1970) assumed that the epidemic in February 1538 was typhus fever.

Even Shrewsbury did not think that most of these epidemics were bubonic. There is no evidence that they were outbreaks of typhus and we suggest that, in the absence of additional information, most were of haemorrhagic plague.

6.2.2 Did plague become endemic in London during the second half of the 16th century?

Parish registers began in England after 1540 and a law was passed that plague burials should be identified therein, so that much better information is available for the second half of the century. Figure 8.1 shows the number of plague deaths recorded in the London Bills of Mortality from 1578 to 1680 and it can be seen that the pestilence grumbled on and was virtually endemic in the city after 1603, with some plague deaths in almost every year. The enormous epidemics produced during the following 60 years were not necessarily introduced from overseas because they could have flared up from one of the isolated cases in the City or have been brought in from the provinces by an infected traveller. This progression towards the endemic state after 1578, with prolonged epidemics, is shown in Fig. 8.1, and Table 6.1 shows the annual number of plague burials in 11 selected London parishes over the period 1557–99; in this table we have analysed all the outbreaks by family reconstitution and some examples are given below. In addition, there are reported plague deaths in the parish of St Pancras, Soper Lane, as follows: 1542, 2; 1543, 18; 1547, 6; 1548, 14; 1549, 1. We have also examined for this period the registers of the parishes of: St Michael Bassishaw; St Mary Somerset; St Bene't, Paul's Wharf; St Antholin, Budge Row; St Peter, Paul's Wharf; St Paul, Covent Garden; and Our Lady in Aldermanbury. There are no recorded plague burials. Although the total mortality in the metropolis in some years was high, as shown in Fig. 8.1, deaths were few in the individual parishes studied.

Nevertheless, our analyses revealed the same pattern in each epidemic with a latent period of 10–12 days and an infectious period of 25–27 days, i.e. *within* households the latent period could be below the normal 12 days. As usual, the analyses suggest (although it is impossible to prove the point) that people were at their most infectious during the first 10 days of the infectious period. They were probably less infectious once the symptoms had appeared, except, perhaps, to those who were nursing them.

With a few exceptions, these epidemics were minor affairs, indicative of a low household and interhousehold R_0 and hence they usually died out quickly. And yet these people would have been living in close proximity in London in crowded dwelling houses. We tentatively suggest that some or many of the population in these London parishes that experienced regular, but minor, epidemics were resistant or immune to plague, thereby reducing the density of susceptibles and hence the effective R_0.

Another feature of the plague epidemics in London in both the 16th and 17th centuries (see Chapter 8) revealed by our analyses is that a high proportion of the deaths were of apprentices and servants (both male and female). This is probably because they were young people brought in from the provinces who lacked any form of resistance or immunity. Rappaport (1989) showed that between 1540 and 1589 about 15% of indentured apprentice carpenters died before they could complete their service. Immigrant apprentices came to London in large numbers and the total number in 1600 has been estimated as 32 000 to 40 000 (Kitch, 1986).

6.2.3 *Epidemics in London, 1542–1600*

(i) Plague broke out in the parish of St Pancras, Soper Lane, in the autumn of 1542, with two deaths in the Yourke household: a daughter on 24 October and a servant on 17 December, an interval of 54 days implying a reinfection in this family from another parish because no other plague burials are recorded in the St Pancras registers until the following summer. Antony Gressa was infected as a new primary case on 27 April 1543 and was buried on 3 June (Fig. 6.1). John Westawe was next infected about 18 June, after Gressa had died, and so he must have been infected from outside the parish and was a new primary case. He died on 25 July and the plague then followed its usual course but with a low infective contact rate. The primary case infected 2–5 secondaries, all in separate households ($R_0 = 2$ to 5), who then infected 7–11 tertiaries, again, with one exception, in separate families, indicating a very low household contact rate (overall $R_0 \approx 2$ to 5). Only 2–3

Table 6.1. *Annual plague deaths in parishes in London in the second half of the 16th century*

Year	Parish A	B	C	D	E	F	G	H	I	J	K
1557	4	—	—	—	—	—	—	—	—	—	—
1558	67	—	—	—	—	—	—	—	—	—	—
1559	49	—	—	—	—	—	—	—	—	—	—
1560	—	—	—	—	—	—	—	—	—	—	—
1561	—	—	—	—	—	—	—	—	—	—	—
1562	—	—	—	—	—	—	—	—	—	—	—
1563	129	—	81	—	16	(a)	(b)	—	—	—	23
1564	14	—	—	—	—	—	—	—	—	—	—
1565	—	—	—	—	—	—	—	—	—	—	—
1566	—	—	—	—	—	—	—	—	—	—	—
1567	—	—	—	—	—	—	—	—	—	—	—
1568	—	—	—	—	—	—	—	—	—	—	—
1569	—	—	6	—	2	8	—	—	—	—	1
1570	—	—	11	—	—	7	—	—	—	—	—
1571	—	—	—	—	—	—	—	—	—	—	—
1572	—	—	4	—	—	—	—	—	—	—	—
1573	—	—	2	—	—	—	—	—	—	—	—
1574	—	—	29	—	—	—	—	—	—	—	—
1575	>32	—	8	—	—	—	—	—	—	—	2
1576	9	—	3	—	—	—	—	—	—	—	—
1577	14	—	3	—	—	—	—	—	—	—	—
1578	—	—	36	—	2	17	—	—	—	—	1
1579	—	—	15	—	—	—	—	—	—	—	—
1580	—	—	—	—	—	—	—	—	—	—	—
1581	—	—	20	—	—	—	—	—	3	—	—
1582	—	7	20	—	4	14	—	—	—	—	1
1583	—	2	7	—	—	20	—	—	—	6	3
1584	—	1	—	—	—	—	—	—	—	10	—

1585	—	2	—	—	—	—	—	—	—	—	—
1586	—	—	—	—	—	—	—	—	—	—	—
1587	—	—	—	—	—	—	—	—	—	—	—
1588	—	—	—	—	—	—	—	—	—	—	—
1589	—	—	—	—	—	—	—	—	—	—	—
1590	—	—	—	—	—	—	—	—	—	—	—
1591	—	—	—	—	—	—	—	—	—	—	—
1592	22	—	—	—	—	—	—	—	—	—	—
1593	164	8	65	5	11	53	>1(c)	15(d)	—	—	—
1594	2	—	4	—	—	4	—	—	—	—	—
1595	—	—	—	—	—	—	—	—	—	—	—
1596	—	—	—	—	—	—	—	—	—	—	—
1597	—	—	1	—	—	—	—	—	—	—	—
1598	—	—	—	—	—	—	—	—	—	—	—
1599	—	—	—	—	—	—	—	—	—	—	—

Abbreviations: A, St Martin-in-the-Fields; B, St Peter, Cornhill; C, St Olave, Hart Street; D, All Hallows, Bread Street; E, St Pancras, Soper Lane; F, St Michael, Cornhill; G, St Dionis, Backchurch; H, Our Lady in Aldermanbury; I, Kensington; J, Christ Church, Newgate Street; K, All Hallows, Honey Lane.

Notes:

(a) There is an entry in the registers between 27 March and 11 June 1563 as follows: 'The beginning of the plague in this p'ishe'.

(b) Footnote at 2 August 1563 says 'The numerous burials during this and the two following months were owing to the prevalence of the Plague of this year'.

(c) There is an entry recording the burial of a plague victim in 1593 and there are many burials in the following months.

(d) There is a break in recording the victims of plague between 5 August and 9 October 1593, although there are a large number of burials during this interval.

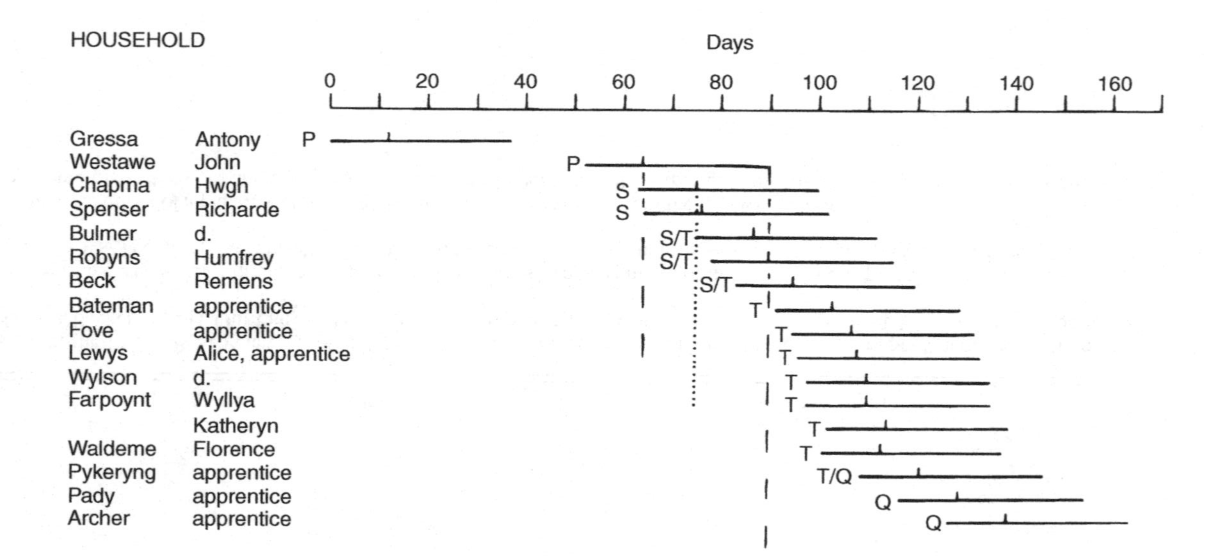

Fig. 6.1. Suggested sequence of infections in the plague epidemic in the parish of St Pancras, Soper Lane, London, in 1543. There were two separate primary cases (see text). Scale: days after 27 April 1543. For further details and abbreviations, see Fig. 5.6.

quaternaries, all apprentices, were infected so that $R_0 < 1$ and the epidemic died. Figure 6.1 shows that, although only 17 people died, this outbreak in 1543 lasted over 5 months because of the low R_0. Wylla Farpoynt did not infect Katheryn Farpoynt so that there were apparently no household infections in this outbreak.

(ii) Plague was said to have broken out in the summer of 1547 and Creighton (1894) recorded that the occupiers of infected houses were ordered to mark them with a cross on their street doors. It reappeared in the spring of 1548 but with only three plague deaths in the parish of St Pancras. Shrewsbury (1970) quoted records that it 'raigned sore in London with great death of people'.

(iii) After a dozen years in which London was free from plague, the disease returned with high mortality in 1563. A deadly plague epidemic was raging at the English garrison at Havre in France and the disease may have been imported from there. It is generally agreed that this outbreak in 1563 was the worst in the 16th century. Figure 6.2 shows the weekly plague deaths in the city and liberties, but excluding the out parishes, from June 1563 to January 1564; it had started earlier in the year and was epidemic by 12 June but it did not reach its peak until September–October. Thereafter it declined and Creighton recorded only 13 plague deaths in the week beginning 21 January 1564. The outbreak clearly followed typical Reed and Frost dynamics, lasting over 8 months, indicative of the usual long serial generation time and a low R_0.

This epidemic not only grumbled on through the winter, rising to 40 deaths in 1 week in February, as Shrewsbury (1970) showed, but continued through March–June 1564, albeit with fewer than 10 weekly deaths. It was, therefore, an atypical type (i) epidemic and is indicative of how plagues now persisted in London. Shrewsbury estimated from the data available that 20 000 people in London died in this plague, at least 25% of the population.

Creighton (1894) gave the following quotation from Dr John Jones

> 'His other observation is interesting as proving the possibility of repeated attacks of the buboes in the same person, an observation abundantly confirmed, as we shall see, in the London plagues of 1603 and 1665: "Here now, gentel readers, I think good to admonish all such as have had the plague, that they flie the trust of ignoraunt persons, who use to saye that he who hath once had the plague shal not neded to feare the havinge of it anye more: the whych by this example whyche foloweth (that chaunced to a certayne Bakers wife without Tempel barre in London, Anno Do. 1563) you

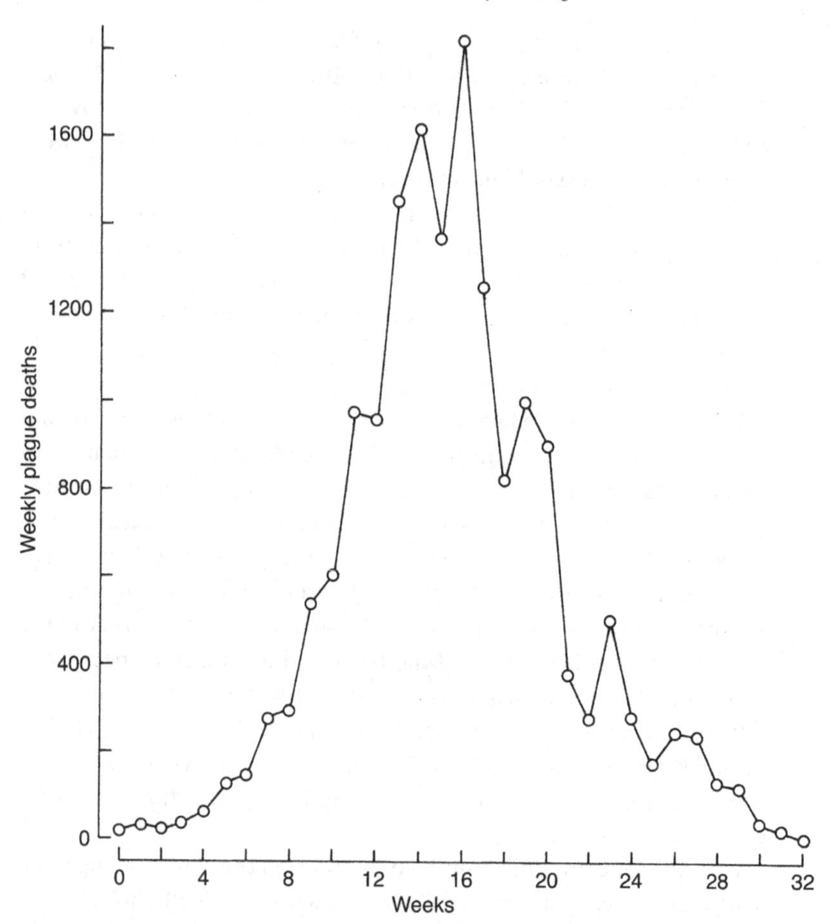

Fig. 6.2. Weekly plague deaths in London from June 1563 to January 1564. Abscissa: weeks after 12 June 1563. Data source: Creighton (1894).

> shall find to be worthelye to be repeated: this sayde wyfe had the plage at Midsommer and at Bartholomewtide, and at Michaelmas, and the first time it brake, the second time it brake, but ran littell, the thirde time it appeared and brake not: but she died, notwythstanding she was twyce afore healed."'

We have analysed the mortality of 1563 in the parish of St Dionis, Backchurch, where the epidemic is not officially recorded in the registers but a footnote records that 'the numerous burials during this and the following two months were owing to the prevalence of the plague of this year' and are typical of haemorrhagic plague. The start of the outbreak with seven co-primaries is illustrated in Fig. 6.3. Probably all

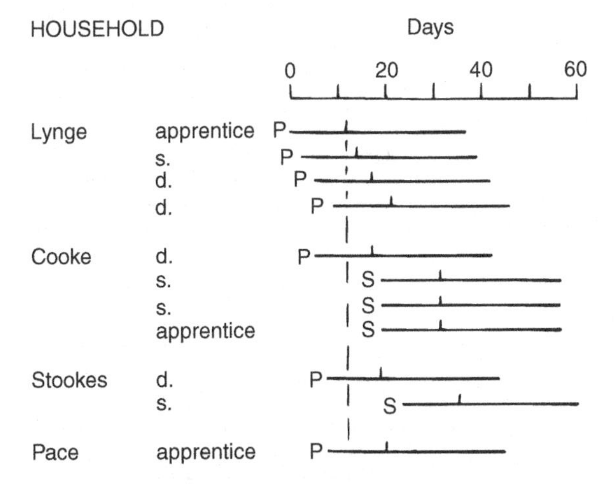

Fig. 6.3. Suggested sequence of infections at the start of the plague epidemic in the parish of St Dionis, Backchurch, London in 1563. Scale: days after 26 June 1563. For further details and abbreviations, see Fig. 5.6.

four victims in the Lynge household were primaries, beginning with the apprentice. The daughter of the Cooke household infected three secondaries, her two brothers and the apprentice. The daughter of the Stookes household infected her brother. The apprentice (the last primary) was the only victim in the Pace household.

The prolonged infection (over 100 days) in Mr Sears' house in the parish of St Dionis began with the apprentice and then, sequentially with three servants before the death of Rychard Sears (Fig. 6.4A). A child and an apprentice were initially infected in Mr Revelle's house and then, sequentially, another apprentice and three servants (Fig. 6.4B).

(iv) There was a small epidemic in the parish of St Martin-in-the-Fields in 1575 that followed the standard pattern and is illustrated in Fig. 6.5. The infection began at a favourable time of year at the start of April but, nevertheless, the effective interhousehold contact rate was low and only 23 people died even though the outbreak persisted for over 4 months. The servant in the Grayes household was the single primary case who infected 4–8 secondaries but these infected only 4–11 tertiaries ($R_0 \approx 1.5$ to 2.5) with few of them in new households. Consequently, there were only 3–8 quaternaries ($R_0 \approx 1$) and only 2–4 fifth generation infectives with an $R_0 < 1$, and so the epidemic died out.

(v) There were smaller epidemics in 1578 and 1582 and the complete outbreak in the parish of St Michael, Cornhill, in 1582 is illustrated in

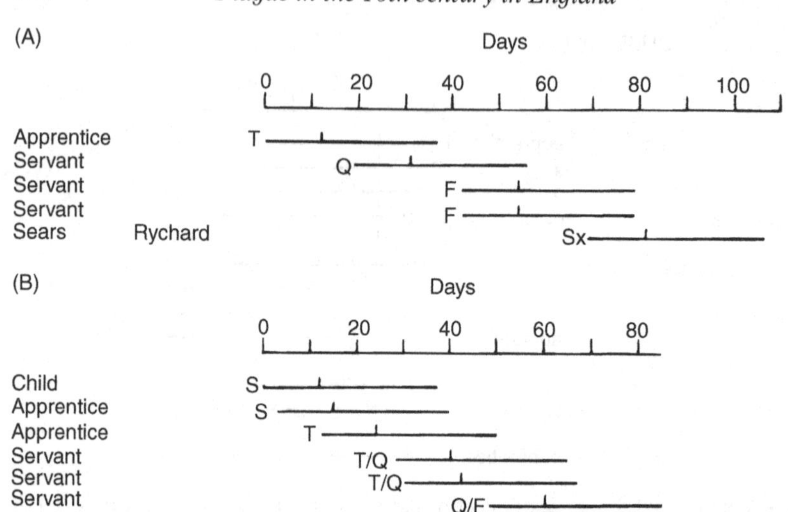

Fig. 6.4. The epidemic in two selected households in the parish of St Dionis, Backchurch, in 1563. (A) Mr Sears' (= Sayrs) household where the infection began with a tertiary infection. Scale: days after 25 July 1563. (B) Mr Revelle's household where the infection began with a secondary case. Scale: days after 3 August 1563. Note the high number of servants and apprentices dying and the slow spread of the plague through the households. For further details and abbreviations, see Fig. 5.6. Sx, sixth generation infection.

Fig. 6.6. Only 34 people died but the epidemic lasted from September 1582 to September 1583, indicative of a very low R_0 within the parish. Thus, the infection began in the autumn, typical of a type (ii) epidemic (upper part of Fig. 6.6) with two co-primaries in two households. The servant introduced the plague into the Poole family, infecting two members plus another servant. The two primaries infected 5–6 secondaries ($R_0 \approx 3$) but the third, fourth, fifth, sixth and seventh generations of the infection showed a fall in the contact rate, typical of winter conditions (R_0 successively 0.4, 2, 0.4, 1, 0.3) and hence the epidemic died out, the last victim dying on 13 January. This type (ii) epidemic did not survive through the winter, but 43 days later a boy in the Cockson family was infected as a new primary, presumably from outside the parish, so initiating a small-scale type (i) epidemic, with deaths occurring from April to September. The lower part of Fig. 6.6 illustrates the continuing low R_0.

(vi) The next major outbreak in London (the last of the 16th century) was in 1593. This epidemic began in the autumn of 1592 and is said to have caused 2000 deaths before the end of the year. On 7 September,

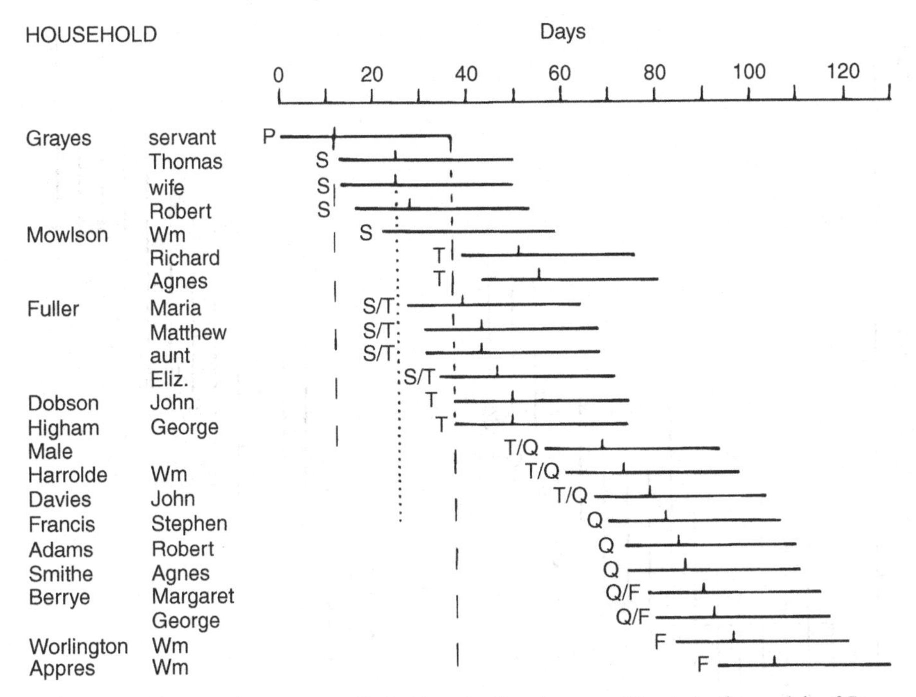

Fig. 6.5. Suggested sequence of infections in the plague epidemic in the parish of St Martin-in-the-Fields, London, in 1575. Scale: days after 1 April 1575. For further details and abbreviations, see Fig. 5.6.

soldiers from the north on their way to Southampton to embark for foreign parts had to pass round London 'to avoid the infection which is much spread abroad' in the city. On 16 September, the spoil of a great Spanish carrack at Dartmouth could be brought no further than Greenwich, on account of the contagion in London. It is noteworthy that the infection lasted through the winter; even in mid-winter people were leaving London: 'the plague is so sore that none of worthy stay about these places' (Creighton, 1894). On 21 January it was officially noted that, after diminishing for some weeks, plague was increasing again in the city, and its continuing increase provoked the prohibition on the 28 January of plays, bear-baitings, bowlings, bullfights, and all sports and like assemblies within the Lord Mayor's jurisdiction. Simultaneously the prohibition was extended to cover the outer Liberties in Middlesex and Surrey within 7 miles of the city.

The plague was evidently rampant in the spring of 1593 and reached its peak in the third week of August. Creighton (1894) identified Fleet Ditch as the most infected part of the city, whereas Shrewsbury (1970)

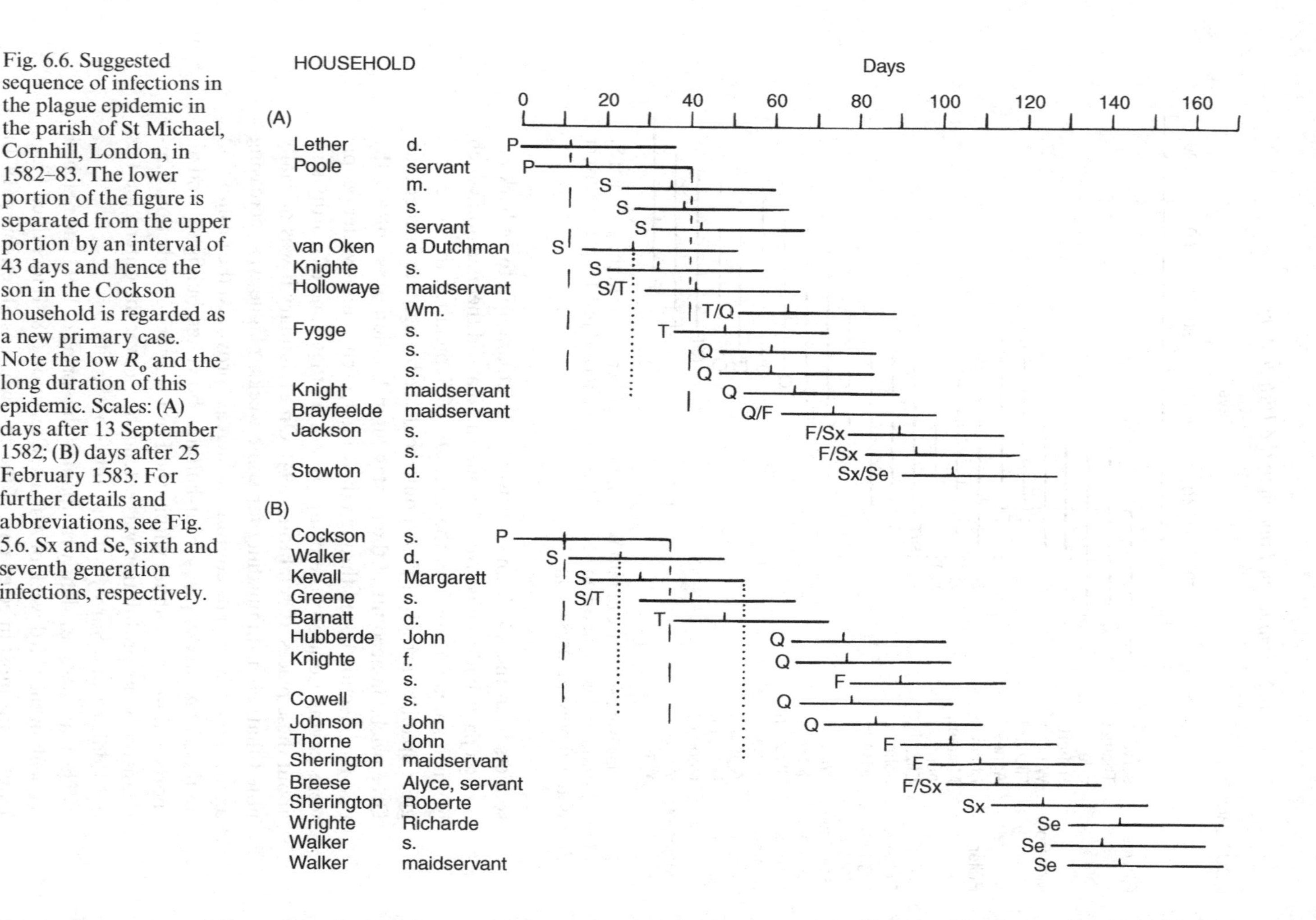

Fig. 6.6. Suggested sequence of infections in the plague epidemic in the parish of St Michael, Cornhill, London, in 1582–83. The lower portion of the figure is separated from the upper portion by an interval of 43 days and hence the son in the Cockson household is regarded as a new primary case. Note the low R_o and the long duration of this epidemic. Scales: (A) days after 13 September 1582; (B) days after 25 February 1583. For further details and abbreviations, see Fig. 5.6. Sx and Se, sixth and seventh generation infections, respectively.

affirmed that St Katharine by the Tower was the parish where the epidemic raged most violently; 525 burials were recorded in the registers in 1593 and Shrewsbury believed that the plague mortality rate was at least 63%. By September, the disease was epidemic in Greenwich and it radiated outwards through Middlesex into Essex, Hertfordshire and Buckinghamshire. Mortality in London from the plague of 1593 has been estimated differently, but probably over 17 000 people died, i.e. 60% of the total of deaths from all causes.

There is little information concerning the symptoms shown by those contracting the disease, but Creighton (1894) quoted a letter written by Richard Stapes on 3 November 1593 commenting on the plague: 'my son-in-law has buried his servant; but I cannot say his was the sickness because the visitors reported that the tokens did not appear on him as on the other'. These tokens were probably not buboes but were the characteristic marks on the body by which victims of the disease were identified. Signs and tokens are discussed more fully in section 13.12; unfortunately, we have no further information concerning the epidemic of 1593 but evidently not everyone dying at that time displayed them.

We see from this summary that the plagues of 1563 and 1593 in London had features in common that were also characteristic of the major plagues: an explosive epidemic with Reed and Frost dynamics that reached its peak in late summer, clear continuation through the winter months at a low level, appearance of the diagnostic tokens, infection radiating outwards to other communities and a high percentage mortality.

6.3 Plagues in central and southern England during the 16th century

During the first half of the 16th century there are numerous anecdotal accounts of scattered epidemics throughout England and, after 1550, many parish registers record deaths from 'the pestilence' or 'the plague'. In general, these outbreaks do not appear to have had their origins in a London focus. In addition, many have searched the registers for years of crisis mortality and some have ascribed these to outbreaks of plague, although they were almost certainly the result of other factors (see section 1.5). Plagues in northern England in this century seem to have had different origins and, in some cases, their rapid spread can be traced with some accuracy. They are described in Chapter 7.

The universities of Oxford and Cambridge seem to have been remarkably unhealthy places, perhaps because the universities' archives provide more written evidence than in other towns, but there are repeated reports of terms being curtailed, of the colleges being closed or even of the universities moving temporarily to another location. Sometimes these were precautionary measures.

The following is a brief, chronological summary of these various outbreaks of lethal infectious diseases in central and southern provincial England, discounting epidemics of Sweating Sickness or of suspected typhus.

1503 Plague at Oxford and Exeter, discounted as bubonic by Shrewsbury (1970).

1509–10 Probably the European pandemic of influenza.

1513 A grace to dispense with lectures at Cambridge because of plague that was apparently widespread in this year. It was also in London and on the Continent and Shrewsbury suggested that this may have been 'black' or haemorrhagic measles, another example of a lethal epidemic that was not bubonic plague.

1517 Plague reported as active this year in Nottingham and Oxford (Creighton, 1894).

1524 Fear of plague caused Cambridge University to postpone the Michaelmas term.

1525 'Vehement plague' at Oxford.

1526 The Easter term at Cambridge was prorogued and the disease may have persisted for 3 years.

1529 Shrewsbury attacked by disease.

1532 Plague ravaging Kent spread to London; Cambridge University term postponed although there is no record of an epidemic there. Plague reported in the parish register of Much Wenlock; the infective that brought the disease must have travelled a considerable distance.

1536 Plague active in Shrewsbury, Somerset, Devonshire, Cornwall, Doncaster (in October when the weather was so cold that even Shrewsbury discounted bubonic plague) and Oxford (where the university was again dispersed).

1537–38 A severe epidemic in a number of widespread places in England including Hull, Reading, near Buckingham, Towcester, Launceston, Kingston-upon-Thames, Portsmouth, Devonshire and

Chester. Term at Cambridge University was adjourned but there is no evidence of an epidemic disease in 1537. It is noteworthy that the pestilence was severe in the January and February of 1538 not only in Devonshire but in the colder, eastern counties of Huntingdonshire and Derbyshire, spreading to Norfolk and York in March and to Cambridge in April. Plague was evidently widespread throughout the metapopulation in this outbreak.

1540 There was an alarming amount of sickness and pestilence throughout England in this hot and dry summer; Liverpool was badly affected and Watford recorded 40 plague burials out of a total of 47 burials between July and September.

1544 There were several scattered, localised outbreaks of plague reported in southern ports and on the southeast coast of England in this year, for example in Bristol, Dover, Rye and Canterbury; plague was also prevalent in and around Calais. Plague was also widespread around Portsmouth in the following summer and 11 of the 34 ships in the fleet were infected and from them 903 plague-sick sailors and marines were removed. Shrewsbury (1970) also reported that some soldiers at Portsmouth were dropping dead, 'full of the marks' of plague as they paraded to receive their pay.

1545 Plague was reported in a belt across the east and Midlands at Marlborough, Loughborough, Oundle and at Cambridge, where the university was dissolved.

1546 Cambridge again attacked and plague erupted at several places in Northamptonshire, Devonshire and Shrewsbury.

1547 An epidemic in Cornwall at Stratton and Camborne; the register at Stratton recorded 155 plague burials.

1551 Plague raged throughout the summer at Bristol.

1557 High mortality (possibly related to a disease) was reported in four widely separated places: Lincoln, Evesham, Colyton and Solihull.

1558 Plague said to be bad in Loughborough and Stratford-upon-Avon.

1559 Severe plague recorded in the registers of Oswestry, with plague deaths persisting through December and January of the following year. Plague was also reported as widely prevalent in Cheshire, Staffordshire and Nottinghamshire.

1563–65 Shrewsbury described the plague of 1563 (which he believed to be bubonic) as probably the severest national outbreak in the English provinces in the 16th century. It broke out in the English possession at Havre in France and spread to the Channel Islands about midsummer 1563. There was also high mortality at Rye and Hastings (191 burials) and outbreaks during 1563–64 at Derby, Leicester (only four plague burials), Stratford-upon-Avon (see section 6.4), Lichfield, Canterbury, Cambridge, Shrewsbury, Bristol and even as far north as Helmsley in Yorkshire where it was said to be 'most hot and fearful so that many died and fled'. Apart from a concentration in the Midlands the other localities are widely separated and we have no evidence of how the epidemic might have spread but there were probably at least two separate entry points via the ports.

1570 Plague reported in Exeter and Northampton.

1574 Plague reported at Chester, Cambridge and Peterborough.

1575 Severe plague at Stamford, together with 50 plague burials at Holy Trinity, Hull. Slack (1977b) also reported plague in Bristol and, using the 'Easter Books' of Christ Church parish, he identified the affected households and these are shown schematically in Fig. 6.7. Wine Street and Broad Street were two of the main highways of the town and contained several large households with servants and apprentices; many of the occupants fled. The Pithay was a poor, overcrowded alley leading to a workhouse. It can be seen that the plague spread erratically and that there was apparently a low contact rate in many households with only one death reported.

1577 Plague reported at Cambridge, Rye and Dover.

1578 Plague caused great mortality at Bury St Edmonds and ravaged East Anglia, Essex, Hertfordshire, Cambridgeshire, but was also reported in Cornwall and the Midlands. Probably two separate entry points at the ports.

1579 Plague reported in Wolverhampton and Evesham.

1580 Widespread epidemics at Norwich, Plymouth, Herefordshire, Gloucestershire, Bury St Edmonds and Rye.

1582 Epidemics in a number of widely separated places but the mortality seems to have been slight.

1583 Heavy plague mortality recorded at Norwich and Southampton.

1585 Plague reported in East Anglia.

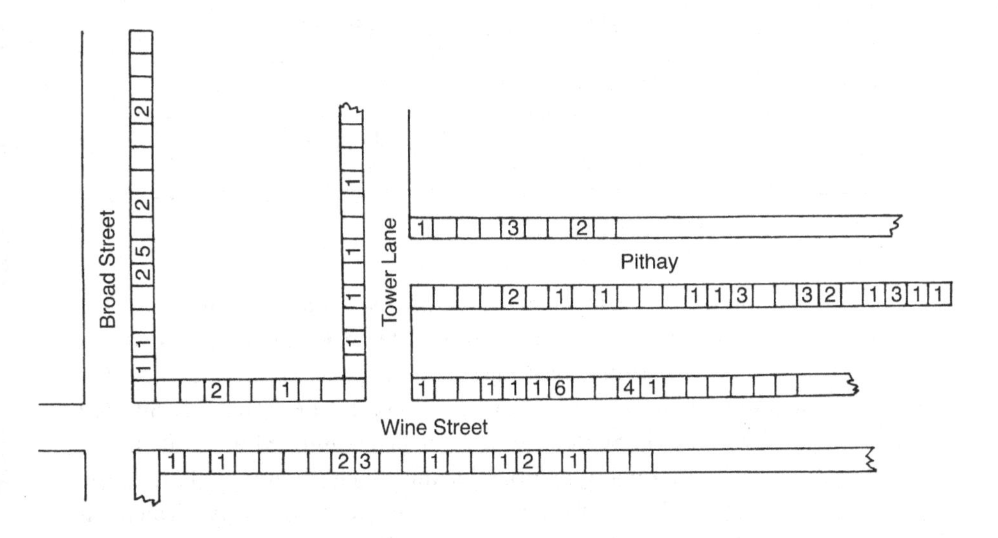

Fig. 6.7. Spread of the epidemic at Bristol in 1575. Schematic layout of streets in which each square represents one house with the number of victims indicated. Data from Slack (1977b).

1586–87 Calamitous visitations of plague reported at Chesterfield and Lincoln.

1590–91 Severe plague was reported in Plymouth, Exeter, Tiverton, Dartmouth and Totnes, and Somerset was also grievously infected. Exeter's population of 9100 recorded 1030 burials, although by 1600 the population numbers had recovered to 8900. The epidemic was apparently most severe in December 1590 and burials continued at a high level through the winter (Table 6.2). In the outbreak at Totnes, although recorded plague deaths peaked in August 1590, they continued through until February 1591 and in December and January the only burials were because of plague.

1592 Plague and 'a pestelent burning ague' reported in Derby, Lichfield, Bewdley, Worcester, Gloucester, Tewksbury, although Shrewsbury suspected that the mortality was not great. In Holt in Norfolk a 'great plague' erupted on 4 August 1592, according to the burials register and continued through until 26 February 1593 (probably a type (ii) epidemic that did persist through to the spring).

1593 This was a year of a major epidemic in London (section 6.2.3) and plague was also reported quite widely in the provinces,

Table 6.2. *Totals of monthly burials at Exeter, 1590–91*

1590	1591										
D	J	F	M	A	M	J	J	A	S	O	N
94	62	55	47	43	42	24	18	47	26	15	18

Data source: Shrewsbury (1970).

mainly from a focus in the Midlands. It was grievous in Tewksbury with 560 plague burials registered; Creighton (1894) stated that Canterbury, Nottingham and Lincoln were attacked; the university at Cambridge was dispersed; Leicester was affected by plague in September 1593, which continued until late in the spring of 1594 (type (ii) epidemic which did not explode in the second phase); plague was said to be particularly virulent in Derby, lasting from October 1592 through the winter until October 1593 (type (ii) epidemic), where 'there was not two houses together free from it' and yet it is reported that it never entered the house of a tanner, a tobacconist or a shoemaker (see Shrewsbury, 1970). Creighton (1894) affirmed that Lichfield sustained a plague death-roll of 1100 and 174 of the inhabitants of Bishop's Castle, Shropshire, died of plague over a 21-month period with a peak in August 1593. The plague also struck savagely at Presteigne, some 16 miles south of Bishop's Castle, where 300 died, and at Gloucester where 81 plague deaths were recorded in the registers of St Nicholas (see Shrewsbury, 1970).

6.4 Case study of the plague at Stratford-upon-Avon, 1564

Mean monthly burials at Stratford-upon-Avon in 1564 were about five when the plague struck on 11 July and the registers recorded 'Hic incepit pestis'. The rapid rise in monthly deaths is shown in Fig. 6.8; the epidemic lasted 6 months, from July to December with a peak in September (84 deaths) and exhibited the characteristic Reed and Frost dynamics for the plague (type (i) epidemic). A total of 237 burials were recorded during this time but, since deaths from other causes were included, plague mortality can be estimated at about 220. This was a major outbreak, but with a lower mortality than that, say, at Penrith in 1597–98.

We have reconstructed the spread of the plague through the families at Stratford-upon-Avon and a pattern of events identical with those observed

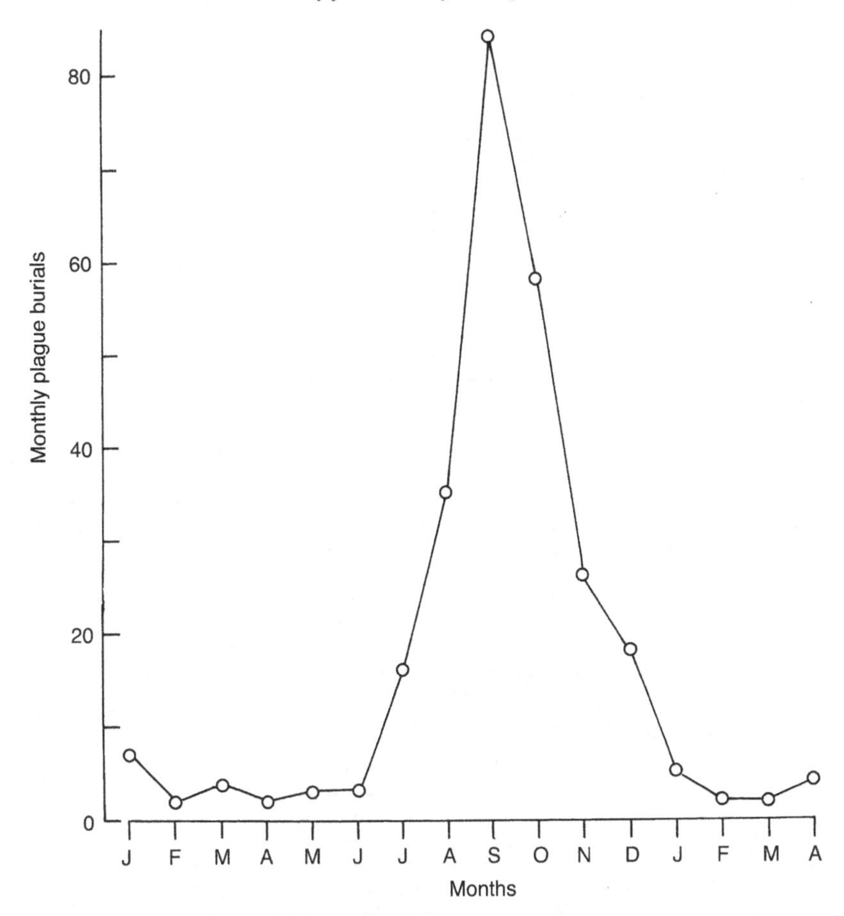

Fig. 6.8. Monthly burials at Stratford-upon-Avon, January 1564 to April 1565.

in other plague epidemics is seen, there being a slow build-up of mortality with the deaths confined within the households in the initial stages before the explosion of the epidemic. As usual, it is easiest to dissect out the epidemiology during the early stages of the outbreak before events become completely confused when the pestilence was at its peak and in Fig. 6.9 we show how the infection was passed through and between the first 22 families. This is an example of a major epidemic that began in mid-summer.

The first person to die of the plague at Stratford-upon-Avon was Oliver Gunne an apprentice of Thomas Degge (= Deeg, alias Gethen) and he is included in the Degge family in Fig. 6.9 because Joanna, the wife of Thomas Degge, was buried 9 days later. It is unlikely that she was infected by Gunne and they were probably co-primary cases and possibly they had

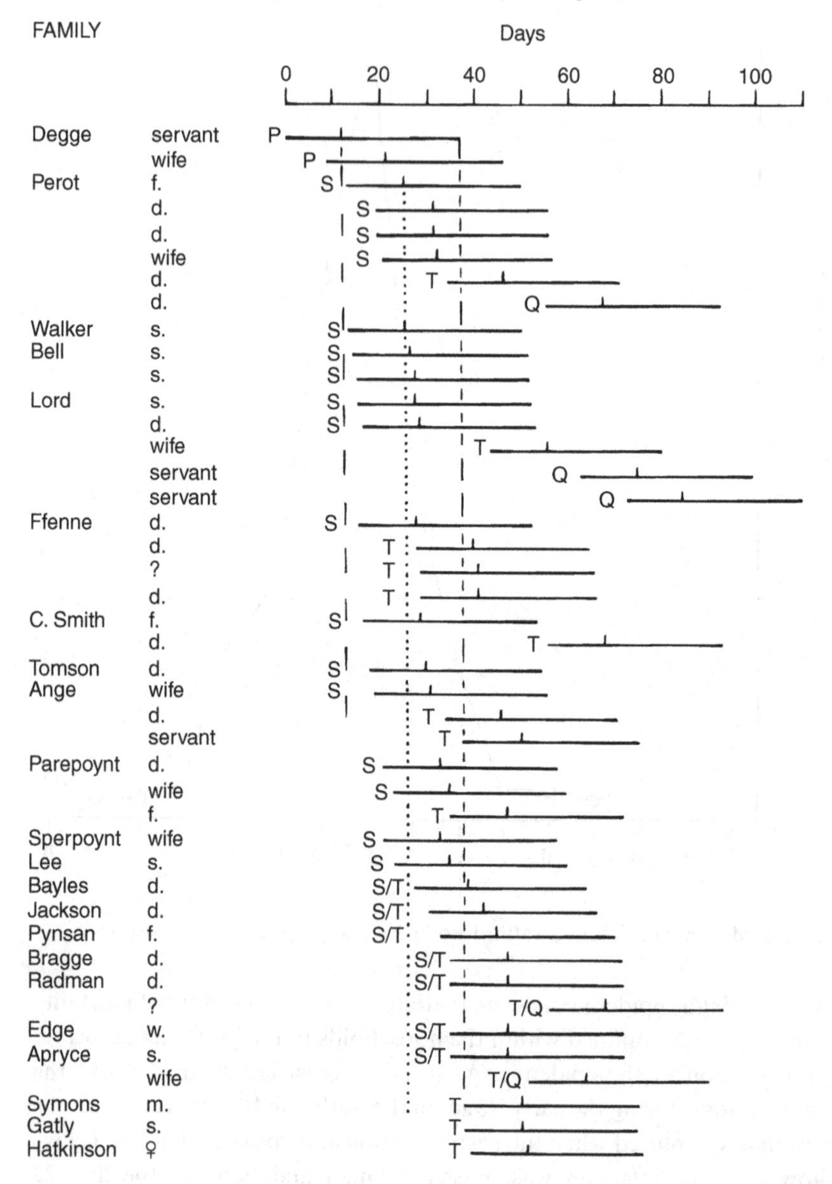

Fig. 6.9. Suggested sequence of infections at the start of the plague epidemic at Stratford-upon-Avon, 1564. Note the high R_0. Scale: days after 4 June 1564. For further details and abbreviations, see Fig. 5.6. ♀, female.

visited a town where the plague was just breaking out (plague was reported at Lichfield and Bristol in 1564; Creighton, 1894; Shrewsbury, 1970) and returned to Stratford carrying the infection. Joanna Degge would not have become infective until $t = 21$ (25 June), but by that time Gunne (who would not have been showing symptoms) had already infected secondary cases in the Perot, Walker, Bell, Lord and Ffenne families. Figure 6.9 illustrates how the epidemic slowly got under way and spread through these families with tertiary infections. Continued analysis of the burials in July and August suggests that the two co-primary cases may have infected 17–24 secondaries, with Oliver Gunne being responsible for the majority ($R_o = 8$ to 12).

6.5 Conclusions

We can draw few conclusions concerning these many diverse outbreaks of lethal infectious diseases that were continually recorded in southern and central provincial England during the 16th century. They seem to be largely, but not entirely, confined to medium-sized and larger towns, but this may be because we lack written evidence concerning the smaller parishes and villages. However, if substantial outbreaks were confined to populations above a critical size (i.e. haemorrhagic plague was density dependent), as in the plague of northern England in 1597–98 (see Chapter 7), the epidemiology would be the opposite of authentic outbreaks of bubonic plague in India, where it is essentially a rural disease of village communities. The epidemics were generally explosive and probably followed Reed and Frost dynamics, lasting for less than 12 months, with a peak mortality usually about late summer.

As yet, we have few details of how fast each of the epidemics spread through the metapopulation nor which were the foci where the outbreaks were initiated. Since, in many instances, the epidemics peaked simultaneously in late summer over a wide area (as in the Midlands in 1593), we conclude that the disease must have spread with great rapidity and often over substantial distances. In some instances mortality may have been slight, perhaps some three to four times the seasonal average, but in some towns perhaps 40% of the population died, with the bulk of the deaths occurring within a 3-month period; this is not the epidemiology of bubonic plague. There seem to be few studies of the age-specific mortalities in these plagues. Were the deaths mainly in infants or in children or in the aged? With this information it might be possible to eliminate weanling diarrhoea or smallpox as the cause. However, we conclude that there is sufficient circumstantial evidence to suggest that the major outbreaks described above had features in common with the Black Death and with the serious plagues of the 17th century about which we have more detailed information. This conclusion is backed-up by the case study at Stratford-upon-Avon where the epidemiological details that can be deduced correspond closely with our other case studies of haemorrhagic plagues.

7

Plagues in the 16th century in northern England: a metapopulation study

Plagues in northern England in the 16th and 17th centuries appear to have different patterns and dynamics from the wave-like spread of the Black Death and from the radial, or from the apparently erratic and unpredictable, movement of the epidemics in central and southern England that were caused by the movement of infectives over substantial distances. Usually the spread of the infection in northern England can be monitored from the records and, frequently, the infections appear to move southwards along well-defined corridors in the northeast from the Scottish borders and Northumberland. The Pennines, which form the backbone of England, effectively divided the Northern Province into eastern (Northumberland and Yorkshire) and western (Cumberland, Westmorland and Lancashire) halves, each with different terrains, and an epidemic disease was brought across by infectives travelling on the roads through the gaps.

Thus, the Northern Territory acted as a separate metapopulation, semi-isolated from Scotland and the rest of England. As we have seen, the inhabitants of northern England had been much occupied in defending themselves from raids and cattle stealing by the Scots, and Carlisle in the northwest and Durham in the northeast were major defensive centres. The terrain, particularly to the west of the Pennines, was very different from that of central and southern England and this was an important determinant of the dynamics of the epidemics. The counties of Cumberland and Westmorland have been described as backward and impoverished (Appleby, 1975) and the area 'remote from large industrial and trading centres; much of it was inaccessible to travellers, and all of it regarded with repulsion by outsiders' (Thirsk, 1967); in addition, these counties were left almost untouched by the various agricultural revolutions that spread across the rest of the country during the 16th and 17th centuries. The economy of the north was more like that of Scotland (where mortality crises and harvest failures persisted up to 1690), Ireland and parts of the Continent, rather than that of lowland England. The region suffered mortality crises in 1587 and 1596–97 because of the synchrony of high wheat prices and low wool prices, which led to extreme hardship and famine (Scott & Duncan, 1998). The population in the extreme northwest, therefore, already subsisting on an inadequate diet, were further weakened and malnourished by the famine of 1596–97 when the plague struck.

To the east of the Pennines, communications and conditions for farming were better in the coastal plains of Northumberland and Durham, in spite of the northern latitudes, cold winter climate and raids by the Scots. The more intensive farming on the eastern side of the Northern Province was accompanied by attempts to stabilise the distribution of land; common fields were associated with village settlements and it appears that the land was used fairly intensively (Thirsk, 1967), in contrast with the situation in Cumbria. To the south lay the vast area of Yorkshire, the bulk of which lay in the lowlands and much of the husbandry and many of the communities resembled those of the Midlands and southern England (Thirsk, 1967). Plague occurred much more frequently along this eastern corridor than it did in the wilder country to the west of the Pennines.

7.1 The first half of the 16th century

The majority of pestilences and epidemics in the Northern Province in the first half of the 16th century appear to have been confined to the northeast, from York to Berwick-upon-Tweed on the Scottish borders, but a virulent plague broke out in Lancashire (Axon, 1894; Sharpe France, 1939) and Liverpool was nearly depopulated in 1540 and a further 250 died there in 1548. Sharpe France believed that the sudden rise in the number of deaths in Croston (some 20 miles to the north of Liverpool) in the months of *January* and *February* (clearly not bubonic plague) can be attributed to the same cause but without more details it is not possible to determine whether these were haemorrhagic plague epidemics. The pestilence was very active in northeastern England in 1538 and Shrewsbury (1970) provided the following description:

> As early as March plague was killing the citizens of York, and by the beginning of April its activity was so fierce that the corporation ordered that all the plague-sick should be removed to certain houses outside the Lathrop gate that had been specially set aside for their reception, that the gate should be closed and that no infected person was to move about in or to enter the city. Later in the year plague spread further north and the Council of the North informed Henry VIII towards the end of August that it was prevalent in Durham and Newcastle-upon-Tyne,

while a second report, early in December, affirmed that it was still active at various places in the two counties.

The parish register of Wragley in Yorkshire revealed that the mean annual burials were 10, which rose to 97 in 1542, of which 60 were in July and August. Berwick-upon-Tweed was the scene of a 'great plague' in the following year, 1543, a further example of an isolated outbreak of an unknown disease.

When plague was prevalent simultaneously in southern England and around Calais in 1544, there was an epidemic that was apparently widespread in the northeast. Newcastle-upon-Tyne was invaded in the summer (the infection presumably entered via the port) and was sorely afflicted by August, with increasing violence in September, so that by October 'all the honest inhabitants' had fled from the stricken town. This plague was said to be raging over most of Northumberland (a desolate county) and 'sundry other places in the North'.

Plague returned to the northeast in the following year, 1545, again following the characteristic seasonal pattern beginning at Berwick-upon-Tweed in April and continuing throughout the summer and, during this time, was reported to have spread over most of Northumberland. It was again active in Newcastle-upon-Tyne in October 1545.

York was affected by an epidemic that was officially recognised by January 1550 when

the corporation ordered all the people, 'dwelling in Laythrop' to evacuate their houses, which were needed for the accommodation of the sick. These houses had been built by the corporation twelve years before for the reception then of the plague-sick and evidently they had been occupied by tenants or squatters in the interval. As the Laythrop accommodation now proved inadequate, the corporation erected two buildings in February on Hob Moor, and later in the month it levied a fund on the four wards of the city for the relief of the sick poor. In the spring of 1551 plague appeared in the city and by the beginning of May it was causing the authorities much trouble. Apparently the buildings at Layerthorpe and on Hob Moor were full of patients by this time, because on 7 May the corporation ordered all the plague-sick to keep in their own houses, and as the disease continued to grow it cancelled the Corpus Christi play on the 18th. After June the outbreak seems surprisingly to have subsisted spontaneously . . . It was during this pestilential period that the city council decreed that every infected house 'shall have Rede Crosse sat uppon the Dower', which appears to be the first use of this colour as a plague-sign in England. Laycock avers that in the two summers of 1550 and 1551 the parish of St Martin cum Gregory lost more than half its population from epidemic disease, and as this was one of the city's healthy parishes in his opinion, he considers that York lost at least half its population at this time.

(Shrewsbury, 1970)

7.2 Plague and pestilence in the Northern Province, 1550–95

The parish registers of Penrith record that 'Plauge was in Penreth and Kendall 1554'; this single line is much-quoted but there appears to be no other account nor any further details of this epidemic and Barnes (1891) commented on the absence of local records of visitations of the plague in Cumberland and Westmorland.

South Lancashire was badly affected by an epidemic in 1558 (Axon, 1894) that was described by Sharpe France (1939) as follows:

> Liverpool in particular was badly hit, about a quarter of the population being killed. The *Town Books* say that the previous year had also been a bad one. There was 'a great plage' in Manchester, and the authorities of Liverpool were very concerned over the danger of its spreading to their town. Unfortunately it did so, and an unlucky Irishman, with the Welsh name of John Hughes, was held to be responsible. He was accused of having been ill when he arrived in the town from Manchester and of having taken his dirty clothes to be washed at the house of a certain Nicholas Braye. A child of Braye contracted the disease and died. Hughes was brought before the mayor and underwent severe questioning, but the accusations could not be proved. However, whether he had carried the plague, or it had come by some other channel, several others in the same house died shortly after, 'and so after that it increased daily and daily to a great number, that died between St. Lawrence's day [10 August] and Martlemas [11 November] then next after, the whole number of 240 and odd persons.' The severity of the outbreak was such that the St. Martin's fair was cancelled and no market was held during a period of three months.

This appears to be another account where a travelling, symptomless infective brought the pestilence from a distant focus (some 30 miles away) but the account is unsatisfactory. Did Hughes contract the disease in a mild form (he was said to be ill) but recovered? How did the members of the Braye family apparently die so quickly? The south Lancashire towns of Ormskirk, Prescot and Standish also recorded a rise in mortality at this time and a 'great plague' visited Preston in 1562.

An epidemic was probably raging from the spring of 1570 in Rotherham and Selby in Yorkshire because in June the corporation of York ordered a continuous day-and-night watch to be kept at the city gates to prevent any suspect persons from either town where the sickness had recently been active from entering. Further north, the plague was 'very sorry' at Newcastle-upon-Tyne in August.

Plague broke out again at Newcastle-upon-Tyne in 1577 and may have spread from there to Hawkshead in the Lake District (Lancashire), crossing the Pennines via the Roman Road alongside Hadrian's Wall. However, the etiology and epidemiology of this outbreak is not clear. After the entry for 17 November 1577 the Hawkshead registers recorded 'entered in November: In this monthe begane the pestelent sicknes in or pishe [our parish] wch was brought in by one George Barwicke' but it was not until 27 December 1577 (40 days later) that the burial of 'George Barwicke wch brought in the sickness' was recorded. It is difficult to reconcile these records with the usual pattern of events in the plague where the incoming infective is the first to die. The first plague burials on 19 and 24 November 1597 were of Elsabeth and Richard Barwicke, respectively, who look like co-primaries, and on 11 and 15 December Edwine and Richard Barwicke, respectively, were buried, apparently co-secondaries. George Barwicke may have been a later secondary or a tertiary infection. The arrival of an infective from outside and the temporal spread of the epidemic at Hawkshead are broadly consistent with plague epidemiology and, possibly, George Barwicke returned to the parish with two of the members of his family who were then already infected (Elsabeth and Richard), whereas George was infected by his relatives and died 6 weeks later.

The plague at Hawkshead began with the two deaths in November 1577 and continued through the winter until 25 February 1578 when the epidemic finished. There was a total of 38 deaths, the majority in December and January, when the outbreak was at its peak, with 16 burials in each month; a type (ii) epidemic which did not flare up again in the spring. Only seven families were affected and, although family reconstitution is not possible (only brief details are given in the registers), analysis suggests that some of these with the same surname were related and visiting one another but were in different households: 12 Tomlinsons died.

The registers also include the note 'Anthony Dixson buried in Langdall the last day of December & taken up agayne & brought to Hauxhead the xjth day of January' and Shrewsbury (1970) gave a quotation that plague was present in the valleys of the Lake District in 1577.

An epidemic erupted again in Newcastle-upon-Tyne in May 1588:

> and in the succeeding months destroyed 1,727 of the townsfolk. The register of St John's parish records that the plague-dead were distributed among 5 burial grounds, to wit, 340 in St John's; 509 at the Chapel; 300 at Allhallows; 400 at St Andrew's, and 103 at St Nicholas. These figures add up to 1,652 burials, leaving 75 corpses which were presumably interred in ground that was afterwards consecrated.

The plague struck at Durham City in the following year, 1589, and, although Creighton related that huts for the accommodation of the plague-sick were erected on Elvet Moor outside the city, the outbreak seems to

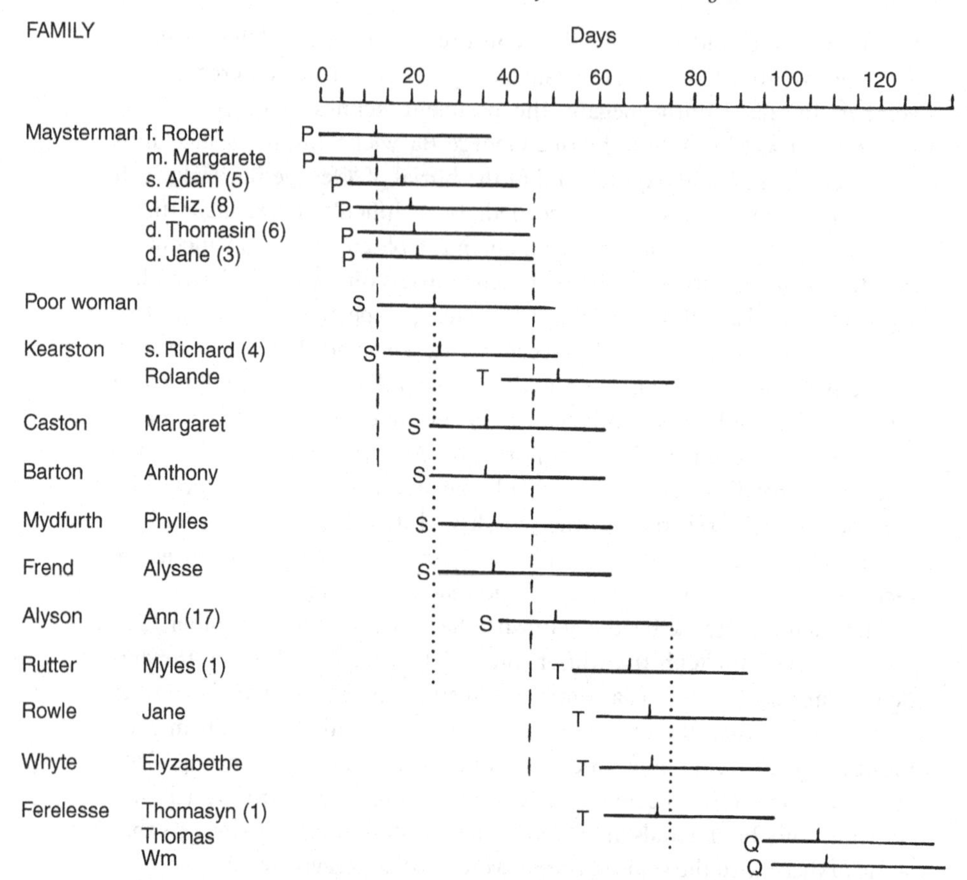

Fig. 7.1. Suggested sequence of infections in the plague epidemic in the parish of St Oswald, Durham, 1589. Scale: days after 14 August 1589. For further details and abbreviations, see Fig. 5.6.

have been mild. The registers for St Mary-le-Bow, Durham, did not record any plague deaths for 1589; indeed only nine burials are given for that year. Only 21 died of the plague in St Oswald, Durham, between 20 September 1589 and 1 January 1590 so that, in this late-starting epidemic, the infection did not continue through the winter. Only 13 families were affected and the progress of the epidemic through the initial families and their probable contacts are shown in Fig. 7.1. Robert Maysterman and his wife were both buried on 20 September, with their son Adam buried on 26 September and four daughters buried between 26 and 29 September. All the members of the Maysterman family were co-primaries, being infected by a common, outside source around 14 August ($R_o = 6$) so that Robert and his wife became infectious around 29 August and so may have been able to carry the infection to the 'poor woman' and the Kearston family (Fig. 7.1), who became the first secondaries. Other co-secondaries were then established in the Caston, Barton, Mydfurth, Frend and Alyson families, so that 6 co-primaries infected 7 secondaries ($R_0 = 1$) who, in turn, infected 5 tertiaries ($R_0 < 1$) (see Fig. 7.1) who in turn infected only 2 quaternaries ($R_0 \ll 1$) and hence the epidemic died out. The estimated parameters for St Oswald were: latent period = 12 days, infectious period = 25 days (including the last 5 days when the victim was showing symptoms).

From the summer of 1589, apart from the usual winter intermissions, there was a progressive expansion of the area of activity of the disease, culminating in the great national outburst of 1593. Thus, in the spring of 1590, Morpeth and Alnwick in Northumberland were said to be infected with 'the sickness of the plague', and in 1593 Durham and York were 'so sore with the plague' that 'none of worth' remained in either city because there was the usual panic exodus of all who had the means to flee.

7.3 The plague of 1597–98 in northern England

The period 1596–97 was one of great hardship and famine in the northwest and there were excessive mortalities (some three to five times greater than the annual average) recorded in many of the parishes. In January 1597, the Dean of Durham wrote to the effect that want and waste had crept into Northumberland, Westmorland and Cumberland, and that the scarcity of food was such that people travelled from Carlisle to Durham, a distance of about 60 miles over some of the worst country in the kingdom, to buy bread. A number of authors have wrongly attributed this increase in burials to the plague simply because the disease was known to be present in market towns by autumn 1597 (Barnes, 1891; Howson, 1961; Shrewsbury, 1970).

The pandemic that raged in the Northern Province apparently began in Newcastle-upon-Tyne and then moved south down the eastern coastal corridor to infect Durham and Darlington. It might have come from Scotland, where it was first recognised at Inveresk in June 1597 and a fortnight later plague was rampant in Edinburgh and the surrounding countryside, or it may have entered via the port at Newcastle (Fig. 7.2). A note in the Penrith parish registers stated 'A sore plague in new castle, durrome & Dernton in the yere of our lod god 1597' and 'A sore plague in Richmond Kendal Penreth Carliell Apulbie & other places in Westmorland and Cumberland in the year of our lord god 1598 of this plague they

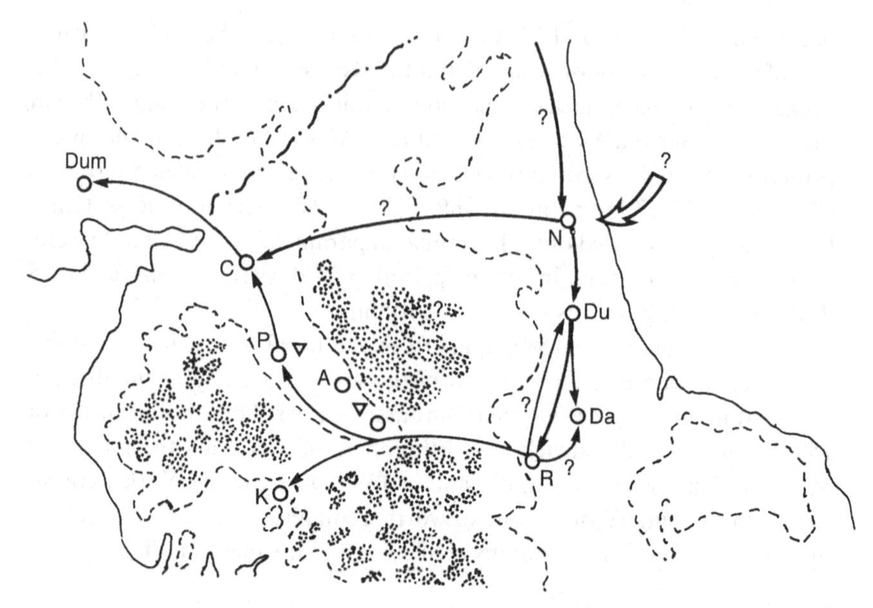

Fig. 7.2. Geographical spread of the plague from northeast to northwest England, 1597–98. Dashed line, land over 650 feet; dotted areas, land over 1640 feet. Circles, market towns; triangles, smaller communities. Note movement of plague across the Stainmore/Bowes Moor gap. The possible introduction of the plague via the port of Newcastle is indicated. A, Appleby; C, Carlisle; Da, Darlington; Du, Durham; Dum, Dumfries; K, Kendal; N, Newcastle; P, Penrith; R, Richmond.

dyed at Kendal . . .'. London, on the other hand, was clear of plague at this time.

On 26 May 1597 the Dean of Durham again complained that there was a great dearth in Durham: on some days, 500 horses are in Newcastle for foreign corn, although that town and Gateshead are dangerously infected. On the 17 September, Lord Burghley, minister of State, is informed that the plague increases at Newcastle, so that the Commissioners cannot yet come thither (the Assizes were not held at all, on account of the plague at Newcastle and Durham); foreign traders were selling corn at a high price until some members of the town council produced a stock of corn for sale at a shilling a bushel less. There are no figures of the plague mortality at Newcastle in 1597, but at Darlington the total deaths by October 17 were 340.

7.4 Durham

Shrewsbury (1970) estimated that the plague burials in Durham in 1597 were as follows:

Elvet	More than 400
St Nicholas	215
St Margaret	200
St Giles	60
St Mary	60
Gaol	24
Total	959+

He regarded these as minimal figures and many of the burials were on the moor. To this total can be added 82 burials recorded in the registers of St Mary-le-Bow during the period July to October 1597, although none of the entries are marked with a 'P', for plague; inspection of the registers confirms that, compared with the preceding and following months, this was an enormous mortality, suggestive of an infectious disease. However, in only two families does more than one person die, which is completely at variance with the high household contact rate often shown in plague epidemics. In general, more adults than children seem to have died, although in the Taylor family the sequence of burials was as follows: wife (husband survived), son aged 5 years, son aged 9 years, sister-in-law, son aged 1 year.

Creighton (1894) stated that the infection broke out again at Darlington and Durham in September 1598.

7.5 The plague at Richmond

Concomitant with this persistent and severe plague in Northumberland and County Durham, there was a severe epidemic in the North Riding of Yorkshire, particularly in the parish of Richmond, where it began in August 1597 when there were 23 deaths, followed by 42 deaths during September. The outbreak coincided with the start of the epidemic at Darlington. The epidemic appears to have reached its height in the summer of 1598, with 93 deaths in May, 99 in June, 182 in July and 194 in August. These figures indicate a grievous calamity in so small a place as Richmond and the stress of the epidemic is shown by the fact that the churchyard was insufficient for the burials, many of the dead having been buried in the Castle Yard and in Clarke's Green. The characteristics of this outbreak closely resembled those observed at Penrith (see Chapter 5): both were market towns and, although the plague arrived in Richmond a month earlier (in August 1597) both began in the late summer, overwintered and re-emerged in the following spring with deaths peaking in summer 1598.

Deaths finally ceased in both by December 1598 (type (ii) epidemic). The populations of both towns suffered grievously, with many victims having to be buried on the fells.

7.6 Plague arrives to the west of the Pennines

From Richmond the disease spread across the Pennines via the Stainmore–Bowes Moor gap, along the trade routes from east to west, to Westmorland and Cumberland (see Fig. 7.2), probably striking first at Penrith and then moving both northwards and southwards.

Events during the plague in the market town of Penrith are described in detail in Chapter 5 and the epidemic spread with remarkable rapidity from this focus to Carlisle, 20 miles to the north, and Kendall, 32 miles to the south (see Figs. 7.2 and 7.3), arriving at both towns on the same day, 3 October 1597, 12 days after the first burial at Penrith.

There are few details of the epidemic at Kendal because the parish registers are not complete but it is evident that the outbreak was severe. During the plague in 1598, the baptisms, weddings and burial sections of register stop in the summer and are not resumed until Christmas, a gap of 5 months. In the burial section for some months before this gap there are some entries with the marginal notation 'p' or 'pla'. Barnes (1891) described how, during the plague, provisions were brought to Coneybeds, a fort situated on Hay Fell, by the country people and left for the inhabitants of Kendal 'which was their only intercourse during that destructive period'.

We have not traced any records of the plague striking at Keswick in the north of the Lake District, but Barnes (1891) included the following paragraph:

> In Keswick there is a tradition that when the plague raged, as no markets were held for fear of the infection, the people of the dales carried their webs and yarns to a large stone, which is very conspicuous on one of the lower elevations of Armboth Fell, and there periodically met and did business with the trades. The stone still goes by the name of the 'web stone'. Mr. J. Fisher Crosthwaite informs me that he has heard old people say that when the plague was in Keswick the country people came to 'Cuddy Beck', but did not cross the little stream. The money was placed in the water and then taken, and the produce was laid on the ground for the Keswickians to take back.

The residents at Penrith adopted similar disinfection tactics during the plague there.

7.7 Carlisle

Again there are no reliable records of the number of people that died of the plague at Carlisle because none of the local registers exist for that period. However, Hughes (1971) has examined a number of documents relating to the period of the plague and he has presented an interesting synthesis that gives a description of events in the town during the epidemic. A census was taken on 20 December 1597 of the city householders with an indication of those families that were visited by the pestilence: the number of households was 323, of which 242 were stricken by plague, which had certainly not run its course by that date. Hughes estimated the population of Carlisle to have been about 1300 at that time.

It is difficult to deduce the lethality of the disease at Carlisle from the documents that Hughes has uncovered. After saying that 242 households were visited he added that 'figures against many of the names may indicate the number of deaths in each household. If this assumption is correct it would give a total of 149 . . .'. Slack (1985) assumed from this that the plague at Carlisle had a low mortality rate but a high morbidity rate 'Only 149 people died . . . but three-quarters of the town's households, 242 out of 323, were infected'. If the disease did have such a low mortality rate, the etiology would be completely different from that at Penrith and would be completely at variance with the tablet in Penrith church where 1196 are described as dying at Carlisle, although this is certainly an exaggerated number; furthermore, this census at Carlisle was taken in December 1597 when the plague had only just begun.

Hughes (1971) described the precautions taken by the civic authorities: a City Council meeting was apparently held on 3 November when it was recorded 'necessarye observations thought meate to be kept in this Cittye, the third day of November 1597: for the avoydinge of further infection of the disease of the plague then suspected there to be, if so it pleace God to blesse there carefull indeavours therein . . .' Infected houses were sealed off; the provisioning of their inhabitants arranged for and orderly arrangements made for the removal and disposal of the dead. Daily visits were to be made by honest experienced men to discover cases of sickness. One of the resolutions laid down was that a weekly collection must be taken in each street for the better relief of every poor person visited; of the total of £209 9s 10d, the amount raised by the citizens themselves was only £14 4s 10d; the largest amount came from the Common Chest and donations were received from several county gentry. The poor were attended without fee or charge for medicine, but those who were in a position to pay were expected to do so. The city gates were placed in charge of honest men whose orders were to prevent the admission of anyone known or suspected of infection or who came from any place where the infection was thought to be. 'Foreigners' and wandering beggars were expelled from the city and during the visitations none from Rickergate, Caldewgate or Botchergate were admitted without a permit from the city bailiff. Movement within the city itself was also restricted. Arrangements were made to pay the stipends of the officers and ministers, of the corpse bearers and the corpse winders and viewers, the latter apparently receiving a flat rate of ten shillings per week. A similar sum was paid to those who cleansed houses where all the inhabitants had died, or had fled to the fields for safety. Help was also given to those of the poor who survived though in daily contact with the sick, and to such as had recovered from the plague.

This account supports the conclusion derived from the Penrith study (Chapter 5) that not all those who came into intimate contact with the infection succumbed and contracted the disease. It also suggests that recovery from this infection was possible. The infectious nature of the disease was fully appreciated and infected houses were marked, as usual, with a red cross 'there to continue until lawfull opening of the same house'. Forty days were considered to be the period of quarantine (as usual), which corresponds well with the estimate of 37 days from infection to death.

It is not known when the practice of removing sufferers to pest-houses began, but other properties were commandeered to deal with the emergency and several isolation hospitals were speedily built outside the city walls. These shelters, variously termed lodges or shields, were situated at Gosling Syke, Stanwix Bank and others directly under the city wall on the Bitts. In each case the shields were sited near a water supply and the choice of a site at Gosling Syke and St Lawrence's Well may have been influenced by the fact that at both these points the limit of the cultivated fields was reached and the moorland waste began. Strict orders were enforced regarding the burial of the dead: special biers were to be provided for carrying the corpses, which had to be buried between 10 a.m. and 4 p.m.; no corpse was to be lifted until the bellman gave word that the grave had been prepared and the beadle had to walk before those carrying the body to give people warning as he came (Hughes, 1971).

7.8 Minor outbreaks of the plague in the Eden Valley

The major epidemics in Cumbria in 1597–98 were confined to the market towns, although the infection appeared in the smaller parishes adjacent to

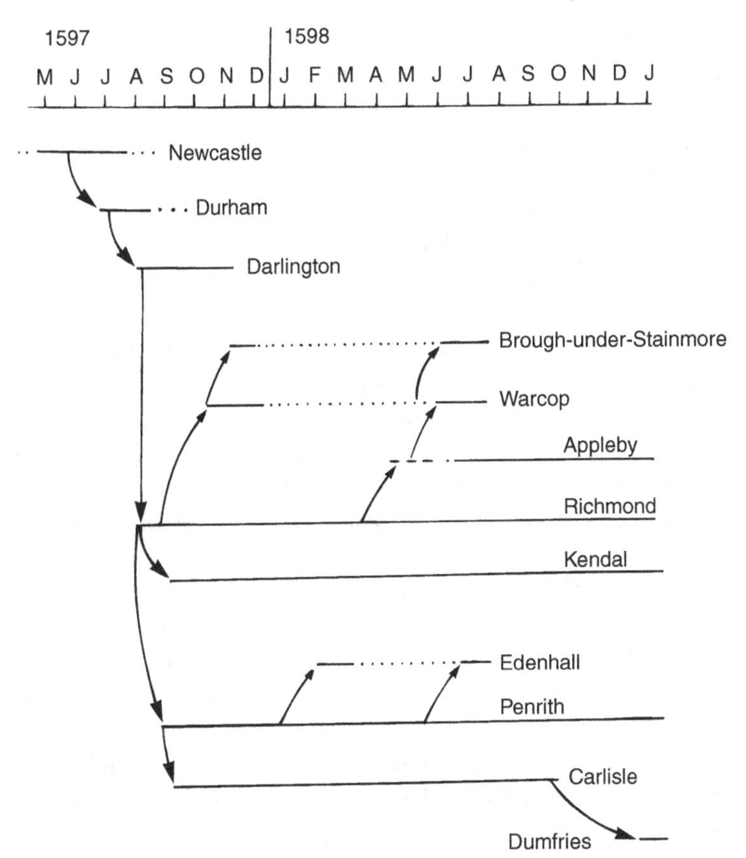

Fig. 7.3. Timing and duration of the plague epidemics, 1597–98. Arrows indicate probable spread of the disease, although Kendal and Carlisle may have received the infection directly from Richmond rather than via Penrith. Solid lines indicate the duration of epidemic.

afflicted market centres (Fig. 7.3). Edenhall, a small parish to the east of Penrith, appears to have suffered a number of plague deaths and the following note appears in the parish register: 'These 4 next following dyed of the plaige, Itm vii . . . Pattrig Rowtlishe was buried wthn Flatts wall neare to his own house being knowne to dye of the plaige.' The death of his wife on 8 March; his servant Elizabeth Thompson on the 11th and his infant son John immediately followed. The first is entered as having been buried 'beside her husband near the said place,' and the last was buried 'beside his father and mother in said place'. This seems to have been an isolated outbreak of the disease. No further deaths from it are recorded until the end of the following July, 1598. A baptism is recorded on 24

March. There was another baptism on 25 April, three burials between this date and August, and then at the head of the next page is this entry: 'The 42 next following dyed of the — [word wanting].'

The first plague death was on 29 July 1598. Some families suffered severely as shown by the following entries: 'Itm First August one child of Andrew Atkinson of the plaige & was buried in flatts cloose. Itm xv & xvi August Andrew Atkinson wiffe iii other children dyed of plaige and were buried their Lodge on Edenhall Fell at a place called Shaddowbourgh.' Twenty deaths occurred in August, and 11 in September. Some were buried in the churchyard and others 'on the backside of their house, on Penrith Fell, or Flatts cloose' (see Barnes, 1891).

Dacre, to the west of Penrith, escaped unscathed, although a note in the register for April 1598 states that they were aware of the plague in Penrith.

At Warcop, the disease appeared to be confined to one part of the parish, and to only two families. Adam Mosse and his two children died of the plague 'as it was thought' on 19 October 1597 (presumably co-primaries). On 4 November 1597 Margaret Mosse and Agnes Lancaster (presumably co-secondaries) were buried in a garth at Blatarne. The next burials occur on 25 May 1598 (presumably a fresh outbreak), when Richard Lancaster and his wife 'died both so daynelye upon the plague as it was thought and were buried in their own yeard at Blatarne'. Another entry on 6 June 1598 stated 'Dyed Thomas son of Richard Lancaster of Bletarne and the barne wherein he died burned and the corps afterwards interred'. Although plague was not given as the cause of death, the fact that purging by fire was deemed necessary would infer a highly infectious and much-feared illness.

The disease did not gain a strong foothold at Brough-under-Stainmore; only eight deaths occurred at first, albeit in 20 days, with seven coming from the family of Abram Wharton. His daughter died of plague on 10 November 1597 and, after 18 days, a son died and another son 2 days later. One day later, Abram, his mother, daughter and maid were buried. There were no further deaths from the plague until between 6 and 30 July of the following year (presumably a fresh infection), when seven more fell victim, of whom five bore the same surname.

The parish registers at Penrith record that the plague also struck at Appleby and the inference is that it was a severe attack to merit inclusion with Kendal and Carlisle. The outbreak apparently occurred late in the pandemic and probably the infection was again spread from Richmond, but might have returned eastwards from Penrith. Between 1 August 1598 and 25 March 1599, 128 persons died at Appleby, Scattergate, Colby and Colby Leathes, and it is inferred that death was because of plague. Appleby market was transferred to Melkinthorpe Woodhouse Farm, between Cliburn and Melkinthorpe, where a field adjoining Gillshaughlin still retained the name 'Little Appleby' in 1894 (Nicholson & Burn, 1777; Furness, 1894).

The plague also broke out again in Darlington and Durham in 1598; possibly being reinfected from Richmond.

These accounts suggest that the pandemic in northern England in 1597–98 was density dependent, i.e. the infection obviously spread to the smaller communities but no full epidemic exploded because of the small size and low density of the population (low R_o; $X < N_T$, section 2.6).

The plague eventually moved northwards from Carlisle and it is noteworthy that it reached Dumfries by the *winter* of 1598, a factor that should have convinced Shrewsbury, by his own criteria, that this could not have been an epidemic of bubonic plague. He says, however, 'there seems to be no doubt about the ravages of bubonic plague in a part of this region in 1598'. The epidemic caused problems for trade and even a scarcity of food; two men sent from Dumfries to Galloway were stopped at Wigton with 38 head of cattle and compensation was sought because the impounded cattle became lean (Creighton, 1894).

7.9 Symptoms of the 1597–98 plague

There are few contemporaneous records of the symptoms of the disease in northwest England in 1597–98, although the accounts of the events during the pestilence in Carlisle (section 7.7) and Penrith (Chapter 5) give a good impression of how the visitation was viewed by the inhabitants. As usual, they believed that the infection could be 'ring-fenced' by preventing the entry of suspected strangers and by isolating those infected in their houses or in specially prepared isolation centres.

Richard Leake 'preacher of the word of God at Killington within the Baronie of Kendall, and in the Countie of Westmerland' delivered his so-called plague sermons either in 1598 or in 1599 and they were printed in London in 1599. He said:

> It pleased God by the space of two yeares together, to giue our country (in the North parts of this land) a taste of his power in iudgement, being prouoked thereunto by our manifold enormious sinnes: he visited us with many and grieuous sicknesses, *as first with the hot feuer, after, with the bloodie issue, and lastly, most fearefully with the extreame disease of the pestilence, inflicted upon many, and shaken at all in our whole countrie* [our italics]. And albeit neither I, nor any of the people

under my charge, were infected therewith, yet had we all of us, the cause thereof within our sinfull hearts, as well as any others.

(see Wilson, 1975)

There is no mention of tokens or of buboes in this account, but there was the usual hot fever which was followed by the 'bloodie issue', although from which source is not clear.

7.10 Conclusions

The plague in northern England in 1597–98 was a lethal, infectious epidemic that spread through the metapopulation in a linear fashion along well-defined corridors because it was constrained by the topography of the region and there were few roads. The pandemic was characterised by the rapidity of its transmission, particularly on the west side of the Pennines. Major epidemics were experienced only in the larger towns and, on this occasion, York escaped the infection. Smaller communities escaped or suffered only minor outbreaks which were confined to a few households; even so, if the infection entered a household, several members of the family would die. Epidemics, therefore, were density dependent and did not explode unless contacts were made with a sufficient number of households (low R_0; $X < N_T$, section 2.6).

Epidemics followed Reed and Frost dynamics but with a long time-course, indicative of a long serial generation time; deaths began in the spring with an initial slow build-up and peaked in the late summer (type (i) epidemics). In several foci the outbreak began in autumn and continued in a few households over winter (low seasonal R_0 but high household R_0) before re-emerging in the spring and initiating a typical summer epidemic (type (ii) epidemics). There is an additional epidemic pattern that we have observed in which the plague began in the autumn and grumbled through the winter but did not re-emerge in the spring ($R_0 < 1$), a variant on type (ii) epidemics. An example is St Oswald, Durham, in 1589 and possibly Hawkshead in 1577–78 (section 7.2). The Black Death, which clearly hit a completely naive population, did not fit into these epidemics: although it was more evident in the warmer months, a full epidemic exploded (with Reed and Frost dynamics) even in mid-winter if an infective arrived.

The major epidemics were, therefore, strongly seasonal, although it is not clear whether this was because of a change in the infectivity of the causative agent in early spring or a change in the behaviour of the inhabitants who started making more effective contacts with their fellow citizens in the warmer weather.

Most towns instituted sensible preventative and quarantine measures, particularly the closure of the entry ports and the attempted prohibition of ingress and egress. Fairs were cancelled. They realised that people showing the dreaded signs in the terminal stages were infectious and isolated these houses as far as they were able and they also seem to have appreciated that the plague could be brought in by apparently healthy travellers. They may have contained the epidemic somewhat by these actions but they were largely defeated by the long incubation period of the disease, during which symptomless infectives could move round the town in summer, particularly in the market, spreading the infection.

8

Plagues in London in the 17th century

We saw in Chapter 6 that plague broke out sporadically in London throughout the 16th century, each epidemic lasting about 9 months during the spring, summer and autumn but it was persistent from 1578 (when there were 3568 plague deaths) to 1582 (2976 plague deaths) and then exploded again at the end of the century in 1593 (10 662 plague deaths) an outbreak that persisted through 1594 and 1595. Does this persistence of the epidemics for 3 to 5 years indicate a change in the epidemiology of the disease or had a new infection or mutation arrived? Some of these epidemics were confined to London, but some spread along the River Thames from Greenwich to Windsor and expanded northwards into the Home Counties, usually within a radius of 20 miles. There were also major epidemics in provincial England in the 16th century, some coincident with the outbreaks in London but, as a generalisation, it does not seem that London usually acted as a focus for these provincial outbreaks, although it frequently spread out to the Home Counties and along the Thames.

Figure 8.1 shows the plague deaths in London from 1578 until 1680 and the three terrible epidemics of the 17th century can be seen. Figure 8.1 is deceptive because the scale on the ordinate necessary to cover the enormous mortality of 1665 disguises the fact that the plague was almost endemic, with deaths in London in most years from 1603 to 1679, although at a low level; this is revealed more clearly in Fig. 8.2 where the annual plague deaths are plotted on a logarithmic scale.

In this chapter we describe the plagues in London in the 17th century. There are many more detailed documents, such as Bills of Mortality and parish registers extant for this period and, in particular, the epidemic of 1665 has been described in many historical studies. We have concentrated here on the aetiological and epidemiological features of the epidemics, including seasonality, spread of the disease within the city, movement out

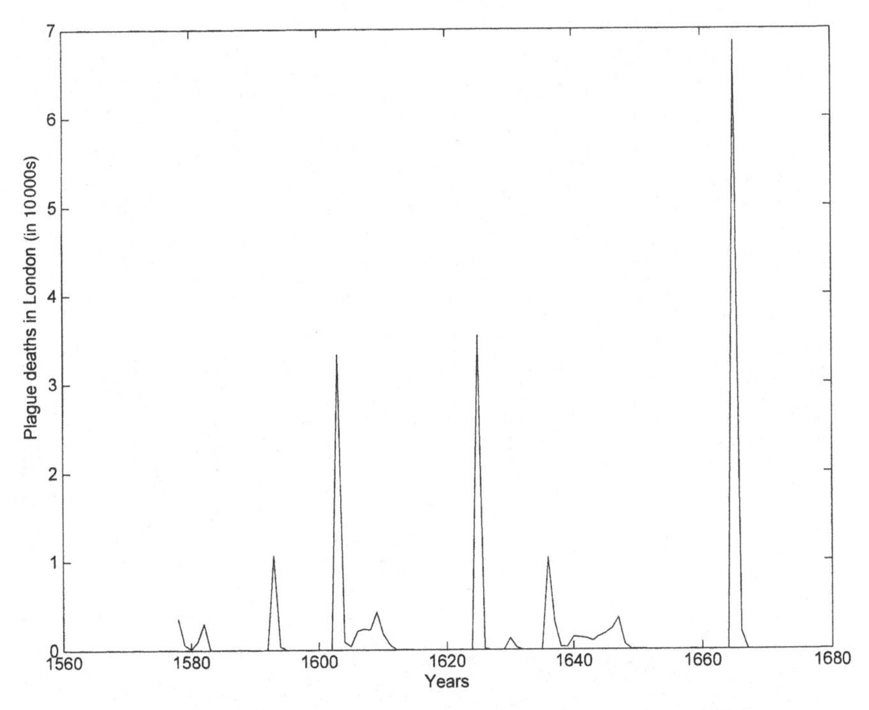

Fig. 8.1. Plague deaths in London, 1578–1680. Data from Creighton (1894).

of London, symptoms, point of outbreak and origins. In short, those details that may help to clarify the nature of the causative organism and which may be compared with the outbreaks in the provinces where considerable information has been collated.

8.1 The outbreak in London in 1603

Plague was active in a number of English towns in 1602 and was also raging in the Low Countries; Shrewsbury (1970) believed that the disease was imported into London from Amsterdam in 1602. The plague probably began in March 1603, although there are records of a few deaths in some weeks in January and February. Creighton (1894) believed, in spite of the lack of direct evidence, that the plague of 1603 began in the east end of London, in the parish of Stepney, where the first burial was recorded on 25 March. Thereafter, the epidemic followed its usual seasonal course with a very slow start in March and April and weekly burials rising sharply during June, peaking in August and September before falling in the autumn and winter (Fig. 8.3); a typical type (i) epidemic of haemorrhagic plague.

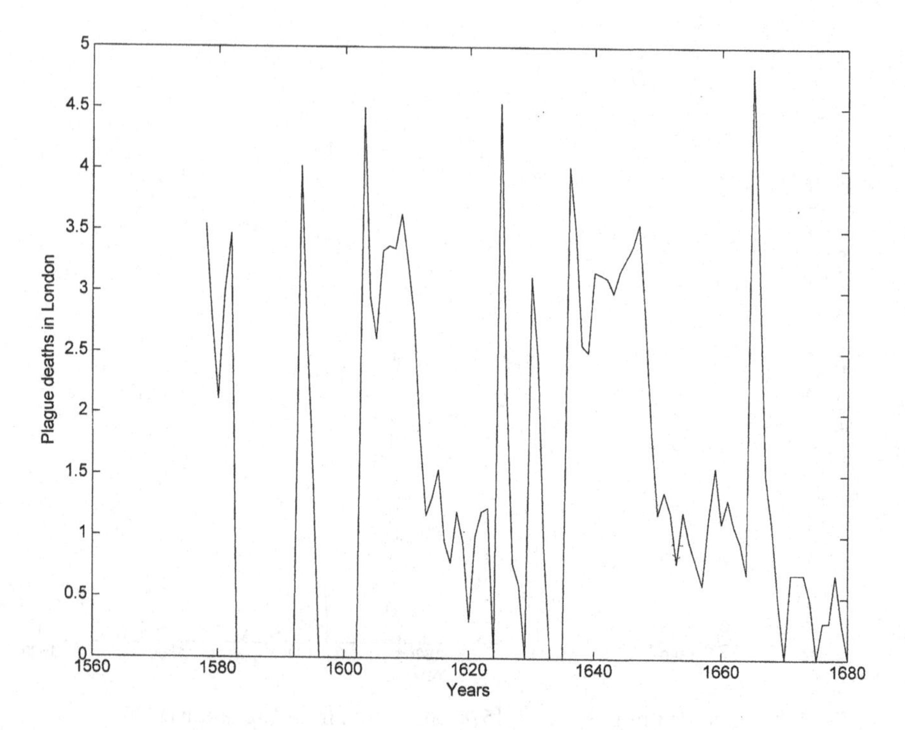

Fig. 8.2. Plague deaths in London, 1578–1680. Ordinate: $\log_{10}$ of the number of deaths. Data from Creighton (1894).

The distribution and intensity of the 1603 epidemic were uneven in the metropolitan parishes, with a wide variation in mortality rates that ranged from 19% in St Antholin, Budge Row, to 94% in St Pancras, Soper Lane. Shrewsbury also calculated the plague mortality rates for Stepney (95%), Newington-Butts (90%), Islington (85%), Lambeth (97%) and Hackney (88%) and concluded that the epidemic was much more deadly in these overcrowded London suburbs than in the city itself. We have analysed the epidemic of 1603 in the parish of St Helen, Bishopsgate, where the first plague victim was recorded in the registers on 25 June and the last on 1 February 1604 and during this time 103 people were buried. The start of the outbreak in the first 32 households is shown in Fig. 8.4: once again, the servants formed a high proportion of the deaths. There were three co-primaries, with the daughter in the Marshall family being infected first on 19 May 1603. The spread of the epidemic was slow at first: the three co-primaries probably infected six secondaries ($R_0 = 2$) who, in turn infected 13 tertiaries ($R_0 = 2$). Thereafter the epidemic suddenly exploded

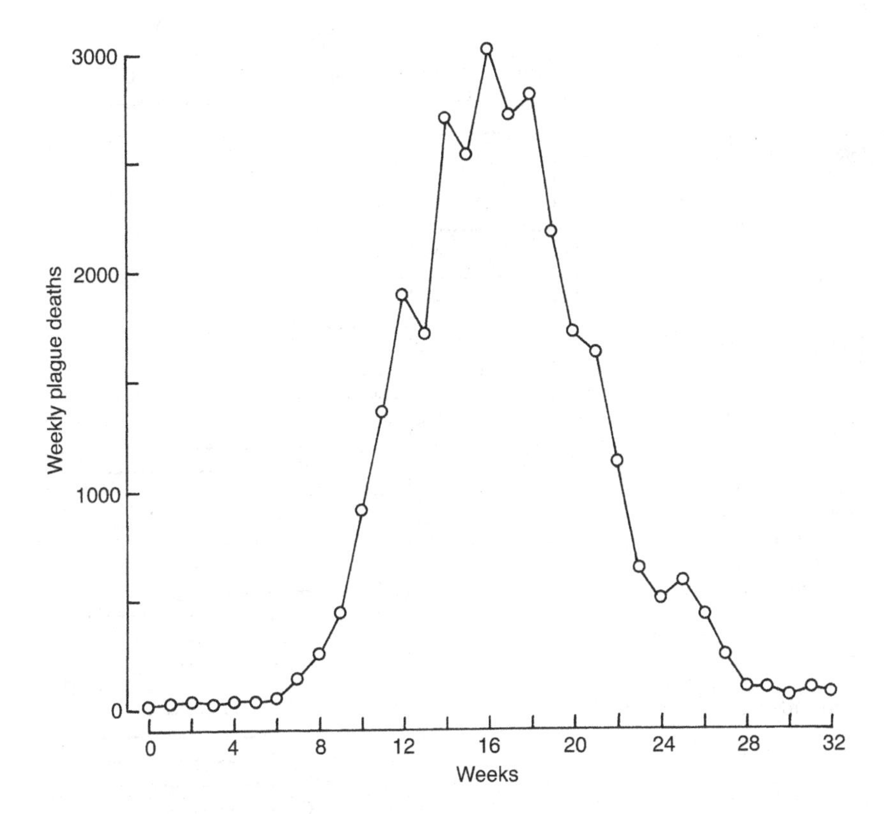

Fig. 8.3. Weekly plague mortality in London, May to December 1603. Abscissa: weeks after 12 May 1603. Data from Creighton (1894).

towards the end of August with an estimated total of 46 quaternaries ($R_0 = 3.5$). As usual, there were multiple deaths in some households (the Harvey family lost three of its members and three servants over 18 days in September) but there was only one victim in the majority. The last quaternary case is judged to have died on 26 September (the son of the Fenner family, shown in Fig. 8.4, died on 3 September).

Surprisingly, after this, the epidemic fizzled out over the winter: a servant in the Sturgeon family (buried 9 October) infected the son, who died 28 days later on 6 November. He, in turn, at the start of his infectious period infected the son of the Furnis family, who died 11 days later on 17 November. The next recorded burial, Thomas Richardson, was buried after a 42-day interval on 29 December. This is explicable in three ways. (i) An infection from outside the parish. (ii) An infection from someone in the parish who contracted the plague but did not die of it because of the

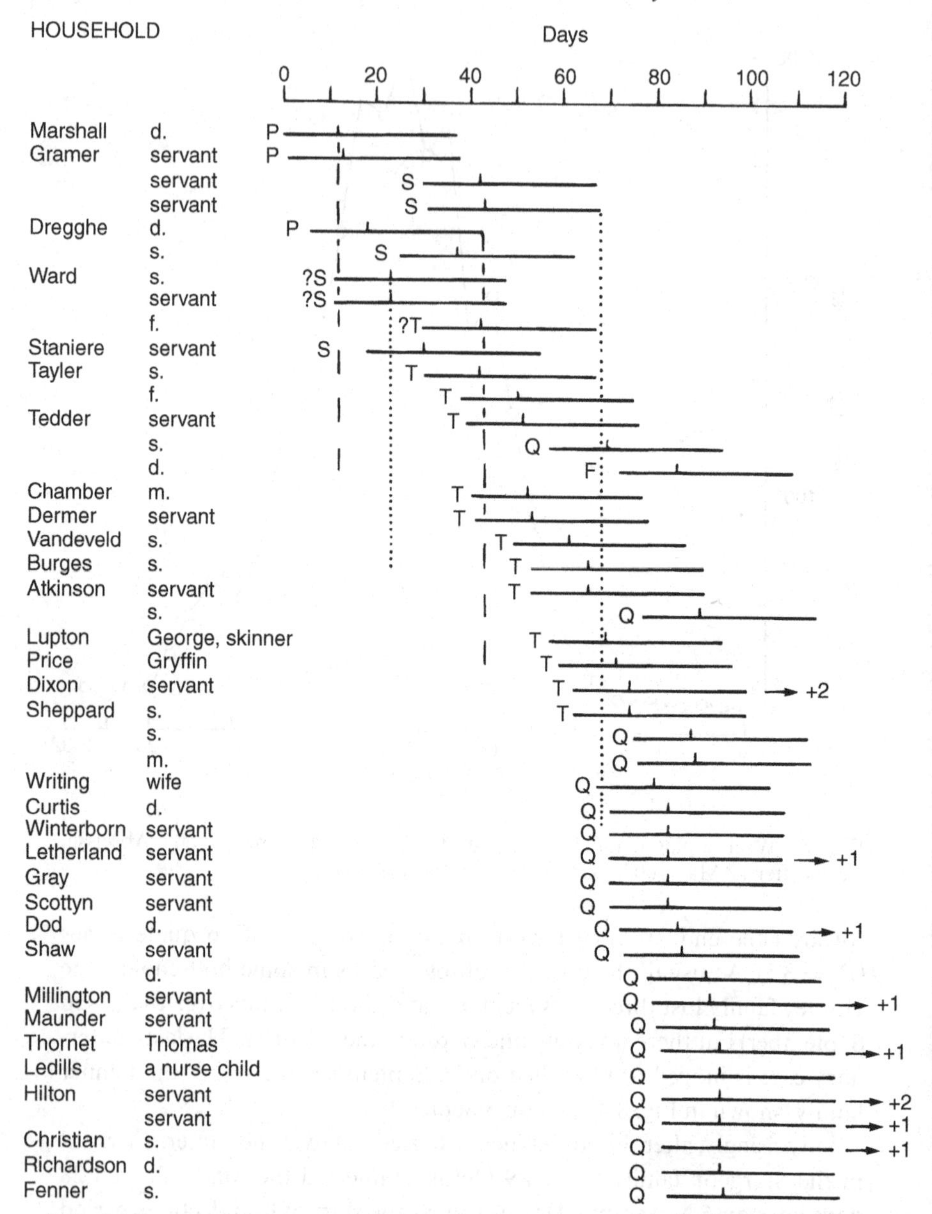

Fig. 8.4. Suggested sequence of infections at the start of the plague epidemic in the parish of St Helen, Bishopsgate, London in 1603. Scale: days after 19 May 1603. For further details and abbreviations, see Fig. 5.6.

exceptionally cold winter conditions in December 1603 (see section 13.10). (iii) An extension of the latent and infectious periods in cold winter conditions; we have other indications of this effect in our analyses of London epidemics and in Penrith (Chapter 5). Finally, Thomas Richardson infected his servant just before he died and the servant, in turn, died 34 days later on 1 February 1604.

Of the symptoms of the disease in this epidemic we have discovered only the following notes. A dying man was visited by a friendly neighbour, who promised to order the coffin, but he died himself an hour before this infected friend. A churchwarden in Thames Street, on being asked for space in the churchyard answered mockingly that he wanted it for himself and he did occupy it 3 days later (Creighton, 1894). Evidently, the disease could follow the same rapid course as that described in other outbreaks of plague in other situations in other centuries; 3 days is a very typical time-course from the first detection of symptoms to death. Creighton also quoted from *A New Treatise of the Pestilence, etc. the Like Not Before This Time Published, and Therefore Necessarie for all Manner of Persons in this Time of Contagion* by 'S. H. Studious in Phisicke' (published in 1603), who stated the theory of the plague bubo: it was a way made by nature to expel the 'venomous and corrupt matter which is noisesome unto it'. He advised incising the bubo and helping it to suppurate 'which was the treatment in the Black Death'. The epidemic of 1603 in London evidently was supposed to have features in common with the Great Pestilence 250 years previously.

Dogs were slaughtered during this plague in London. Shrewsbury said that the Venetian secretary recorded that the weather in May 1603 was unusually hot, which aroused the fear that the plague would spread, especially as the authorities had taken no action against it 'except to kill the dogs and mark the houses'. The churchwardens of St Margaret's parish paid for the slaughter of 502 dogs at 1d each.

Unlike many of the plagues in London in the 16th century, the epidemic of 1603 spread widely. On 8 August 1603, St Bartholomew and Sturbridge Fairs and all others within 50 miles of London were cancelled. Creighton (1894) stated that many of the country parishes nearest to London had plague burials in 1603 that he believed to be because of Londoners fleeing from the epidemic (always the standard response of the more wealthy citizens to an outbreak of the pestilence) and in the Croydon register there is a note that 'many died in the highways near the city'. It is interesting (and curious) that the infection in the country near London had been attracting notice before the plague in the capital caused any alarm: the Lord Mayor wrote to the Privy Council concerning the steps that had been taken 'to prevent the spread of the plague in the counties of Middlesex and Surrey'. The King levied a special rate on 10 villages in Kent on 20 July 1603 to relieve the sufferers in a grievous plague; Creighton contended that such rates were usually levied when an epidemic was nearly over and concluded that the outbreak in Kent must have been at least as early as that in London. Perhaps the plague in 1603 was not imported directly into the ports of London but came from Kent.

Thus Kent became very generally infected and by October 1603 few towns on the road from London to Dover were free from it. It spread elsewhere in the Home Counties: Surrey and Sussex were affected and in Essex it 'swept off great numbers' in Colchester. The plague spread westwards along the Thames Valley to Oxford and also moved southwest to Wiltshire where an order directed that a watch be kept in every town and village in the county for the arrest of all vagrants who were to be summarily ejected if they were suspected to be 'dangerous of infection'.

Plague was very active in England during the period 1602–6, with major foci in the Midlands and the northern counties, as we describe in section 9.1. It is probable that the disease exhibited changes in its epidemiology after 1600; not only was the mortality very much greater than before, as in London in 1603, but isolated epidemics in towns and cities changed to pandemics in the metapopulation, as presaged by the pandemic of northern England in 1597–98 (section 7.3). These pandemics persisted over winter for 3 or more years and their spread can be monitored along the communication routes, particularly the roads (see section 13.9).

8.2 Plague in London after the epidemic of 1603

There were only 910 and 410 plague deaths reported in the Bills of Mortality of London for 1604 and 1605, respectively (when there were severe epidemics in the provinces, section 9.1), but the pestilence became endemic for 5 years thereafter. The total plague deaths in each year were as follows: 1606, 2124; 1607, 2352; 1608, 2262; 1609, 4240; 1610, 1803.

Figure 8.5 shows that the seasonal pattern of mortality was the same in each of these 5 years, peaking with great regularity in the months of September and October. Once again, when we have detailed weekly or monthly data available, we see the characteristic seasonal pattern, emerging in late spring and peaking in late summer/early autumn.

An analysis of most of the epidemic at the London parish of St Mary Somerset in 1606, when only 38 died, is shown in Fig. 8.6. There were four co-primaries infected over 4 days in March but this epidemic struggled to

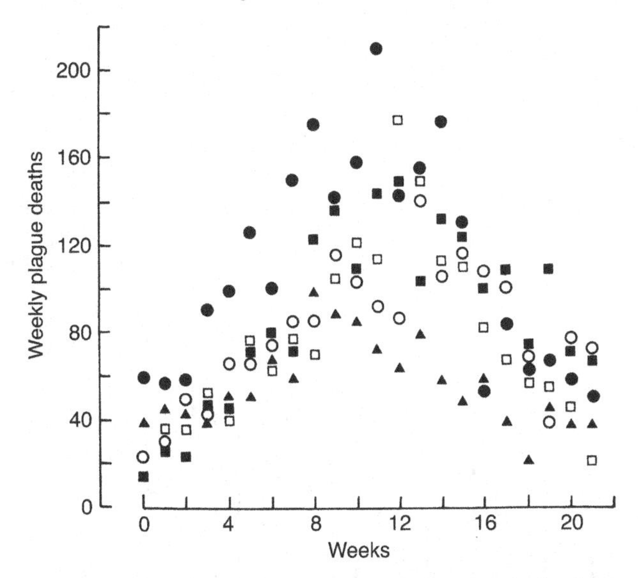

Fig. 8.5. Seasonality (July to November) of weekly plague deaths in London, 1606–10. Abscissa: weeks after the first week in July. Note the peaks in the first week of September. Open circles, 1606; open squares, 1607; closed squares, 1608; closed circles, 1609; closed triangles, 1610. Data from Creighton (1894).

continue because of the low R_0. Only one secondary was infected (within the Kinge family) producing an effective contact rate below 1 (0.25). Only one tertiary case is recorded ($R_0 = 1$) but she (a daughter in the Hoare family) infected seven quaternaries ($R_0 = 7$) who, in turn, infected eight fifth generation victims ($R_0 = 1$), the last dying on 31 August. Thereafter, the epidemic fizzled out quite quickly because of the low R_0. The servant in the Ricardsonne household was buried on 15 October but he had infected a servant in the Nuame household who died on 31 October (not shown in Fig. 8.6). Three months later, the maid servant in the Smithe household was recorded as a plague burial on 1 February 1607 and must have been a reinfection. She probably infected the servant in the Halye family who died exactly 37 days later. Again, the plague struck particularly fiercely at servants.

There were also four plague burials in St Mary Somerset in autumn 1607 (one primary, two secondaries, one tertiary).

8.3 Plague in London in 1625

The weather in the run-up to the plague of 1625, when over 35 000 died, was unusual: the summer of 1624 was unusually hot and dry; October was

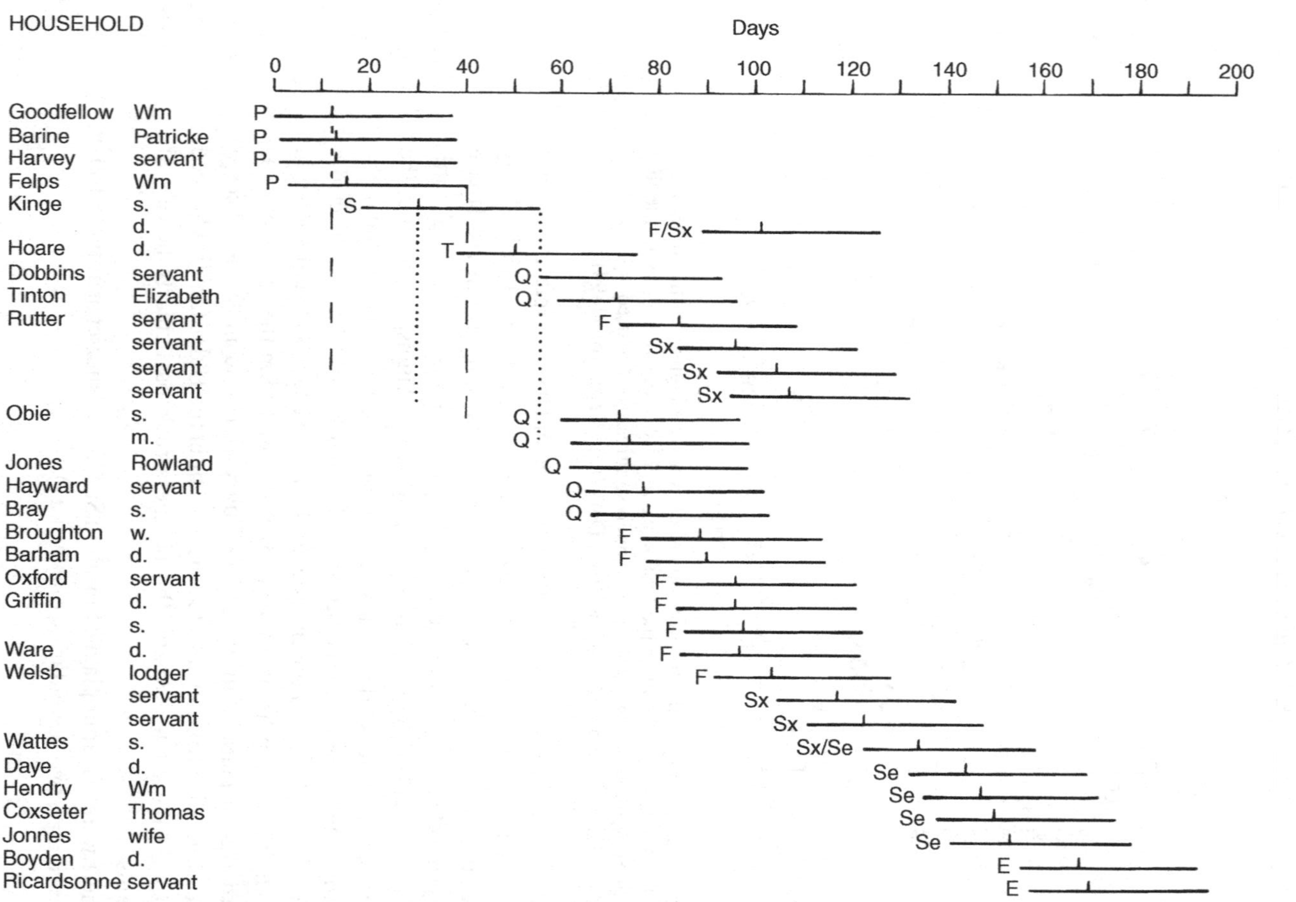

Fig. 8.6. Suggested sequence of infections in the parish of St Mary Somerset, London, in 1606. Scale: days after 21 March 1606. For further details and abbreviations, see Fig. 5.6. Sx, Se and E, sixth, seventh and eighth generation infections, respectively.

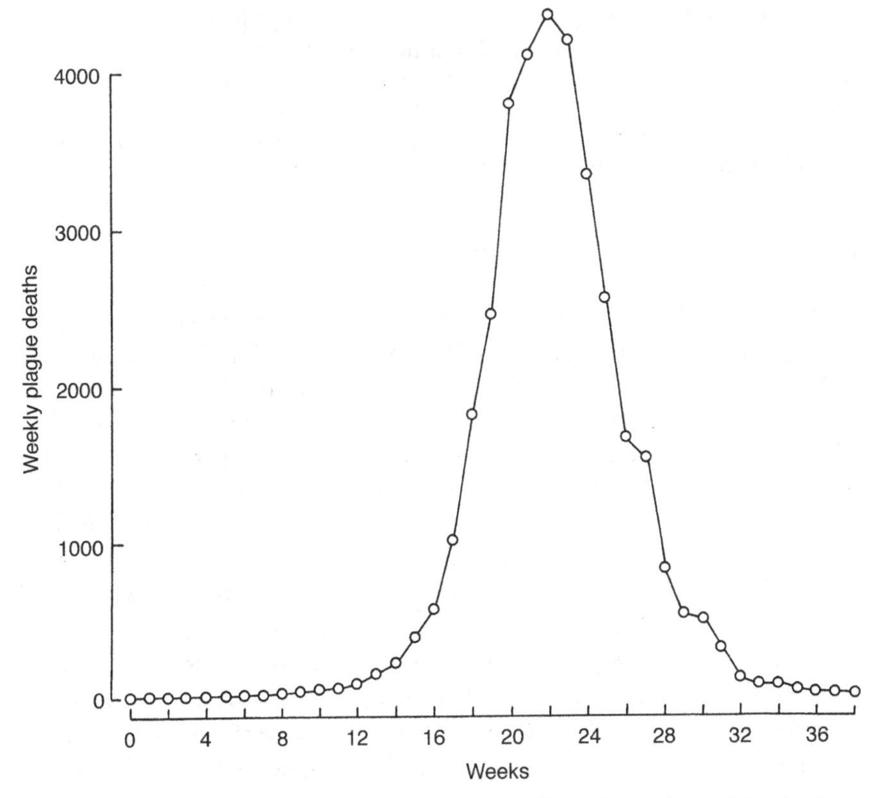

Fig. 8.7. Weekly plague deaths in London, March to December 1625. Abscissa: weeks after 17 March 1625. Data from Creighton (1894).

exceptionally fine; January 1625 was warm and mild; spring was described as 'wholesome' but the early summer was extremely cold. June 1625 was a month of ceaseless rain and both the hay harvest and corn harvest were spoilt.

Figure 8.7 shows the weekly plague deaths in 1625 that Creighton (1894) believed demonstrated how the epidemic increased after the rains in June, but the data suggest that the epidemic followed the usual seasonal pattern: there was a very slow build-up with 2 plague deaths in January and 12 in February and weekly plague deaths thereafter did not exceed 30 until May 12 (Fig. 8.7) when the epidemic exploded and reached its peak in August when over 4400 deaths were recorded in one week.

The Lord Mayor received the following reprimand from the Privy Council in March 1625, which suggests that plague was already a serious problem this early in the year:

We understand that the plague doth daylie encrease in the citty and that ther dyed this last weeke seven of it in one parrish and although it hath beene thus encreasing divers weekes yet wee cannot heare that any good course hath beene taken for preventing it either by carrying the infected persons to the pesthouse or setting watch upon them or by burning of the stuffe of the deceased which being of little or noe value might easily be recompensed.

Seven deaths in one parish suggests a greater death rate than that given in the Bills of Mortality (Fig. 8.7).

Shrewsbury (1970) has extracted from eight parish registers in London details of the spread of the disease in 1625 in 17 selected families with a total of 80 burials, and inspection of the data shows clearly that its progress follows exactly the pattern that we have demonstrated in our analyses of communities in London and the provinces, with a latent period of 10–12 days.

The analysis of the start of the remarkable and severe epidemic at St Martin-in-the-Fields in 1625, when some 191 persons died, is shown in Fig. 8.8. The epidemic began here explosively and 179 of the victims died between 4 July and 8 August. We have identified at least 34 apparent co-primaries (Fig. 8.8), presumably the result of multiple infections from adjacent parishes. This epidemic at St Martin-in-the-Fields was also characterised by a low *household* R_0, since there were few secondary infections within the families. The outbreak disappeared as quickly as it had begun and there were no plague burials after 17 August.

Of the symptoms of the 1625 plague in London, Shrewsbury (1970) quoted a correspondent: 'The physicians do in a manner agree that this sickness is not directly the plague, as *not having any sore* [our italics] or any such like accident, but only contagious in blood or kindred'; he deduced that the victims did not show buboes and suggested that their absence 'was due in all probability to the extremely high virulence of the responsible strain of *Pasteurella* [= *Yersinia*] *pestis* at the start of the 1625 epidemic.'

A story tells of a woman who fled to Croydon and looking back on Streatham Hill said '"farewell plague" but soon after was taken sick, *had these tokens on her breast* [our italics] . . .' (see section 13.12).

The Privy Council on the 21 October 1625 censured 'the undiscreet and

---

Fig. 8.8. (*opposite*) Suggested sequence of infections at the start of the plague epidemic in the parish of St Martin-in-the-Fields, London, in 1625. Scale: days after 28 May 1625. For further details and abbreviations, see Fig. 5.6.

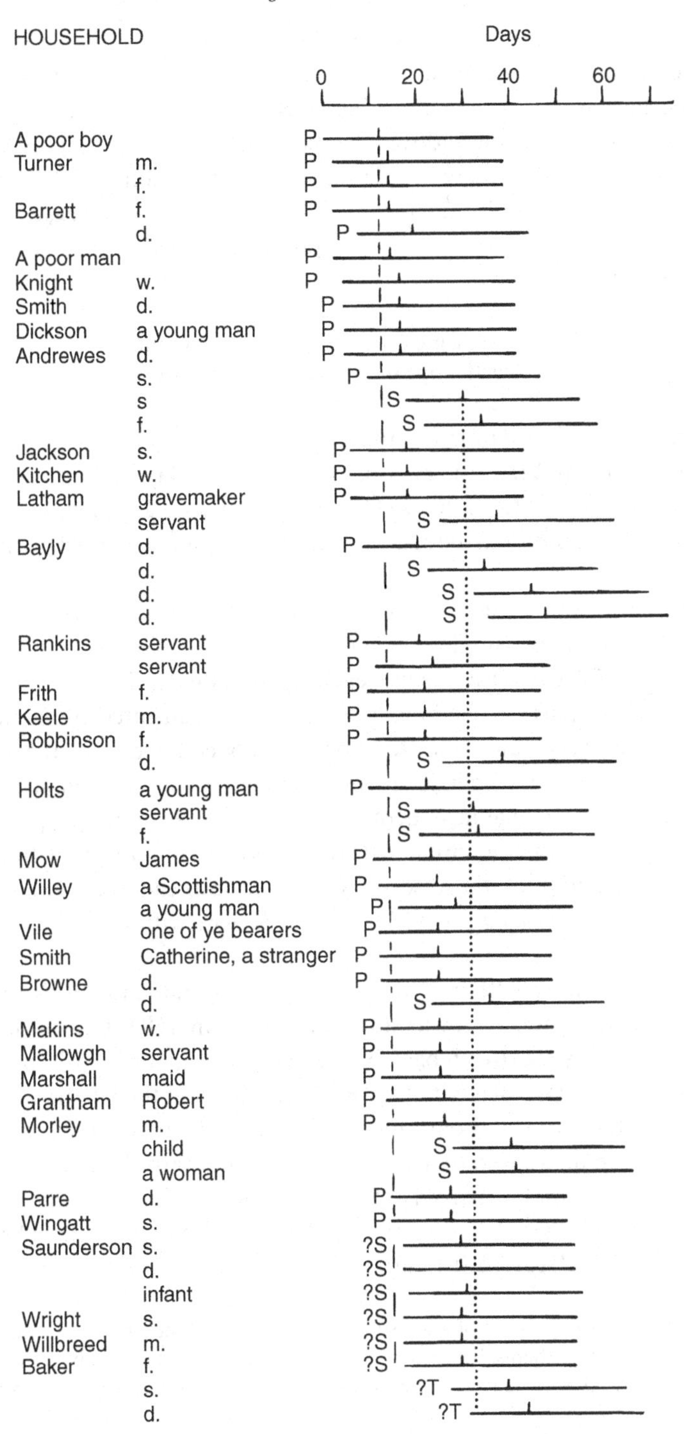
HOUSEHOLD
Days
0
20
40
60
A poor boy
Turner m.
f.
Barrett f.
d.
A poor man
Knight w.
Smith d.
Dickson a young man
Andrewes d.
s.
s
f.
Jackson s.
Kitchen w.
Latham gravemaker
servant
Bayly d.
d.
d.
d.
Rankins servant
servant
Frith f.
Keele m.
Robbinson f.
d.
Holts a young man
servant
f.
Mow James
Willey a Scottishman
a young man
Vile one of ye bearers
Smith Catherine, a stranger
Browne d.
d.
Makins w.
Mallowgh servant
Marshall maid
Grantham Robert
Morley m.
child
a woman
Parre d.
Wingatt s.
Saunderson s.
d.
infant
Wright s.
Willbreed m.
Baker f.
s.
d.
P
S
?S
?T

unruly caryage of the inhabitants of Westminster of whom those who have the sore running upon them goe as freely abroade conversing promiscuously with others as if they were not infected'. Shrewsbury assumed that these people were convalescent and we conclude, firstly, that it was possible to recover from the infection and, secondly, that at least some of the victims displayed open sores during the recovery stage of the disease. These conclusions are supported by a contemporaneous poem, which is quoted by Creighton:

> Some with their carbuncles and sores new burst
> Are fled with hope they have escaped the worst

Creighton concluded from this that the buboes and boils might come out more than once and that the best chance of survival lay in their suppuration. In June 1625, Lord Russell 'being to go to Parliament had his shoemaker to pull on this boots, who fell down dead of the plague in his presence'.

Again, dogs were destroyed: the parish of St Margaret, Westminster, paid £2 17s 8d for the slaughter of 466 dogs.

By July 1625 the disease was universally distributed in London and had spread to other parts of the kingdom, probably augmented by the flight into the country of all those Londoners who could do so, the usual response to a major epidemic. In mid-September the Tuscan Resident reported from his refuge near Bedford that almost as many people were dying of plague 'within the circuit of three miles from London' as within the capital itself, and that the disease was so widespread throughout the kingdom that it was impossible to go anywhere without a danger of contracting it (Shrewsbury, 1970).

However, the people fleeing from the plague in London met with a poor reception in the country towns and villages, as in 1603. Creighton described how 'They are driven back by men with bills and halberds, passing through village after village in disgrace until they end their journey; they sleep in stables, barns and outhouses, or even by the roadsides in ditches and in the open fields. And that was the lot of comparatively wealthy men'. A stranger from London arrived at Southampton on 27 August and died in the fields; he had a 'good store of money about him which was taken before he was cold.' The Dean of St Paul's wrote:

> The citizens fled away as out of a house on fire, and stuffed their pockets with their best ware, and threw themselves into the highways, and were not received so much as into barns, and perished so: some of them with more money about them than

would have bought the village where they died. A justice of the peace told me of one that died so with £1400 about him.

(Creighton, 1894)

Sir John Coke sent a report to Lord Brooke on 18 October 1625 that seems to imply that by then the disease was losing its virulence: 'We are full of hope that God beginneth to stay his hand, because now in London the tenth person dieth not of those that are sick and generally the plague seems changed into an ague'.

The plague spread throughout the Home Counties and also through the counties bordering the English Channel, Kent, Sussex and Hampshire, by July. Shrewsbury averred that the plague moved from London to the West Country via Wiltshire, where it had already arrived by 5 August. It broke out in Oxford by the end of July and in East Anglia by the end of June. Indeed, as we describe in section 9.3, plague seems to have been widespread in England in 1625 and this continued into the following year, although Creighton (1894) rather belittled this provincial pandemic. The general belief seems to be that the epidemic in London spread widely and remarkably rapidly from this focus.

8.4 Recovery of the population of London after 1625

There were 54 000 deaths registered in London in 1625, 35 000 of them because of plague, but these gaps were rapidly filled, presumably largely because of immigration. By 1627 baptisms were again at 8408, having been 8299 in the year before the plague. In 1629, baptisms exceeded burials by more than 1000 and continued to be slightly in excess until the next plague of 1636. London in 1625, therefore, is another example of a population that recovered remarkably quickly after a mortality crisis.

The population of London in 1625 was of the order of 300 000 (Creighton, 1894; Shrewsbury, 1970) so that the 35 000 deaths represent approximately a 12% plague mortality, although with so many fleeing from the capital and possibly dying elsewhere, the proportion of those staying behind who died would be higher. Nevertheless, the percentage plague mortality in London would have been much less than in some outbreaks in the provinces. For example, in Penrith in the 1597–98 epidemic some 45% of the population died (section 5.2), although a smaller proportion of the citizens of Penrith may have had the opportunity or the means to have fled from the town. Does this suggest that the non-immigrant population of London had some degree of immunity because of repeated exposure to the disease?

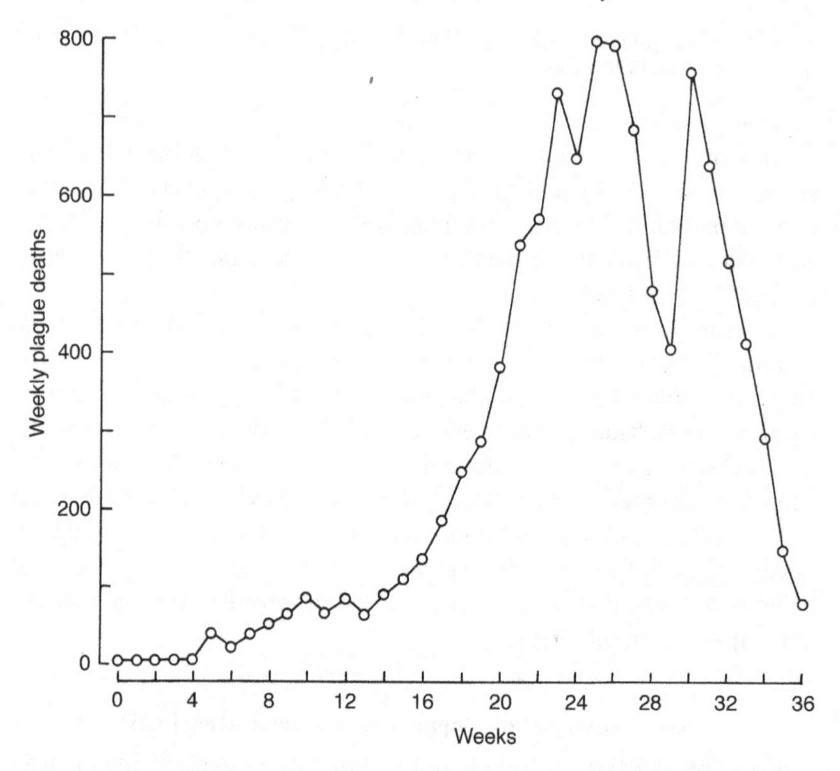

Fig. 8.9. Weekly plague deaths in London, April to December 1636. Abscissa: weeks after 7 April 1636. Data from Creighton (1894).

8.5 Plague in London in 1636

There was a violent outbreak of plague in the northern counties in 1635 (section 9.4), a year when the infection was quiet in London. An epidemic erupted, however, in 1636 in the capital, the third in the 17th century. It was on a smaller scale, however, with only 10 400 being 'buried of plague' (see Fig. 8.1). Although the mortality followed the normal overall seasonal pattern, with the characteristically slow build-up, the peak of the epidemic was a little later than usual with the maximum weekly burials being recorded in the last week of September and the first week of October (Fig. 8.9). The epidemic did not subside until December and it continued to smoulder through the winter and broke out again in the spring of the following year, 1637.

Like the plagues of 1603 and 1625, the outbreak of 1636 in London again began in the eastern suburbs; deaths from plague were reported in the plague-houses in Stepney and Whitechapel in mid-April, some 3 weeks before the epidemic inside the walled city, which eventually bore the brunt of the attack. Shrewsbury (1970) believed that the appearance of the plague in the eastern suburbs was consistent with its marine importation, probably from Holland.

An analysis of the outbreak in the parish of St Mary Somerset, August to December 1636, is shown in Fig. 8.10: only 33 died because of a low R_0. The single primary produced 5–6 secondaries ($R_0 = 5.5$) who infected only 2–3 tertiaries ($R_0 \approx 0.5$), producing 5 quaternaries ($R_0 = 2$), 10 fifth generation ($R_0 = 2$), 4 sixth generation ($R_0 = 0.4$) and 5 seventh generation ($R_0 \approx 1$) infections. Although only 23 households were affected, this epidemic lasted 170 days because the infection spread slowly to other families in the autumn and winter before being extinguished at the end of December.

The plague travelled westwards along the Thames Valley, as usual, via Westminster to Isleworth and then to Reading. It was also active at Faversham, a river port on the north Kent coast, where 78 plague burials were registered between May and November 1636. Faversham, perhaps because of its role as a port, seems to have suffered from plague quite frequently in previous years and it may have acted as one of the points of entry for the epidemics in Kent. Plague was active in at least two other places in Kent in 1636 and it is uncertain whether the outbreak there spread from London or was initiated independently, perhaps as a different disease. Shrewsbury (1970) gave the following account (taken from the *Sussex Archaeological Collection*) of the importation of plague into the hamlet of Kemsing from Sevenoaks in 1636 that was written by Leonard Gale when he was 67 years old:

> I was born in the parish of Sevenoaks, in Kent, my father, a blacksmith . . . had, by a former wife, two sons, and by my mother three sons and one daughter; and when I was between sixteen and seventeen years of age, my father and mother going to visit a friend at Sensom, in the said county, took the plague, and quickly after they came home my mother fell sick and about six days after died, nobody thinking of such a disease. My father made a great burial for her, and abundance came to it, not fearing anything, and notwithstanding several women layd my mother forth, and no manner of clothes were taken out of the chamber when she died, yet not one person took the distemper; this I set down as a miracle. After her burial, we were all one whole week, and a great many people frequented our house, and we our neighbours' houses, but at the week's end, in two days, fell sick my father, my eldest brother, my sister, and myself; and in three days after this my younger brothers, Edward and John, fell sick, and though I was very ill, my father sent me to market to buy provisions, but before I came home it was noysed abroad that it was the plague, and as soon as I was come in adoors, they charged me to keep in, and set a strong watch over us, yet all this while no one took the distemper of or from us, and about

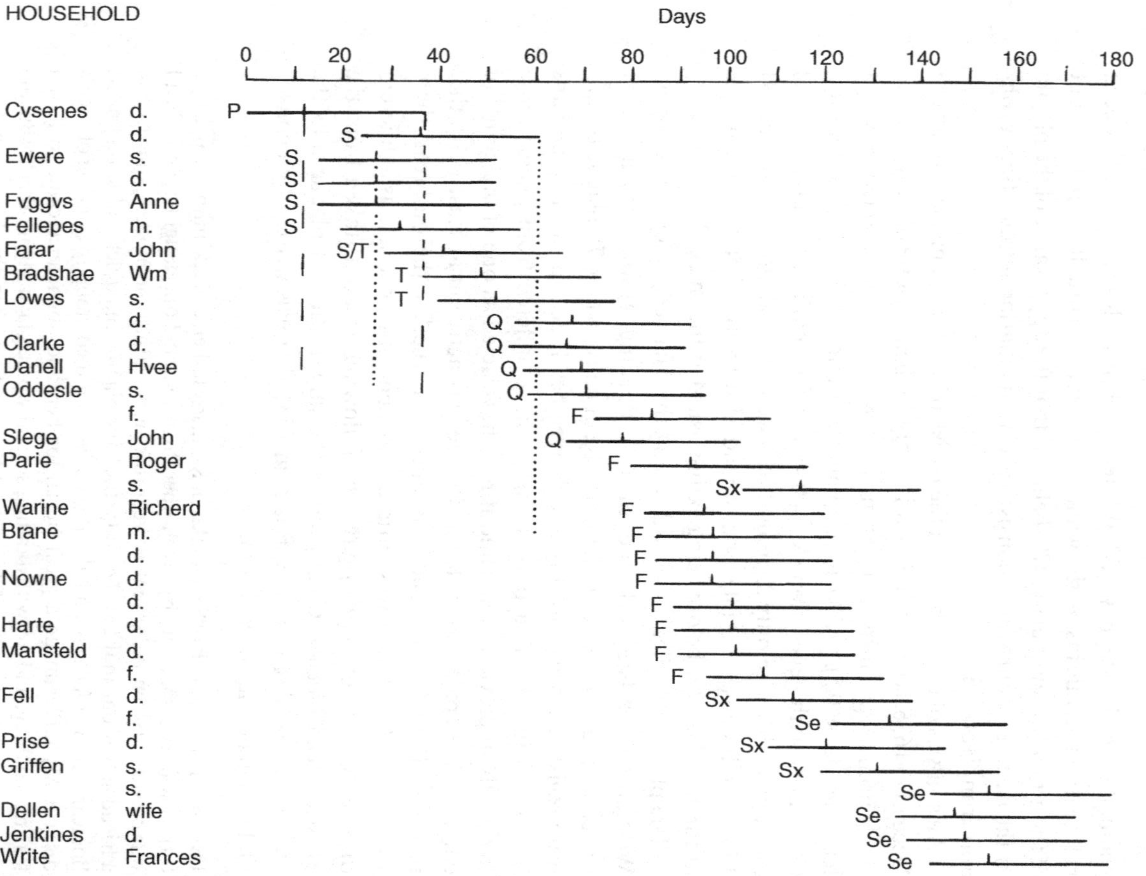

Fig. 8.10. Suggested sequence of infections in the plague epidemic in the parish of St Mary Somerset, London, in 1636. Scale: days after 30 June 1636. For further details and abbreviations, see Fig. 5.6. Sx and Se, sixth and seventh generation infections, respectively.

the sixth day after they were taken, three of them dyed in three hours, one after another, and were all buryed in one grave, and about two days after the two youngest both died together, and were buryed in one grave. All this while I lay sick in another bed, and the tender looked every hour for my death; but it pleased God most miraculously to preserve me, and without any sore breaking, only I had a swelling in my groin, which it was long ere it sunk away, and I have been the worse for it ever since, and when I was recovered, I was shut up with two women, one man, and one child for three months, and neither of them had the distemper.

Shrewsbury (of course) interpreted the foregoing in terms of *Yersinia pestis* and blocked fleas, although he conceded that the strain of the microbe that infected the Gale family was of lowered virulence because it took 6 days to kill Mrs. Gale and because she 'obviously did not exhibit the usual signs of [bubonic] plague'. He concluded that there was no outbreak of plague in the hamlet because the Blacksmith's Forge was at one end of the village so that 'no infected rat left the house'. We suggest that this is another example of where an epidemic of an infectious disease does not explode unless the size and density of a population are above a certain limiting size (see section 13.6; Scott & Duncan, 1998).

The plague continued in London in the following year, 1637, causing 3082 deaths. In 1638 there were only 363 plague deaths but the total mortality, over 13 000, was, as Creighton (1894) pointed out, nearly 2000 more than in the previous year and he suggested that an epidemic of yet another disease may have broken out in London. The four successive years in London, 1640–43, also showed exceptional mortality that was not attributed to plague by Creighton (1894).

We illustrate in Fig. 8.11 the small epidemic in the parish of St Michael Bassishaw where nine families were infected during June to November 1641 and where the plague burials are identified in the registers. The single primary, a servant in the Earith household, infected five secondaries (four of them also servants) in the house in quick succession, beginning on the twelfth day after his initial infection (Fig. 8.11). The disease managed to spread the infection to a member of the Middleton family (a single tertiary) at the very end of the illness of one of the secondaries ($R_0 \ll 1$). Quaternary infections were established in the Gill family either after a very short latent period in the tertiary (Fig. 8.11), or by a fresh infection from outside the parish, or by an infection from someone who did not die of the disease. In any event, a daughter of the Gill family infected her brother (fifth generation) and so the epidemic hung on by the skin of its teeth because he infected a single sixth generation of the epidemic just before he died (Fig. 8.11). This small outbreak was characterised by a low interhousehold

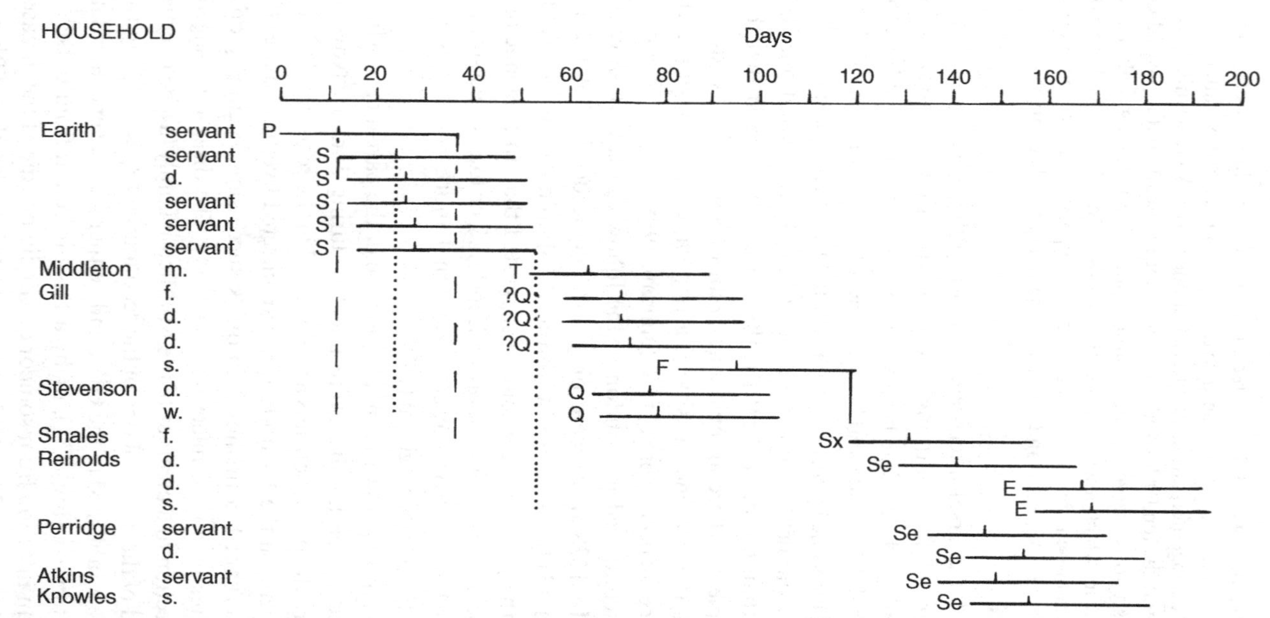

Fig. 8.11. Suggested sequence of infections in the plague epidemic in the parish of St Michael Bassishaw, London, in 1641. Note low R_o and slow spread of this small epidemic, including a 36-day gap between two burials. Scale: days after 28 April 1641. For further details and abbreviations, see Fig. 5.6. Sx, Se and E, sixth, seventh and eighth generation infections, respectively.

contact, even in summer, which extended its duration, and by an apparently short latent period for infections within the family.

8.6 The Great Plague in London in 1665–66

There are many accounts that record graphically the events during the terrible plague of 1665; Pepys gave an eye-witness account in his diary; Defoe (1722) who was only 6 years old at the time reconstructed the story some 55 years after the event; Bell (1924) described his admirable book as the story of the tragedy of the poor because 'in its immensity and in overwhelming proportion it was "the poores Plague"'. This is probably the definitive source for all aspects of the outbreak in London in 1665, although there are other accounts: Harvey (1769) and Leasor (1962). This section covers the aetiological, epidemiological and demographic aspects of the epidemic.

8.6.1 Origins and spread of the epidemic

Although plague was generally endemic in London, there was no major outbreak for almost 30 years between 1636 and 1665. It is surprising that there are relatively few notices of plague throughout England and Wales during the Commonwealth because there are no obvious changes in the social, sanitary or other states to account for this freedom. Shrewsbury (1970) believed that the war with the Netherlands reduced the opportunities for importation of the disease via shipping because Holland was the principal European source for the introduction of *Y. pestis.* Whatever the nature of the infective agent, Holland probably represented the source of some of the importations of plague into England: during the epidemic in the Netherlands in September 1655, the Commissioners of Customs were warned to ensure that no infected refugee was allowed to land in Britain.

The winter of 1664 was severe and the extreme frost did not abate until March 1665 (Bell, 1924). The earth was held in an almost continual black frost from November 1664 and the dry cold continued after the frost broke, producing, it was said, 'an unusual number of cases of pleurisy, pneumonia and angina', the result of the 'direst winter spring and summer that ever man alive knew . . . the grounds were burnt like highways, the meadow ground . . . having but four loads of hay which before bare forty'. There was a death in Long Acre from plague towards the end of December 1664; a bale of silks had come to the house from Holland but had originally been imported there from the Levant and this was sufficient evidence to suggest that *Y. pestis* was introduced by this route. This is most unlikely; the death was a solitary one and there were only five other sporadic and apparently unconnected plague burials in December 1664. Of these six deaths, three were in Whitechapel and the other three were in separate parishes. There was one further death in mid-February until the plague began at the end of April. Harvey (1769) described this latent period as follows:

And being restrained to a house or two, the seeds of it confined themselves to a hard frosty winter of near three months continuance: it lay asleep from Christmas to the middle of February, and then broke out again in the same parish; and after another long rest till April, put forth the malignant quality as soon as the warmth of spring gave sufficient force, and the distemper showed itself again the same place, where it was first: neither can it be proved that these ever met; especially after houses were shut up.

Shrewsbury explained this account as being a description of the epidemiology of bubonic plague (although he attributed the February death to typhus because he could not accept that the fleas could be active in the depths of winter), but it is unnecessary to take any account of these cases, which probably did not presage the epidemic that began in 1665: plague had been endemic in London for much of the 17th century, with a dozen or up to 1000 deaths in most years. However, this account does illustrate how, in non-epidemic situations, the disease could strike in winter in an apparently random and sporadic way, with nobody else dying, although some may have contracted the illness and survived.

Weekly registered plague deaths reached a total of 43 at the beginning of June 1665 and the first official notice of the outbreak in London was a proclamation on 14 June cancelling Barnwell fair 'for fear of spreading the plague'. Pepys wrote in his diary for 15 June 'The town grows very sickly, and people to be afraid of it'.

The plague broke out in the parish of St Giles-in-the-Fields and moved from the western and northern suburbs towards the City, the eastern suburbs and Southwark, the reverse direction to its usual progress. Creighton (1894) quoted Boghurst, an apothecary, who practised in St Giles-in-the-Fields as follows:

'The plague fell first upon the highest ground, for our parish is the highest ground about London, and the best air, and was first infected. Highgate, Hampstead and Acton also all shared in it'. From the west end of the town, Boghurst continues, 'it gradually insinuated and crept down Holborn and the Strand, and then into the City, and at last to the east end of the suburbs, so that it was half a year at the west end of the city [in his experience] before the east end and Stepney was infected, which was about the middle of July. Southwark, being the south suburb, was infected almost as soon as the west end.'

Defoe (1722) amplified this account having also shown

how it began at one end of the town, and proceeded gradually and slowly from one part to another; and like a dark cloud that passes over our heads, which as it thickens and overcasts the air at one end, clears up at the other end: so while the plague went on raging from west to east, as it went forwards east it abated in the west, by which means those parts of the town which were not seized, or who were left, and where it had spent its fury were, as it were, spared to help and assist the other; whereas had the distemper spread itself all over the City and suburbs at once, raging in all places alike, as it has done since in some places abroad, the whole body of the people must have been overwhelmed.

This account (if Defoe is to be believed) gives an insight into the epidemiology of the plague; it spread gradually and inexorably, each parish in turn experienced a flash epidemic that quickly burnt out. Its intensity was nearly over in one place before it had begun in another and Creighton (1894) regarded this as the most interesting epidemiological feature of the epidemic.

The explosive start of the major epidemic in the parish of St Michael Bassishaw in 1665 is shown in Fig. 8.12. Plague burials were recorded in the register from 17 June to 7 January 1666. The single primary case, the daughter in the Chadburne household, infected four other members of her family, some after a short latent period (Fig. 8.12), and 4–15 other secondaries (R_0 = 6 to 17). During this initial period, the contact rates both within and between households were high. The secondaries infected a very large number (perhaps 60) of tertiary cases during the period mid-July to the end of August but, now, the pattern of the epidemic had changed; there were few infections within the household but a high interhousehold contact rate, typical of outbreaks in July and August. The next wave of quaternary cases occurred in September and the epidemic was almost completely extinguished by mid-October.

The plague spread from the metropolis to the townships and parishes within about 25 miles (a day's ride) and also along the Thames Valley (as usual); in Deptford, only 3 miles from London Bridge, mortality was high, but no real spatial pattern emerges because 432 died in Brentford, 20 miles away. Shrewsbury described the events in the market town of Croydon, 10 miles to the south, where the first plague death is believed to be that of a fugitive from London on 12 June 1665 but the first burial entry marked *pestis* in the register is 27 July. Again, the pestilence did not explode, but 39 burials in September in Croydon compared with the monthly average of

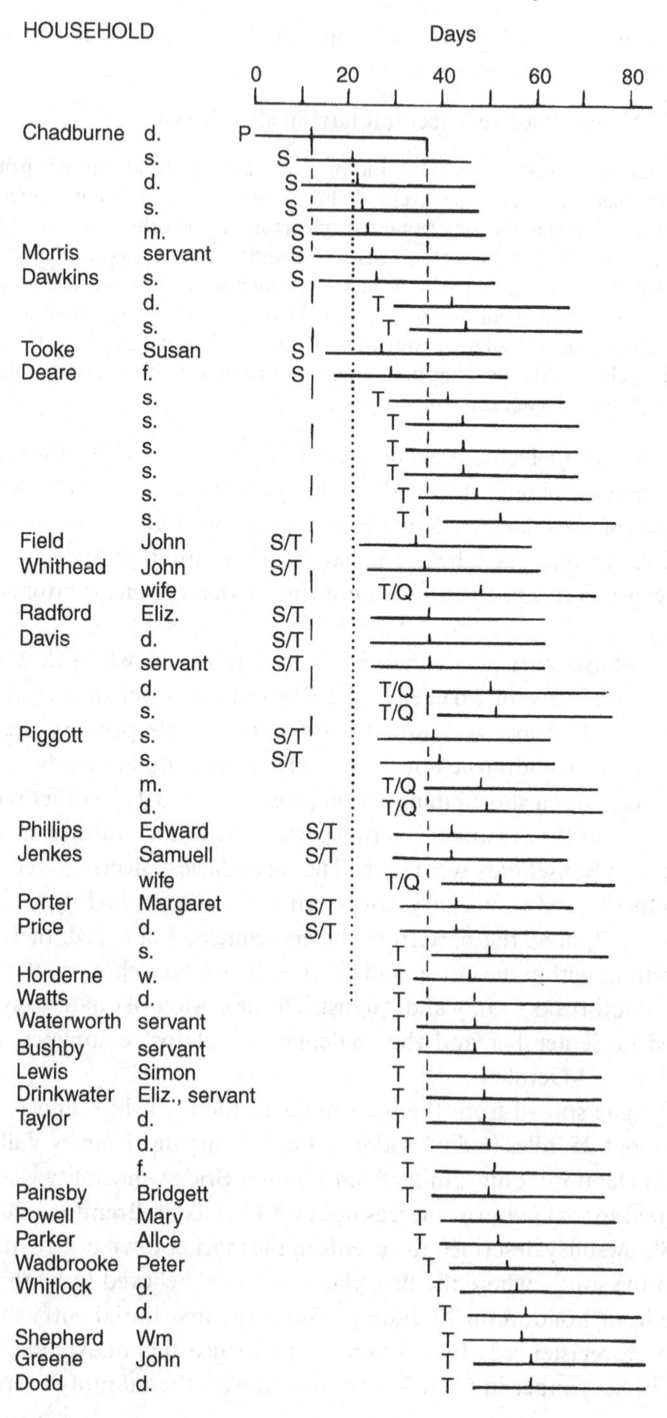
HOUSEHOLD
Days
0
20
40
60
80
Chadburne d.
P
s.
d.
s.
m.
Morris servant
Dawkins s.
d.
s.
Tooke Susan
Deare f.
s.
s.
s.
s.
s.
s.
Field John
Whithead John
wife
Radford Eliz.
Davis d.
servant
d.
s.
Piggott s.
s.
m.
d.
Phillips Edward
Jenkes Samuell
wife
Porter Margaret
Price d.
s.
Horderne w.
Watts d.
Waterworth servant
Bushby servant
Lewis Simon
Drinkwater Eliz., servant
Taylor d.
d.
f.
Painsby Bridgett
Powell Mary
Parker Allce
Wadbrooke Peter
Whitlock d.
d.
Shepherd Wm
Greene John
Dodd d.
S
S/T
T
T/Q

10 suggest the start of an epidemic. Plague burials continued at a steady rate in October (when the weather was dry, cold and frosty), declined by 50% in November and 18 were recorded in December in spite of an exceedingly hard frost on 22 November that heralded a spell of cold weather. Ten members of one family died in December and in the following months. Eight plague burials were recorded in January 1666, eight in February and four in March. Deaths in another family totalled 16 and nearly a quarter of the plague mortality in Croydon was contributed by four families. The epidemiology of the plague in the market town of Croydon was quite unlike that of London; the epidemic never really exploded and was apparently confined to a small number of families. It was probably frequently reinfected by fugitives from the metropolis and continued, but was constrained, through a cold autumn and winter and ceased only in March 1666.

8.6.2 Seasonality and mortality

As in previous plagues, all that could afford to do so fled from London, but the poorer classes in the populous suburbs on both sides of the Thames were left and it was they who suffered most. Their employment and wages mostly ceased when the wealthy left so that malnutrition and starvation were added to the vicissitudes with which they were afflicted and may have exacerbated the lethality of the disease. Their desperate situation led many of them to undertake the dangerous work of the day- and night-watchmen of the shut-up houses, the buriers and the dreaded plague nurses who were appointed by the authorities and who were said to contribute to the deaths of their patients. Shrewsbury gave a quotation 'that the plague stricken were more afraid of the official plague-nurses than of the disease itself'.

The totals of the weekly plague burials in London in 1665 are shown in Fig. 8.13; the epidemic began later in the year than in previous outbreaks, with very few deaths between the end of April and the second week of June. Thereafter, it followed the usual pattern, rising dramatically to a crescendo in mid-September and then falling equally rapidly during the autumn and winter. The total of plague deaths in 1665 according to the Bills of

---

Fig. 8.12. (*opposite*) Suggested sequence of infections at the start of the plague epidemic in the parish of St Michael Bassishaw, London, in 1665. Note the high initial R_0 and the rapid interhousehold spread. Compare with Fig. 8.11. Scale: days after 11 May 1665. For further details and abbreviations, see Fig. 5.6.

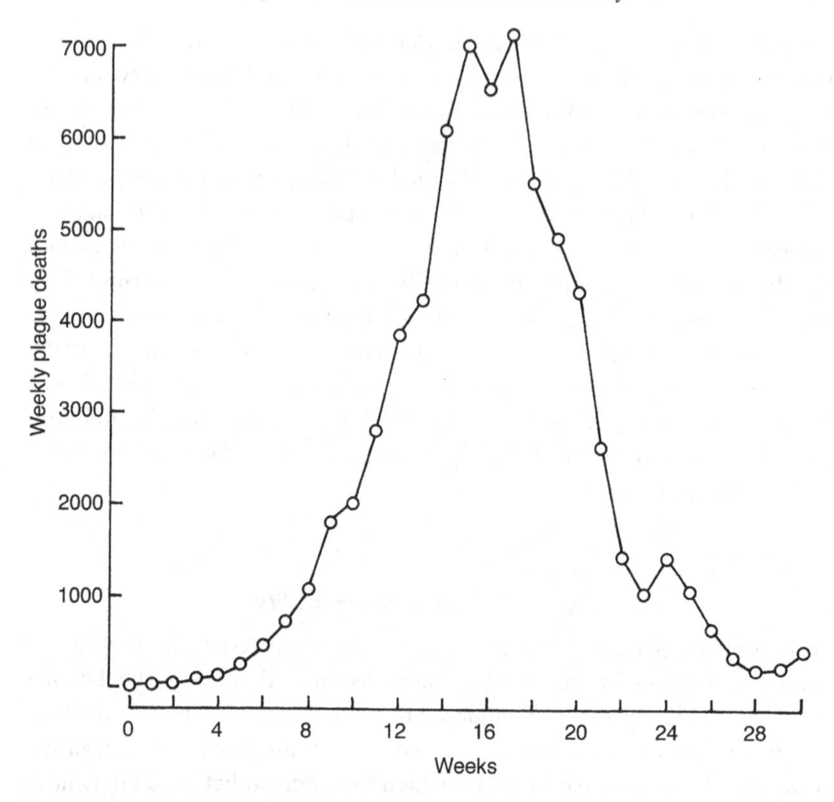

Fig. 8.13. Weekly plague burials in London, May to December 1665. Abscissa: weeks after 23 May 1665. Data from Creighton (1894).

Mortality was 68 595 but there is general agreement that this was an underestimate: Bell (1924) averred that many plague deaths were either deliberately hidden from the women 'searchers' or those officials were bribed, or intimidated to refrain from reporting them. These views are confirmed by the entry in Pepys' diary for 30 August 1665: 'Abroad and met with Hadley, our clerk, who, upon my asking how the plague goes, told me it encreases much, and much in our parish; for, says he, there died nine this week, though I have returned but six: which is a very ill practice, and makes me think it is so in other places; and therefore the plague much greater than people take it to be.' In addition, because the Quakers, Jews and Anabaptists refused to allow the plague deaths among their members to be included in the church returns, the plague mortality among them mostly escaped the bills. Pepys wrote on 31 August 1665, 'In the City died this week 7,496, and of them 6,102 of the plague. But it is feared that the true number of the dead this week is near 10,000; partly from the poor that cannot be taken notice of, through the greatness of the number, and partly from the Quakers and others that will not have any bell rung for them'.

More people in London died of the plague in 1665 than in any other visitation but this should be seen against the sharp rise in the population to about 460 000, giving a plague mortality, according to the official figures, of 15%, which compares to a corresponding value of 13% for the earlier epidemic of 1625.

8.6.3 Signs and symptoms

Because there are more contemporaneous and near-contemporaneous accounts of the epidemic in London in 1665, it is possible to assemble and present for the first time a reasonably reliable description of the symptoms and course of the disease, providing corroboratory evidence where possible. Bell, writing in 1924, gave the following admirable summary which he based on the writing in 1665–66 of Dr Nathaniel Hodges:

> It was rarely longer than two or three days after the attack when the visible signs of Plague appeared on the body. The 'blains,' so-called, were like blisters on the skin, obscurely ringed about. If no worse signs followed, the patients might entertain hope.
>
> 'Buboes' – hence the term, bubonic plague – were tumour-like outgrowths most commonly found under the arms and in the groin, and less frequently behind the ear, two, three, or four in number, and varied greatly in size. They were also called botches. If the growth failed to break naturally, the surgeon opened it by incision. Unhappily the rising of Plague buboes was attended by such severe pain, and feeling as of intolerable burning as the time for suppuration approached, that the sufferers often became raving mad. Incision by the knife and the subsequent cleansing, of course without anaesthetics, were so extremely painful that patients collapsed under it. If the buboes failed to rise and break, there was little expectation of life. As they broke, the fever declined.
>
> With carbuncles, another common eruption in Plague, mortification had always to be dreaded. The surgeon, by cauteries or the lancet, opened the carbuncle, a task necessitating the greatest care, or gangrene destroyed the patient.
>
> These were the common concomitants of the Plague as it developed towards recovery or death. But the sign most feared was that to which the people gave the name of 'the tokens'. The devout and superstitious accepted them as God's sign – 'the heathen are afraid of Thy tokens'. Some called them 'God's marks'. They were the almost certain forerunners of death. Medical observation agreed that very few with these marks upon them recovered health. 'The tokens' were spots upon the skin, breaking out in large numbers, varying in colour, figure, and size. Some, where they had run together, became as broad as a finger nail, others were small as a pin's head, till they enlarged and spread. The colour might be red, with a surrounding

circle inclining towards blue; in others a faint blue, the circle being blackish; others again took a dusky brown tone. Often the flesh was found to be spotted when no discoloration was visible on the skin. No part was immune from these round spots, though the neck, breast, back and thighs were the most common places for them. 'The tokens' sometimes were so numerous as to cover all the body.

It was 'the tokens,' so universally dreaded, that gave to the Plague the name of 'the spotted death'. They customarily appeared after two to four days' progress of the disease, but might rise without any previous warning of infection. A woman, the only one of her family left alive and thinking herself perfectly well, perceived the pestilential spots on her breast and shortly thereafter died. A young man of good constitution, unexpectedly finding 'the tokens' upon him, believed them not to be the genuine marks, he being otherwise in such vigorous health, yet within four hours death confirmed the physician's diagnosis.

Dr. Hodges mentions as being most strange in his first experience that many persons came out of delirium as soon as 'the tokens' appeared, believing that they were in a recovering and hopeful condition. The poor sufferers did not know their fate. He recalls the case of a maid who had no idea that she was attacked by Plague, her pulse being strong and senses perfect, and she complained of no disorder or pain, but on examining her chest he discovered 'the tokens' there. Within two or three hours she was dead. 'The tokens' sometimes first became visible after death.

It is clear from this account that the tokens, spots upon the skin, rather than the buboes were the most important diagnostic feature of the plague (see section 13.12). This description of the signs and symptoms is amplified and corroborated by Creighton's précis of Boghurst's writing:

Of evil omen was 'a white, soft, sudden, puffed up tumour on the neck behind the ears, in the armpit, or in the flank;' also a 'large extended hard tumour under the chin, swelling downwards upon the throat and fetching a great compass' (the brawny swelling of the submaxillary salivary glands and surround tissues). Tokens came out after a violent sweat, which was often induced by purpose of the nurses, who said, 'Cochineal is a fine thing to bring out the tokens.' Nurses often killed their patients by giving them cold drinks . . .

The botches, or buboes (swollen lymph-glands in the neck, armpits or groins), were the most distinctive sign of the plague, having given to it the old name of 'the botch'. Besides these, there were the 'tokens' (specially limited in meaning to livid spots on the skin), carbuncles and blains. Carbuncles, sayd Boghurst, commonly rose upon the most substantial, gross, firm flesh, as the thighs, legs, backside, buttock; they never occurred, that he saw, on the head among the hair, or on the belly. They were not seen until the end of July, were most rife in September and October, commonly in old people, never in children . . .

'Blains are a kind of diminutive carbuncle, but are not so hard, black, and fiery; sometimes there is a little core in them. Generally they are no bigger than a two-penny piece, or a groat at the biggest, with a bladder full of liquor on the top of them, which, if you open but a little, will come out whitish or of a lemon or straw colour.' 'Besides a blain there is a thing you may call a blister, puffing up the skin, long like one's finger in figure, like a blister raised with cantharides; and such usually die.' The following experience is remarkable, but it is doubtful whether Boghurst has not taken it from Diemerbroek: 'Towards the latter end of a plague, many people that stayed, and others that returned, have little angry pustules and blains rising upon them, especially upon the hands, without being sick at all. But such never die, nor infect others.

Creighton continued:

Among the symptoms of a fatal issue, Boghurst mentions the following: Hiccough, continual vomiting, sudden looseness, or two or three stools in succession, shortness of breath, stopping of urine, great inward burning and outward cold, continual great thirst, faltering in the voice, speaking in the throat and occasionally sighing, with a slight pulling-in one side of the mouth when they speak, sleeping with the eyes half-open, trembling of the lips and hands and shaking of the head, staggering in going about rooms, unwillingness to speak, hoarseness preventing speech, cramp in the legs, stiffness of one side of the neck, contraction of the jaws, the vomit running out from the side of the mouth, prolonged bleeding at the nose, the sores decreasing and turning black on a sudden . . . 'some of the infected run about staggering like drunken men, and fall and expire in the streets; while others lie half-dead and comatous . . . Some lie vomiting as if they had drunk poison.'

(compare with the behaviour of the citizens of Chester in 1647, section 9.6.1 and with the plague at Athens, section 1.2.1).

Creighton (1894) gave the following summary of a dissection by Dr George Thompson of a youth who died of plague:

He found what appear to have been infarcts in the lungs; the surface was 'stigmatised with several large ill-favoured marks, much tumified and distended,' from which, on section, their issued 'sanious, dreggy corruption and a pale ichor destitute of any blood.' The stomach contained a black, tenacious matter, like ink. The spleen gave out on section an ichorish matter. The liver was pallid and the kidneys exsanguine. There were 'obscure large marks' on the inner surface of the intestines and stomach. The peritoneal cavity contained a 'virulent ichor or thin liquor, yellowish, or greenish.' There was a decoloured clot in the right ventricle, but 'not one spoonful of that ruddy liquor properly called blood could be obtained in this pestilential body.'

Bell said laconically that the dissection showed that the plague produced far-reaching changes in the internal organs as well as affecting the skin by a multitude of blue or black spots containing congealed blood. 'In fact no organ was found to be free from changes'.

Creighton (1894) summarised Boghurst's account of the epidemiology:

'It usually went through a whole kindred, though living in several places; which was the cause it swept away many whole families . . . In some houses ten out of twelve died, and sixteen out of twenty.' Old people that had many sores upon them, especially carbuncles, almost all died . . . Many people had the spotted fever and the plague both together, and many the French pox and the plague both together, and yet both sorts commonly lived. All sorts died, but more of the good than the bad, more men than women, more of dull complexion than fair. It fell not very thick upon old people till about the middle or slake of the disease, and most in the decrease and declining of the disease. Cats, dogs, cattle, poultry, etc., were free from infection.

Some died in twelve or twenty days, but more in five or six. In summer, about one-half that were sick died; but towards winter, three of four lived. None died suddenly as stricken by lightning: 'I saw none die under twenty or twenty-four hours.' After one rising, or bubo, was broke and run, commonly another and another would rise in several parts of the body, so that *many had the disease upon them half a year; some risings would not break under half a year* [our italics], being so deep in the flesh.

The following account quoted by Creighton gives a good description of the plague spreading through a household and illustrates the medical features described above:

We were eight in the family – three men, three youths, an old woman and a maide; all which came to me, hearing of my stay in town, some to accompany me, others to help me [he was a celebrity in the religious world with a large following]. It was the latter end of September before any of us were touched . . . But at last we were visited . . . At first our maid was smitten; it began with a shivering and trembling in her flesh, and quickly seized on her spirits . . . I came home and the maid was on her death-bed; and another crying out for help, being left alone in a sweating fainting-fit. It was on Monday when the maid was smitten; on Thursday she died full of tokens. On Friday one of the youths had a swelling in his groin, and on the Lord's day died with the marks of the distemper upon him. On the same day another youth did sicken, and on the Wednesday following he died. On the Thursday night his master fell sick of the disease, and within a day or two was full of spots, but strangely recovered . . . The rest were preserved.

The general populace believed that they caught the plague because the infection was in the air of the place – they were exposed to risk if they were living in a plague-ridden spot. Certainly, the practice of shutting up a family in their house when plague first struck must have increased inter-household contact rates.

8.6.4 Changes in virulence

The medical accounts suggest that the course and lethality of the disease changed subtly as the epidemic progressed. Shrewsbury also drew attention to this apparent change in the virulence of the infective organism quoting, in support, 30 recovered people emerging from the pest house. He also quoted Hancock who said that on 10 September 1665 the disease came to its height:

'It now killed in two or three days, and not above one in five recovered; or four in five died . . . but after this period, when the disease was on its decline, it did not kill under eight or ten days, and not above two in five died.' So that it was calculated by Dr Heath that there were not fewer than 60,000 people infected in the last week of September, of whom near 40,000 recovered. For the plague being come to its crisis, its fury began to assuage, and accordingly the Bill decreased almost 2000 that week. 'For, had the mortality been in the same proportion to the numbers infected, as at the height, 50,000 would very probably have been dead instead of 20,000, and 50,000 more would have sickened; for, in a word, the whole mass of people began to sicken, and it looked as if none would escape, as not one house in twenty was uninfected . . . the disease was enervated and the contagion spent. Even the Physicians themselves were surprised: wherever they visited they found their patients better . . . so that in a few days, whole families that expected death every hour, were revived and healed and none died at all out of them.' Yet it appeared that more people fell sick then, when not above one thousand died in a week, than when five or six thousand died in a week.

Boghurst corroborated these conclusions: an epidemic declined in malignity towards the end so that the buboes suppurated and some 60% of patients recovered. Pepys wrote in his diary for 16 October 1665 'Lord! how empty the streets are, and melancholy, so many poor, sick people in the streets full of sores . . .'.

It can also be seen in section 8.6.3 that the signs and pattern of the disease changed during the epidemic: the tokens appeared only rarely until the middle of June and the carbuncles not until the end of July. Shrewsbury declared that 'every English outbreak of bubonic plague in the past ended in this way' and he went to great lengths to explain this change in the pathology as a progressive spontaneous decline in the virulence of *Y. pestis*.

8.6.5 Animals, clothing and wigs

As in previous plagues, animals were slaughtered: multitudes of mice and rats were destroyed by ratsbane and 40 000 dogs and 200 000 cats were killed. The doctors taught that the seeds of the disease could lurk in a bundle of clothes or bedding and that they became 'more virulent through the fermentation that goes on in these circumstances'.

Pepys wrote on 3 September 1665 that he donned his 'new periwigg, bought a good while since, but durst not wear, because the plague was in Westminster when I bought it; and it is a wonder what will be the fashion after the plague is done, as to periwiggs, for nobody will dare to buy any haire for fear of the infection, that it had been cut off people dead of the plague'.

8.6.6 Effect of the Great Fire of London

Popular opinion is that the Great Fire, which broke out on 2 September 1666, was responsible for terminating the Great Plague and also for eliminating bubonic plague for ever, not only from London but from the whole of Britain, probably by destroying the rats in their burrows. This is manifestly incorrect and Shrewsbury (who believed that *Y. pestis* was the infective agent) was at pains to point this out, although he interpreted the facts in terms of the effects on the rat population: the fire was limited to the walled city whereas the liberties and out-parishes had always been the main foci for the epidemics. The epidemic was already fading 10 months before the fire, in the autumn of 1665, accelerated, apparently, by a reduction in virulence.

The plague did not die out in London after the Great Fire, but lingered all through 1666 causing 1998 deaths. Indeed, the plague continued at an endemic level, with a few deaths recorded in most years, for a further 13 years until 1679. The last major epidemic in England was a solitary outbreak at Nottingham in 1667.

But, as we describe in section 3.1, authentic *bubonic plague* came to England a number of times, 250 years later, in the 20th century, one of the unfortunate consequences of swifter steamship travel, but no epidemic ever developed because of the impossibility of establishing an epizootic.

8.6.7 Identity of the first victim in households in plague epidemics in London

We have analysed the major epidemics in London in 1593, 1603, 1625, 1636 and 1665 together with minor outbreaks in the parishes listed in Table 6.1 and have determined the identity of the first victim in each family and the results are shown in Table 8.1. When there were multiple infections in the family, a child was the first victim in 48% of the cases, in contrast with 31% of the families where the infection began with a parent. We suggest that most of the children were susceptible and were more likely to be infected and to bring the disease into the household. When there was only one person infected in the household during the epidemic (the majority of cases), the situation was reversed, with 48% of the victims being adults and 34% children. Many of these adults may have been living in boarding houses, or were travellers or there may have been no children in the household. A high proportion, about 20%, of all the first victims who initiated the plague in their households were servants, suggesting that they were susceptible immigrants.

Table 8.1. *Identities of the first victims in households in epidemics and outbreaks of plague in selected parishes in London*

	Multiple infections				Single infection			
	Adults				Adults			
Year	Male	Female	Child	Servant	Male	Female	Child	Servant
1593	12	9	31	12	43	20	43	29
1603	6	5	20	16	28	15	28	22
1625	17	13	31	16	79	68	80	48
1636	5	6	25	2	32	26	47	8
1665	33	20	76	20	80	76	123	53
Others	24	16	74	45	118	65	136	82
Totals	97	69	257	111	380	270	457	242
	166				650			

Note the difference in the identity of the first victim in multiple and single household infections. See Table 6.1 for parishes analysed.

Analysis of the intervals between the first and second victims in the household in the epidemics of 1603, 1625, 1636 and 1665 ($N = 372$) shows that 99% of them were within 37 days, the estimated interval between the point of infection and the end of the infectious period.

8.7 Dynamics of plague in London

The annual plague mortality in London from 1578 to 1678 is shown in Fig. 8.1, although this is somewhat misleading because the scale hides the fact that the disease was endemic in the interepidemic years (see Fig. 8.2). Time-series analysis of this series provides evidence of a 19-year cycle (not statistically significant), but visual inspection suggests that the major outbreaks were sporadic (Fig. 8.1) and that each of these was followed by rapidly decaying epidemics. These dynamics can be compared with those of smallpox in London after 1800 (see section 2.9), when the endemic level was falling steadily. The two massive outbreaks of smallpox (E_1 and E_2 on Fig. 2.6) were also followed by sharply decaying epidemics. We suggest that the plague in London during the period 1563 to 1665 (as distinct from the towns in rural England) was an *undriven* system that followed simple SEIR dynamics and hence the epidemics decayed (sections 2.6 and 2.7).

In conclusion, the dynamics of the plague in rural England at the end of the 16th and the first half of the 17th centuries were different from those in

London, where the disease was endemic and displayed irregular, catastrophic outbreaks that decayed rapidly, possibly as described by a standard SEIR model. The difference between the dynamics may reflect the different size, population densities, degree of immunity and R_0 of the two different types of population.

9

Plagues in the provinces in the 17th century

As we have seen, the plague in London in the later part of the 16th century broke out sporadically and the epidemics steadily increased in intensity whilst it grumbled on in the interepidemic years (section 6.2), becoming endemic in the 17th century with outbreaks of unparalleled ferocity and mortality (Chapter 8). Meanwhile, there were many reports of widespread outbreaks scattered through the provinces throughout the 16th century (section 6.3). Parish registers were required by law to record burials from pestilence during the Elizabethan period but even though these provide a great deal of invaluable information (much of which has still to be extracted and analysed), it is difficult to discern a pattern in the underlying epidemiology of these outbreaks in the provinces. In general, these provincial outbreaks do not appear to have come from a focus in London. Were there many introductions of the disease via the Kent coast (e.g. in 1532), the south coasts (e.g. in 1544 and 1590) and the East Anglian ports (e.g. in 1585), or was the infection grumbling on through the century with low infectivity ($R_0 \leq 1$) but spread widely by apparently healthy carriers (because of the long incubation period), with epidemics flaring up when and where the conditions were right? Improved communications and much wider travel associated with the wool trade would have exacerbated the spread (see discussion in section 13.9).

Inspection of the results presented in section 6.3 suggest that there might have been two main foci (apart from London) in the metapopulation, one in the southwest and the other in a broad band, running across the centre of the country from East Anglia through the Midlands to the Welsh Marches. The East Anglian ports were important centres for the importation of goods that were then sent onwards via the river systems that they served (see section 13.9) and plague was frequently introduced from continental Europe by this route. In addition, the Northern Province seems to have behaved as a separate metapopulation in the late 16th century, with plague being transmitted along the corridor to the east of the Pennines, with Newcastle acting as the usual point of entry and focus (Chapter 7). In 1597 an epidemic spread across to the corridor to the west of the Pennines.

The evidence suggests that, by the second half of the 16th century, the people of England were able to recognise the symptoms (particularly the tokens) and characteristics of an outbreak of plague and were beginning to put into practice quarantine and other public health measures, including the prevention of people leaving or entering the town (see, for example, Carlisle, section 7.7). These measures were draconian at York in 1604, and strenuous efforts were made; although plague was in the surrounding villages, they prevented the arrival of plague for several months before the epidemic eventually broke out. However, as we have seen, these preventive measures would not have been very effective, firstly, because of the long serial generation time, which allowed the movement of symptomless infectives, and, secondly, as the records at York show, they were much concerned that bales of cloth could carry the infection. We conclude that these were not outbreaks of bubonic plague but that the same unknown infectious agent was responsible for all these major lethal epidemics. Because plague in London became more persistent and the epidemics increased in ferocity during the 17th century, we shall see (section 9.7) that this development was paralleled in the provinces, where there seem to be records of outbreaks (many of which were small in scale) somewhere in the metapopulation in almost every year.

9.1 The years 1603–5

Plague was severe in London in 1603 (Chapter 8) and epidemics occurred widely in the provinces during the period 1603–5; some of these may have been brought by travelling infectives from London but some probably originated from other sources, including entry by different ports.

9.1.1 Spread of the plague in the northeast

There was a severe epidemic at Hull and its environs in 1603 and plague also erupted at Newcastle-upon-Tyne in July 1603; it may have come from the south of Scotland, where it was widespread from June to December 1603, or it may have entered via the ports. The infection spread along the northeast corridor by September 1603 to villages 10 miles southwest of York, which, thereupon, introduced stringent protective measures beginning in July 1603 including closure of markets and the removal of suspect cases to temporary accommodation outside the city walls. All exchange of goods and commodities with Hull, London and Newcastle-upon-Tyne were prohibited but, although the officials scrupulously inspected the bales of materials and clothing coming from these towns, they apparently ignored the travellers, seamen and merchants who brought them and probably carried the infection.

The plague was at several towns in the wapentake (a territorial division of Yorkshire) of Ainstie by October 1603 and, although illegal trading with infected areas continued, the city of York was believed to be still clear of the infection on 29 November 1603. However, the first suspected plague victims were reported on 2 December 1603 (Sakata, 1982). The epidemic probably grumbled on at a very low level through the winter (presumably low R_0), although trading restrictions with London were lifted on 14 February 1604, but it reappeared in mid-March 1604 (when there was an entry in the burial registers of Holy Trinity, Micklegate) and watches were set to ensure that vagrants did not enter the city. By early May, plague had entered York. A 'viewer and clenser' was sought from Newcastle, plague-lodges were erected outside the walls and a number of other public health measures, including the killing of cats and dogs, were enacted. The Council was now unable to control its spread through the City and reported on 5 May that there were infected houses in five parishes, although substantial rises in mortality did not occur until some months later, indicative of the typical, slow build-up of the plague. In the parish of All Saints Pavement, plague burials were specifically identified; the first occurred on 19 April, followed by a group of four between 7 and 10 May, one on 2 June, three on 11 June and it was only from 6 July that plague burials began to occur on an almost daily basis (Galley, 1998). This slow build-up in early summer with the deaths widely spaced in time is typical of the start of an epidemic of haemorrhagic plague with its long serial generation time. The epidemic lasted until November 1604 and 82% of all burials occurred between July and October. Total burials within the city walls numbered 2800 in 1604, 25% to 33% of the population (Galley, 1994, 1995), and the city annals compiled in 1639 reveal that there 'dyed in York 3,512 persons', although it is not possible to give such a precise figure, since the parish register of St Olave recorded that 'The months followinge viz August September October november & December people dyed so fast that they could not be well nombred' (Galley, 1998).

York acted as a focus for the outward dispersion of the plague and the Corporation House Books recorded on 18 April 1604 'the plague is dispersed in diverse townes and place nere vnto this Cittie' and the archbishop reported that his diocese was 'sore visited with the sickness' in October. The parish register of St Giles, Durham, records 18 plague deaths between September 1604 and 25 January 1605 (low winter R_0); evidently the epidemic spread northwards and continued through the following winter but did not explode in the spring, perhaps because of a low R_0.

9.1.2 Chester as a focus

Plague had struck at Chester previously with a severe epidemic in 1517 and with smaller outbreaks in 1558 and 1574 (Axon, 1894) and there were to be repeated attacks during the first 60 years of the 17th century. The city seems to have acted as an entry port for the infection, which probably came from Ireland, and then as a focus for repeated radial dissemination through Cheshire and adjoining counties. The pestilence began in Chester in September 1602 (i.e. before the main London epidemic) in a glover's house in Chester in St John's Lane where 7 died (high household contact rate) and mortality thereupon increased with 60 weekly deaths; 650 died of the plague in 1603. In the parish of Holy Trinity, the epidemic started between 13 December 1603 and 3 January 1604 and the registers record the burial of 'Alice dau to James Hand of Blacon, husbandman bur in church nere the poore man's box: the first that dyed of the plague in the parish but was not known to the parish till she was buried'. There were 986 plague deaths in 1604 (with 55 dying in 1 week) and 812 died between 14 October 1604 and 20 March 1605. After this, another 100 died before the epidemic ceased in early January 1606. Cabins for the victims were erected outside the city at the riverside, close to the New Tower and, on occasion, five or six of the same family died in the course of 2–3 weeks (high household R_0) and it is noteworthy that Creighton (1894) recorded that these multiple deaths were particularly found in the homes of sailors. All fairs were cancelled; some of the justices and a great number of citizens fled into the surrounding countryside.

This plague at Chester was unusual because, although we have no monthly totals of burials, its pattern was different from other provincial epidemics that we have described previously, lasting 40 months, from September 1602 to January 1606, presumably becoming temporarily endemic. Did the epidemic spread through the city from one focus to the next in successive years? The first burial in the parish of Holy and Undivided Trinity, Chester, was followed by only three deaths in separate households in early January and then after a break of 6 weeks the infection began again

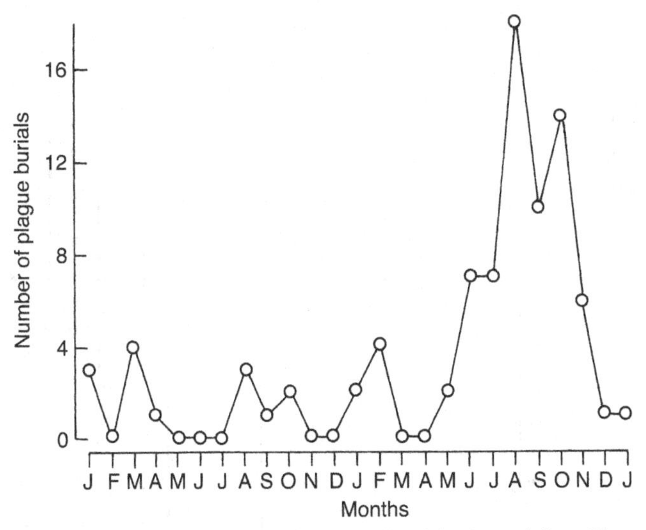

Fig. 9.1. Monthly plague burials in the parish of Holy Trinity, Chester, from January 1604 to January 1606.

on 3 March 1604 with five deaths in three families. Plague burials were not recorded again until a reinfection, when six people in four families died between 18 August and 13 October 1604; these were so spaced out that some must have been infected from an adjacent parish. Another minor outbreak occurred during the winter with six plague burials in three families between 29 January and 5 February 1605 and an isolated death on 24 March 1605. Plague began again 3 months later on 17 May 1605 and grumbled on sporadically for 7 months until 9 January 1606 (the last plague burial), during which time 64 people were buried. Most of the victims had individual surnames and there were few families with multiple deaths. The epidemic in Holy and Undivided Trinity did not follow Reed and Frost dynamics, although there was an increase in deaths over the period August to October (Fig. 9.1) and it is clear that this outbreak followed a pattern different from that seen elsewhere: there were relatively few deaths during the 2-year time period and the epidemic never exploded; it died out several times and was reinfected from a nearby parish. Finally, most frequently, only one person died in a household. We conclude that this was a parish within the city and so was open to infectives entering and that it had experienced plague previously so that many individuals were resistant or immune.

The plague spread radially from Chester for considerable distances into Cheshire, Shropshire and Staffordshire; the infection was probably not

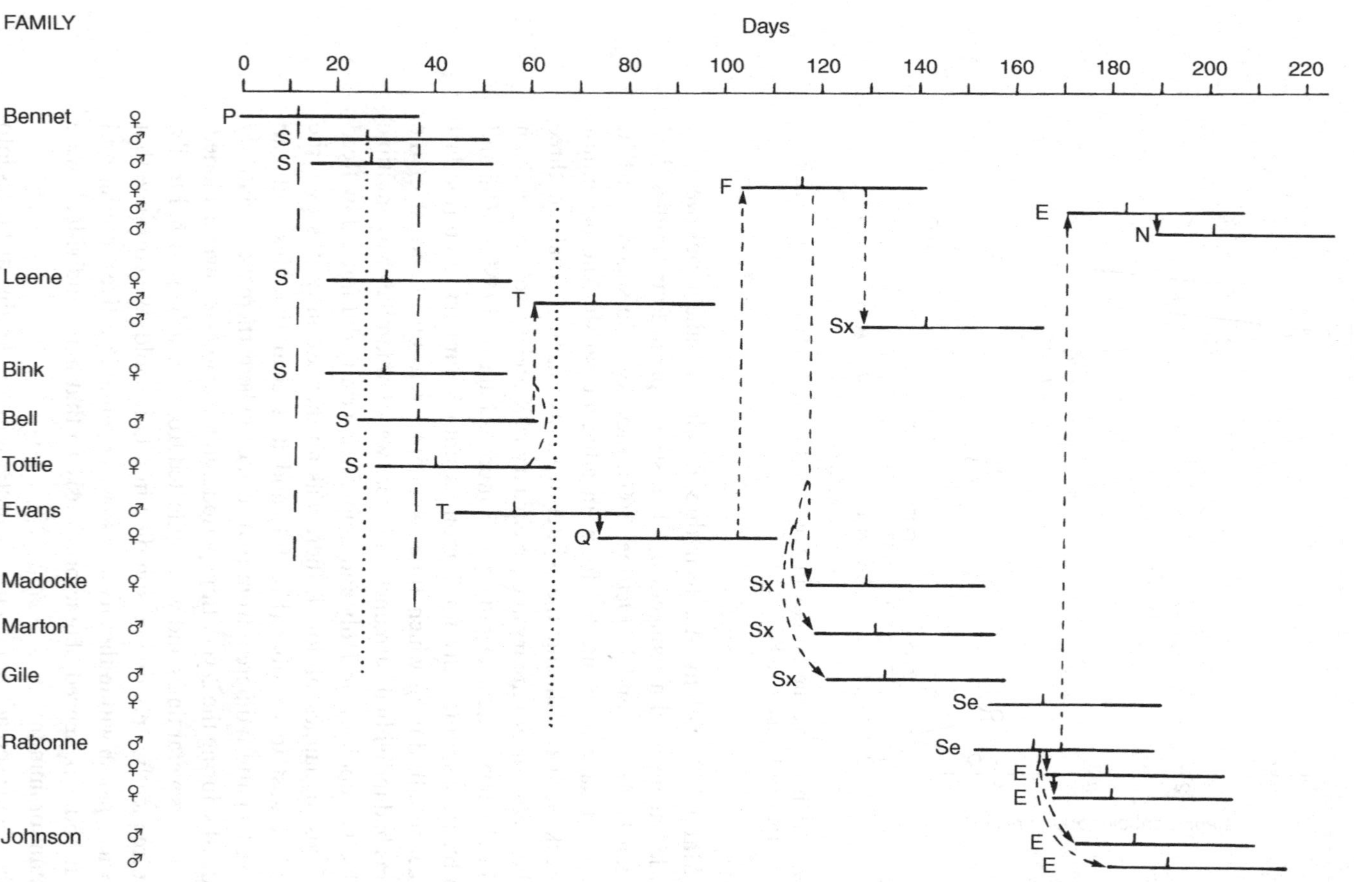

Fig. 9.2. Suggested sequence of infections in the plague epidemic in the parish of West Kirby, Cheshire, in 1604. Note low R_o. Scale: days after 14 April 1604. For further details, see Fig. 5.6. Abbreviations: Q, F, Sx, Se, E and N, quaternary, fifth, sixth, seventh, eighth and ninth generation infections, respectively; ♂, male; ♀, female.

spread from place to place in the majority of these outbreaks, rather the pestilence was carried out separately from the focus. Most of these secondary epidemics were in small, scattered parishes where the mortality was slight.

(i) West Kirby, northwest Wirral, 17 miles from Chester. Plague began here on 21 May 1604, lasted 7 months and finished on 25 December 1604 during which time 14 families were infected and 26 people died. The spread of the epidemic through the families is shown in Fig. 9.2. The epidemic broke out in the Bennett family on 21 May 1604 with one female death and this was followed by two co-secondaries on 4 and 5 June. The primary case infected a further three or four contacts in the Leene, Bink, Bell and Tottie families. The further spread of the disease at West Kirby is shown in Fig. 9.2: the tertiary infection of a male in the Evans family is of particular interest because he subsequently infected a quaternary (a female) in his family. If this quaternary case had not infected anyone else the epidemic would have ceased, but she had a single contact with a female in the Bennett family, which was then reinfected and the outbreak was maintained and continued until the end of the year when it disappeared. The epidemic lasted over some nine generations, although few people died because of the low R_0, estimated values for which are: primary to secondaries = 6; secondaries to tertiaries = 0.33; tertiaries to quaternaries = 1; fifth to sixth generations = 4; sixth to seventh generations = 0.5; seventh to eighth generations = 2. Estimated parameters: latent period = 12 days, infectious period (including symptoms) = 25 days; a type (i) epidemic.

(ii) Malpas, Cheshire, 12 miles southeast of Chester. The plague was a minor outbreak as the registers show:

Σ 28 June 1604: Dominick a gentlemen that died at Mrs. Maude Brereton's of Edge by the plague and was brought to the church on a drag or sledge with a horse.

Σ 29 June 1604. Thos. Plymlose, servant to Mrs. Maude Brereton of Edge who died of the same infection.

(iii) Whitchurch, Cheshire–Shropshire border, 16 miles from Chester. Plague reported here on 26 August 1604 in the Malpas registers (Richards, 1947). It may have been the source of the infection in Malpas, the adjoining parish, and the plague may have spread from here to Mucklestone in Staffordshire and to Alderley in Shropshire.

(iv) Nantwich, Cheshire, 17 miles southeast of Chester. Said to be invaded from Chester, with the epidemic beginning in June 1604 whereupon it was abandoned by those inhabitants that could afford to do so (Axon, 1894). Mortality was heavy with 430 deaths registered from June 1604 to March 1605, a persistent type (i) epidemic.

(v) Weaverham, Cheshire, 12 miles east of Chester. Only three plague deaths in two families in September 1604.

(vi) Plague was also reported to be at Macclesfield and Congleton in Cheshire in 1604, both about 30 miles east of Chester. There were over 70 burials recorded at Macclesfield between 3 September and 3 October and the epidemic spread characteristically through whole families but was apparently largely confined to certain streets (Axon, 1894).

(vii) Plague broke out at Shrewsbury (Shropshire), probably carried there along the corridor from Chester which lay 35 miles to the north. There were four parishes in Shrewsbury but records are available for only two. The plague struck predominantly in St Mary's but the registers record the deaths of three children that came from the parish of St Alkmond and the registers there show a small rise in burials in August and September. It is claimed that 677 persons died in Shrewsbury between June 1604 and April 1605.

The epidemic began at St Mary's on 2 June 1604 and died down in November but did not disappear until March 1605 (persistent type (i) epidemic) by which time 119 burials had been recorded. The summer outbreak did not show a clear Reed and Frost peak mortality but the early spread of the epidemic within and between families, shown in Fig. 9.3, is completely consistent with the epidemiology of haemorrhagic plague. Six individuals in six separate families were initially infected and are regarded as co-primaries (although they may have been co-secondaries from an unknown primary); these introduced 17–22 secondary infections into their own and other families ($R_0 \approx 6$). Tertiary infections began at the end of July. However, the epidemic at St Mary's was characterised by a relatively low spread of the infection *through families* during July to September 1604, thereby reducing the peak of the epidemic. This pattern suggests that some of the population might have been resistant because of prior exposure to the infection; plague had previously visited Shrewsbury in 1536 and 1576.

Shrewsbury appeared to act as a focus for the plague in Shropshire in 1604: four plague burials in one family in the parish of Myddle (= Middle); 16 plague burials in Condover beginning in August

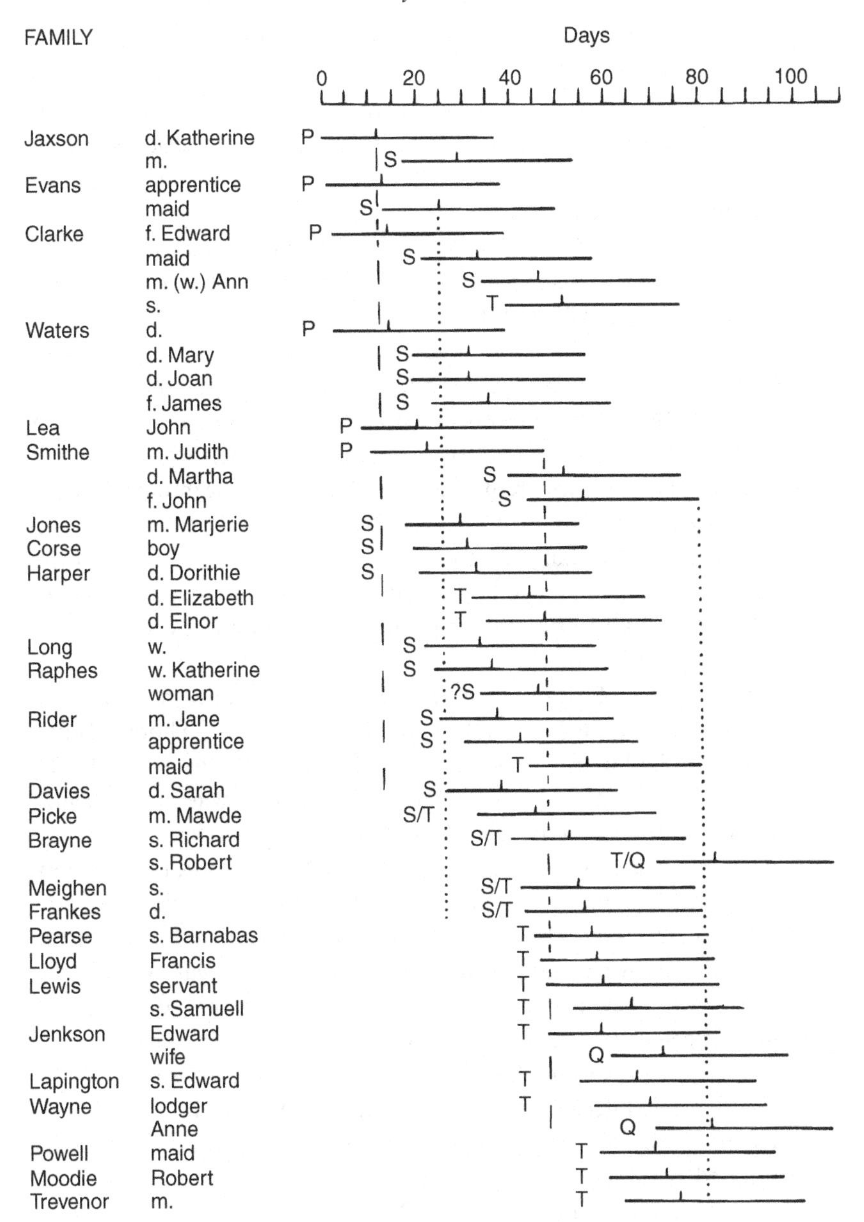

Fig. 9.3. Suggested sequence of infections at the start of the plague epidemic in the parish of St Mary, Shrewsbury, in 1604. Scale: days after 2 May 1604. For further details and abbreviations, see Fig. 5.6.

1604, together with five plague burials in 1605; plague was reported at Hodnet in 1604.

(viii) Plague was also reported to have spread from Chester to Staffordshire in 1604, being reported at Stone and Alton parishes (some 40 miles southeast of Chester) and at Penkridge (14 miles south of Stone).

(ix) Plague was recorded at Backford (a parish adjacent to Chester), with three burials in December 1604 and a separate outbreak in September–November 1605 with 10 burials in four families.

(x) Woodchurch, north Wirral, Cheshire, lies 14 miles north of Chester and a small outbreak of plague began there on 3 April 1605, with five deaths in the Belie family, one death each in the Bennett, Warmingian and Hodgeon families and two deaths in the Sherlocke family, all within 9 days and apparently resulting from the same initial infection. These led to a further three secondary infections and deaths in late April and two tertiary infections in late May 1605, whereupon the epidemic finished only to begin again (probably by a reinfection from Chester) in July 1605 with one plague death followed by five secondaries, all in the Leonarde family.

(xi) The registers of Heswall, Wirral, Cheshire, 12 miles northwest of Chester record on 19 May 1605 the plague death of Thomas Hamnett, who had infected his wife Alles Hamnett, who was buried on 9 June 1605 (evidently a secondary). The parish was subsequently reinfected (presumably again from Chester where the plague was in its later stages) with a plague death on 25 August 1605, i.e. relatively late in the summer. There were 24 deaths in this outbreak, the last on 3 December 1605. The suggested spread of this second phase of the epidemic between and through the families at Heswall is shown in Fig. 9.4: the new primary case was Wyllyam Thorntune who probably infected six secondaries in five families ($R_0 = 6$). Tertiary infections probably began about the 25 to 26 August 1605 and, in all, 11 families were infected ($R_0 = 1$). Subsequent values for R_0 are tertiaries to quaternaries = 1.3, quaternaries to fifth generation = 0.5. Estimated parameters are: latent period = 12 days, infectious period (including showing symptoms) = 25 days.

9.1.3 Plague at Manchester

Plague was said to have struck Manchester in 1558 (although there is no record in the local annals); there was a 'sore sickness' in 1565 and the death

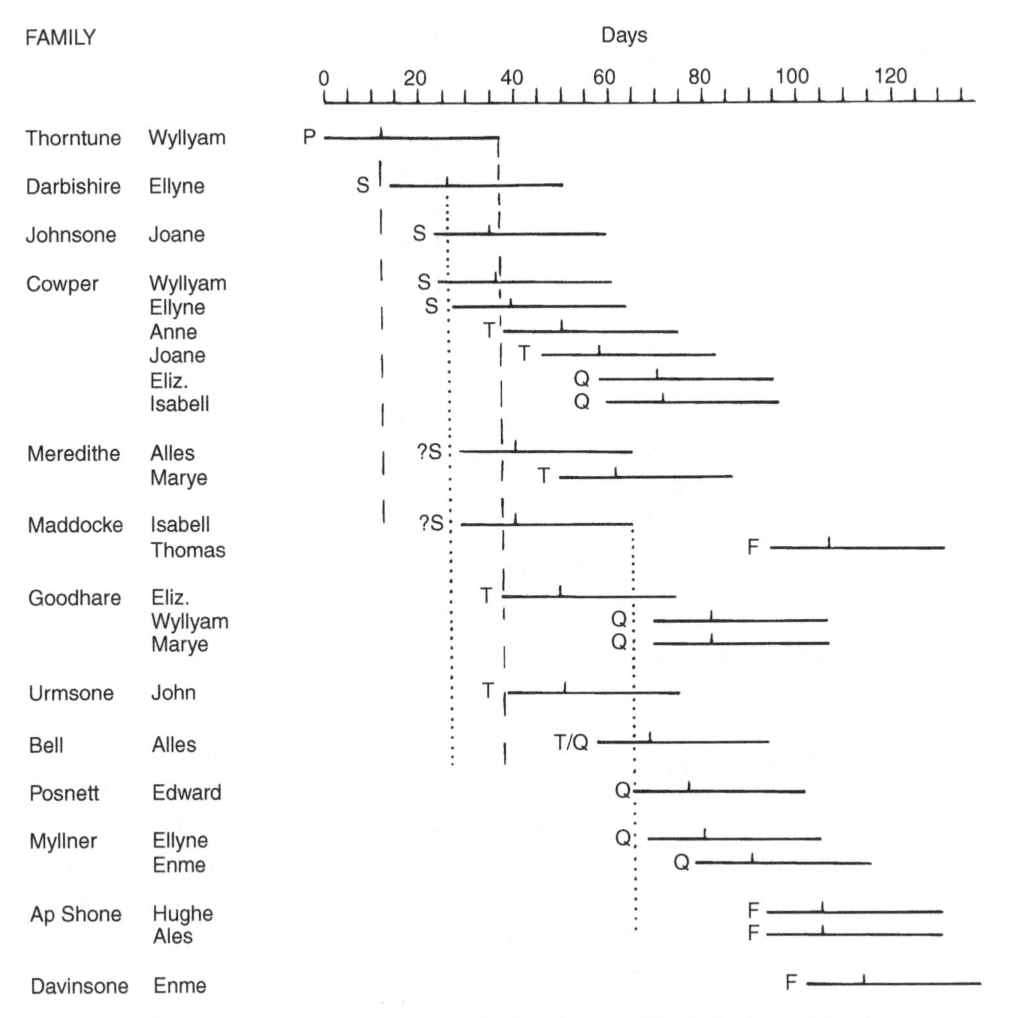

Fig. 9.4. Suggested sequence of infections in the plague epidemic in the parish of Heswall in 1605. Scale: days after 19 July 1605. For further details and abbreviations, see Fig. 5.6.

of 70 parishioners in April 1588 (Axon, 1894). On 7 April 1594 the baptism registers record: 'May d. to Rauffe Cloughe of Faylsworth being xviij weeks old before it came to baptysme for yt the plague was then very contagious in Cloughs house'. However, the burial registers for March to May 1594 show that mortality was low at this time.

The burial registers of the cathedral church of Manchester show a heavy mortality (at least double the mean annual rate) in 1598 but inspection of

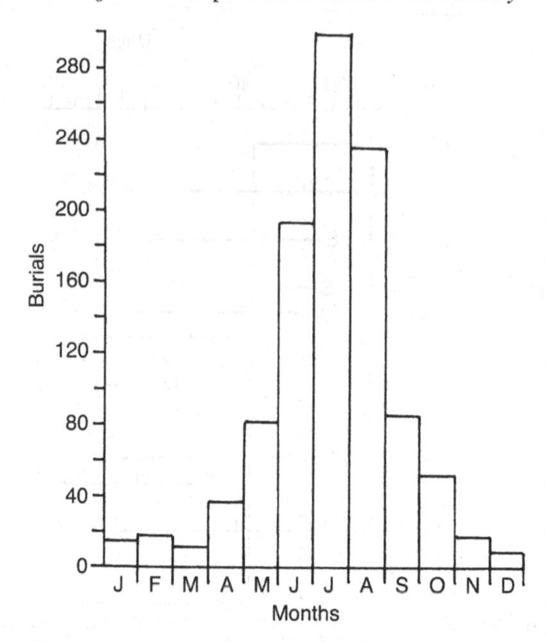

Fig. 9.5. Burials recorded in Manchester parish register in 1605. After Willan (1983).

the registers suggests that this was not the result of a plague epidemic. Plague burials were not recorded in the registers in 1605 but it is evident from contemporary accounts that this was a major epidemic of the pestilence. There were no weddings during June to August and no baptisms in August because of the 'extremitye of the sickenes'. The usual preventative measures were introduced: a tax was levied for the infected; the Court Leet was not held; people coming into Manchester were required to produce a certificate that they were free from infection; plague-cabins were established on Collyhurste (Willan, 1983).

The plague of 1605 began in April and followed the characteristic Reed and Frost dynamics, gathering momentum slowly, deaths peaking in July and probably ending in December 1605 (see Fig. 9.5), although Willan (1983) states that sporadic cases occurred until February 1606, i.e. a persistent type (i) epidemic.

A total of 1053 burials was recorded in Manchester in 1605 but the absence of any specific marking of plague victims in the registers makes any analysis of the epidemic difficult. Nevertheless, the epidemiology appears to be different from other outbreaks that we have described. We assume that the pestilence began in early to mid April and by 24 April 1605 the daughter in the Salle family was buried (the primary) and she was followed 16 days later (10 May) by a male Salle death and then on 14 May (maidservant and a child), on 18 May by the widow and on 23 May by a brother, all secondaries. This fits with approximate latent and infectious periods of 12 and 25 days, respectively, and with the typical progress of haemorrhagic plague in families. However, before that date, on the following days in April, the burials in the Leeze household were recorded within the space of 5 days: 19th (female), 23rd (female), 23rd (daughter), 24th (male), 24th (wife), 24th (infant). These must all have been household co-primaries, i.e. infected from a common source, a most unusual pattern at the start of an epidemic. However, the appearance of multiple primary infections within a family was repeated many times in the plague at Manchester; 74% of the second burials occurred within 12 days of the first, suggesting that *within families* the latent period may have been shorter than usual. Only a further 18% were clear household secondaries, dying within 37 days of the first burial. This high cumulative percentage dying within 12 days is similar to that recorded between the second and third burials and between the third and fourth burials. Perhaps crowded living conditions or the social behaviour (including going to market *en famille*) of the inhabitants of Manchester may have accounted for the multiple initial infections in a household.

Manchester also had an epidemic in the late autumn of 1606 and through the winter of 1607 that Axon (1894) recorded as plague, although the registers do not show an excess mortality at this time. Patients were sent to the pest-house and a tax was levied for the relief of the stricken.

The 1605 plague in Manchester may have been carried from Chester (35 miles) where the disease was raging. It then spread southwards to Stockport (5 miles) and Macclesfield (14 miles). Plague appeared in Stockport in the autumn, on 9 October 1605, and persisted through to 14 August 1606, during which time 51 people died (type (ii) epidemic, low mortality, low R_0). At Macclesfield, the epidemic also broke out in the autumn, on 3 September 1605 and 70 burials were recorded before 3 October 1605, spreading particularly through families (high household contact rate), see Axon (1894).

9.1.4 The Midlands and East Anglia

After the London plague of 1603, the pestilence was widespread in a broad band stretching from East Anglia across the Midlands to Bristol in the west during the period 1603–5. The reports are scattered and often anecdotal, but there is no doubt that many places were affected, although most of them did not experience a major mortality crisis. Shrewsbury (1970) has assembled the evidence but it is not possible to determine the spatial components of this outbreak. It also spread westwards from London along the Thames Valley to Oxford and thence southwestwards into Wiltshire, reaching Devizes, Marlborough and Salisbury by early autumn 1603. Plague was already raging in Bristol by this time, having begun before July 1603 with, it is estimated, more than 3000 deaths before January 1604 (type (i) epidemic). Plague erupted in Bath in May 1604 and peaked in July but with only 50 deaths during 14 weeks of the summer. Bristol was said to be reinfected in July 1604 and it may have acted as a secondary entry port for the infection.

Plague was also reported to be active on the opposite side of the country in East Anglia. Lowestoft and Norwich may have acted as entry ports for this area because there was heavy mortality in both these towns in 1602–3. Although a sick house was established at Ipswich in May 1603 and bailiffs were appointed to supervise the infected houses and all ships were prohibited from coming to the town, only four persons were reported as dying from plague. Only 10 plague burials were reported at Cambridge.

Thus plague could have spread to the Midlands by infectives travelling from a number of foci: (i) directly from London, (ii) northwards from Wiltshire, (iii) via ports on the east coast, (iv) from the port of Bristol, (v) from the port of Chester and thence via the focus at Shrewsbury or via Staffordshire. Plague was reported scattered at this time at Northampton (231 burials), Boston (where the market was closed), Birmingham, Congleton, Leicester (a mild epidemic), Tewkesbury (23 plague burials), Walsall (16 plague burials) and Stamford in Lincolnshire (nearly 600 deaths).

In 1605, plague broke out in Oxford (where temporary pest-houses were established), Godmanchester in Huntingdonshire (30 plague burials), Northampton and Grendon in Northamptonshire (12 victims) and Cambridge (Shrewsbury, 1970).

It can be seen that the pattern of the epidemics across central England changed in the early part of the 17th century; henceforth plague became persistent with outbreaks scattered over the metapopulation and the pestilence was probably reinforced by infectives regularly coming in through the ports from overseas. In general the reported mortality became less severe, perhaps because the towns had experienced the pestilence previously. Even if better information were available it would probably be difficult to define the spatial component of the epidemics. Much of the movement along the roads in the Midlands at that time would have been associated with the wool trade and, since the disease had a long incubation period, the epidemics would appear to spread almost randomly.

9.1.5 Southern England, 1603–5

The Home Counties were struck by plague in 1603, probably initiated by infectives coming from London; Colchester in Essex was badly affected and Kent was said to be generally infected. Outbreaks were recorded in Surrey (108 burials at Dorking) and Sussex. In 1604, plague was active in the counties south of the Thames from Sussex to Devonshire, although with varying intensity: only five victims in one family were recorded in Horsham, Sussex, whereas the port of Southampton and Salisbury were said to be ravaged. Cranbourne in Dorset recorded 71 plague deaths. In Exeter (where extensive preventative measures were taken) deaths from plague were only 26 in 1603–4.

9.2 The years 1609–11

Plague continued to erupt sporadically over much of rural England and, again, it is difficult to determine a spatial pattern. It was widespread but scattered in the Midlands in 1609–11: Gloucestershire (22 plague burials in the hamlet of Tredington with three further deaths in the following April after overwintering), Worcestershire, Wiltshire (a severe outbreak in Chippenham in 1611), Shropshire (after overwintering at Ludlow, there were 96 plague burials spread over the summer with no clear peak, perhaps because of the earlier outbreaks there), Derbyshire (53 deaths over the summer in Belper), Lincolnshire and East Anglia (the plague mortality was low at Norwich but the river was watched in case infected persons should come from Yarmouth) with a serious outbreak at Cambridge in 1610. In Leicestershire, Loughborough, which had experienced high mortalities previously in 1545, 1558 and 1602–3, was struck by a plague epidemic in August 1609 which lasted 18 months and ended on 19 February 1611, having twice overwintered. The death toll was 452 and cabins were erected on the outskirts of the town.

Chester, again, had a severe outbreak in 1610 and the pestilence may have been carried back and forth between the city and Liverpool, which was attacked in 1609 (when 28 cases were confined in the cabins) and 1610. No epidemics have been traced in the smaller parishes of Cheshire.

Plague also broke out in Newcastle-upon-Tyne in 1609, when 20 burials were reported, followed by 160 plague deaths in the following year, 1610, between April and December. The epidemic does not appear to have spread southwards along the northeast corridor.

9.3 The plague of 1625–26

The plague continued with sporadic and widely scattered outbreaks during the period 1617–24, but usually the mortality was low, although it was said to be in a virulent form at Leicester in 1623. Scarborough (north Yorkshire coast) had an outbreak in the summer of 1624 and Yarmouth (on the Norfolk coast) and 'some places of the country thereabouts' were reported to be suffering (unusually) from a winter plague in February 1624.

Concomitant with the plague in London, the disease was widespread in England in 1625–26, only the northwest corridor being spared. Again, it is difficult to determine the spatial components of the pandemic. The pestilence broke out in the southern counties bordering the English Channel (Kent, Sussex, Hampshire, Devon), particularly in the ports (by which it may have gained multiple entry) during the summer of 1625. The mortality was probably lower than in previous epidemics. It spread widely in Devon and thence to Somerset. It arrived in Wiltshire (perhaps from London) by August 1625 and became widespread. There were 62 deaths in Winchester, Hampshire, between August and October 1625, not a heavy mortality. Plague was widespread again in Cornwall, Devon, Dorset and Somerset in 1626 and reappeared the following year. Ashburton (Devon) recorded 366 plague deaths where plague was already active in January and February 1626, having presumably begun in the preceding late autumn, and reached its peak by April and May (Fig. 5.2C), a variant of a type (ii) epidemic, with a very early peak.

In East Anglia, the plague broke out in Norwich in June 1625 and is said to have been imported from Yarmouth on the coast. It reached its usual, late summer peak and then overwintered and broke out again in March 1626 and continued through a second summer, with a plague mortality over 16 months of 1431; this appears to have been a greatly extended type (i) epidemic with two clear peaks, but may have been the result of a reinfection in spring 1626. The pestilence was also recorded at Grantham in Lincolnshire and at several localities in the east Midlands, including Leicester again, where it appeared in spring 1625 and was believed to have come from London in spite of precautions to stop the transport of wares or passengers. A haberdasher who had collected a hamper of hats and other commodities from London ignored these orders but fell ill and died of plague about 25 miles south of Leicester; this is an example of an infective

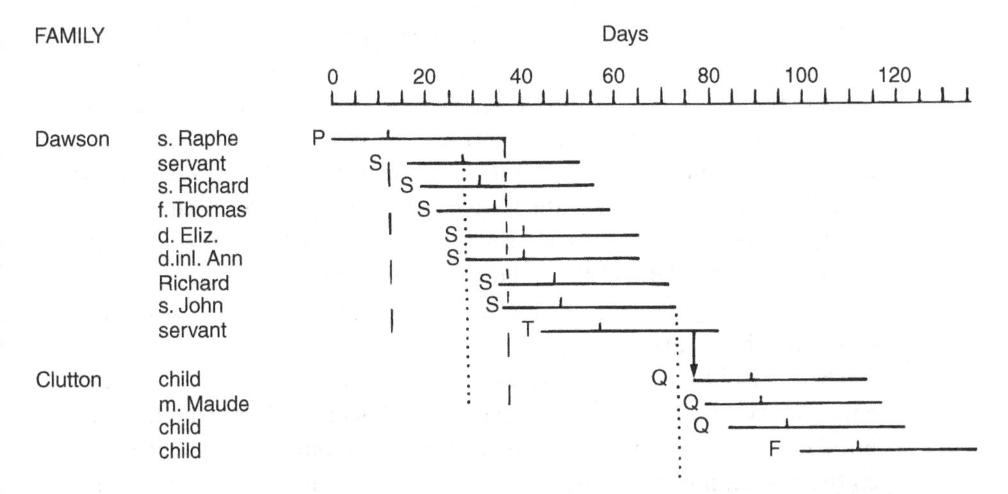

Fig. 9.6. Suggested sequence of infections in the Dawson and Clutton families at Malpas, Cheshire, in 1625. Scale: days after 18 June 1625. For further details and abbreviations, see Fig. 5.6. d.inl., daughter-in-law.

travelling a considerable distance and bringing the pestilence from London to the Midlands. This epidemic was probably minor, with less than 50 plague deaths.

Although there is no record of plague at Chester in 1625, the registers of two parishes in Cheshire contain references to very small, contained outbreaks of plague. The registers of Malpas are not explicit because members of the first household infected were buried near the house and not in the churchyard. Apparently Raffe Dawson returned from London (where the plague was raging), a distance of some 185 miles, to his father's house in the hamlet of Bradley. He had been infected with the pestilence, as the registers record, and was probably infectious by early July 1625. Raffe Dawson displayed the characteristic symptoms and died about 25 July 1625. There was a high household contact rate and all the members died and were buried near to the house. There were at least five and possibly seven co-secondaries (Fig. 9.6; $R_0 = 7$). When Richard Dawson (brother to the head of the family, Thomas Dawson) 'being sick of the plague and perceyving he must die at yt time, arose out of his bed, and made his grave and casuing his nefew, John Dawson, to cast some straw into the grave, wich was not farre from the house, and went and layed him down in the said grave, and cased clothes to be layed uppon, and soe dep'ted out of this world; this he did, because he was a strong man, and heavier then his said nefew, and another wench were able to burye.'

The Clutton family were then also infected by the servant to the Dawson family, a tertiary (Fig. 9.6) and these victims were buried by their relatives in the churchyard. The story of the epidemic at Malpas shows how, because of the long latent and infectious periods, the pestilence could be carried over a great distance from London, a journey that must have taken some time. The high household contact rate ensured the death of apparently the whole household, who had no resistance, whereas the low population density in the hamlet and the isolation measures that are indicated by the relatives burying the dead around the house and not in the churchyard, contained the outbreak.

There was also a small outbreak of plague coincident with the later events at Malpas in the adjacent parish of Shocklack. John Handley was buried 3 September 1625 and, with hindsight, is thought to have introduced the plague, possibly from a visit to Malpas, because his children, John (15 years old) and Elizabeth (19 years old) both fell sick on 23 September. John died 2 days later 'And by reason he died so suddenly, and having Red-specks [the Tokens] found upon him, he was supposed to have died of the Plague. And therefore was carried to the Church upon a Dragge by his Mother, Elleyn Handley and Randle gylbert his half brother. And was buried in the Church Yard, at the Steeple end out of the alley; without Service, ringing or any other ceremonies of the Church'. Elizabeth Handley died on 27 September 'And because she had Red-specks found upon her and some sore under her arm, she was likewise suspected to have died of the Plague. And therefore she was buried in her Mother's Croft, near the Orchard upon Wed 28 Sep. And upon Monday the 3rd day of Oct., the aforesaid Eliz. Handley was taken up again out of her grave and brought to the Church, upon a Dragge by her half brother Randle gylbert, aforesaid, which buried her near to her brother John Handley the younger without any ceremonies of the Church'. Elleyn Handley (the mother) was buried 9 October 1625 'And because her two children had died with Red-specks upon them therefore she was suspected to die of the Plague. Wherefore she was carried to the Church upon a dragge by her son, Ran Gylbert, and laid by [the bodies of] her two children, John and Eliz. Handley'. Randle Gylbert (aged 28 years), who assisted with these family burials and who was 'supposed to have brought these troubles and sickness into the Parish of Shocklach, was confined to keep his Mother's house, and there kept in by watch and ward, night and day, himself alone a long time. And upon Sunday 2 Oct. he was stripped and viewed by certain neighbours and Parishioners, and then had no sign of any sore found upon him'. He escaped the plague and was married the following year (Richards, 1947). It is noteworthy that the red spots or tokens were the characteristic symptoms of the plague.

Plague was also very active again in the northeast corridor in 1625–26, appearing in various parishes: in the North Riding of Yorkshire (the city of York escaped); in Gateshead, Barnard Castle, Whickham (a single household) and Sunderland in County Durham; and in Newcastle-upon-Tyne in Northumberland in both 1625 and 1626 where the epidemic was probably not severe. Sunderland was said to be dangerously infected and there were 89 plague burials at Gateshead.

9.4 The years 1630–37

Cambridge was a major focus of the plague in the provinces in 1630, with 214 victims; it began with seven deaths in April and it was claimed that it had been imported by a soldier 'who had a sore upon him' (Shrewsbury, 1970) a type (i) epidemic initiated by a stranger. The plague was very widely dispersed in 1630: Canterbury (Kent), Aylesford (Kent), Norwich (Norfolk), Shrewsbury (Shropshire), Preston (Lancashire) with 1100 burials registered between November 1630 and November 1631 (a type (ii) epidemic) and Bedfordshire.

Plague broke out in Alford (Lincolnshire), the first death being on 22 July 1630, with 94 plague deaths recorded before 9 November 1630, a type (i) epidemic. The early stages of the epidemic are shown in Fig. 9.7 and the striking feature of this outbreak is that there were 22 burials in the first 10 days, all apparently co-primaries. In reality, there were probably one or two genuine primaries who introduced the infection into the parish earlier and who were undetected and hence these first 22 victims were probably co-secondaries. The effective household contact rate was not high during the opening stages of this outbreak. The epidemic peaked in August 1630 with 45 plague deaths and with only 10 exceptions, the remaining plague deaths were single events in the families, a common feature of plague mortality in summer.

Plague was widespread in the Midlands in 1631, penetrating into East Anglia. Loughborough recorded 135 but Norwich only 20–30 plague deaths. Louth, in Lincolnshire, only 10 miles northwest of Alford, had a major, typical epidemic that began in April 1631 with 754 plague deaths of which nearly 500 occurred in July–August (Creighton, 1894), probably the result of the pestilence arriving in a naive population; a type (i) epidemic.

Plague appeared elsewhere in Lincolnshire in 1631 and the parish registers of St Margaret's, Lincoln, record 'Here the sickness began' on 25 July

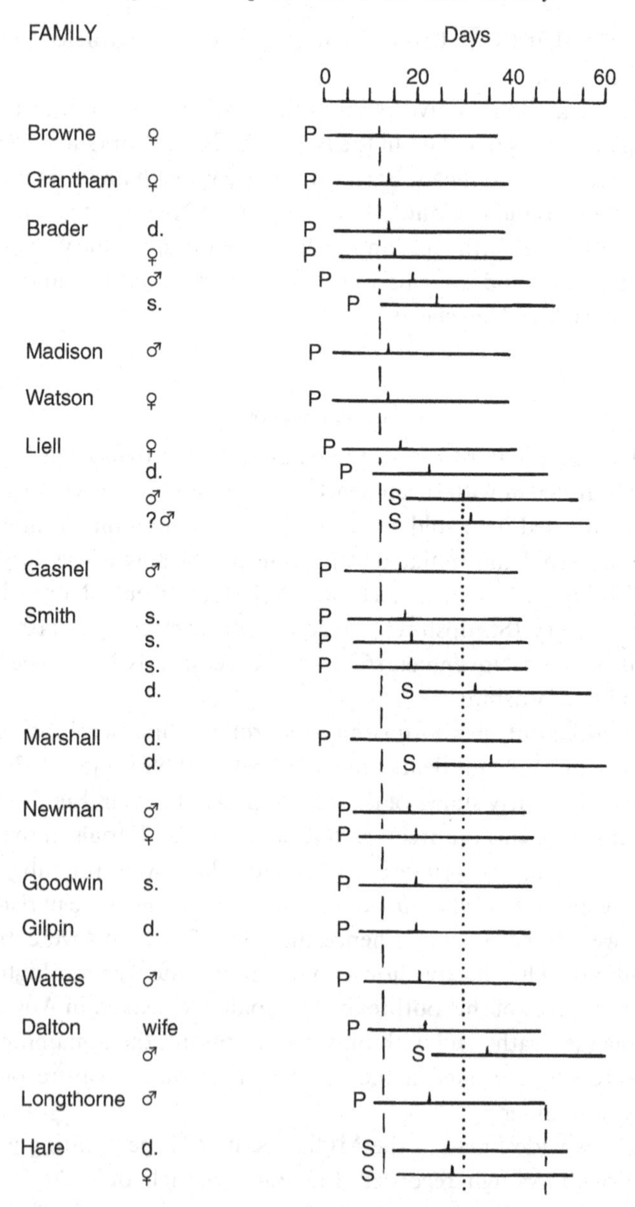

Fig. 9.7. Suggested sequence of infections at the start of the plague epidemic in the parish of Alford, Lincolnshire, in 1630. For abbreviations, see Fig. 5.6. ♀, female; ♂, male. Scale: days after 15 June 1630.

1631. Although plague deaths are not specifically recorded, probably about 43 died before the 'last that died of the sickness' on 4 December 1631; a type (i) epidemic. Only two families were stricken in another parish, St Peter's, in Lincoln. Many of the burials are single events in the families and there were probably 10 co-primaries. This pattern of isolated cases in the families, coupled with a low death rate in the city, suggests that Lincoln may have experienced outbreaks of plague before and there is evidence of epidemics there in 1557, 1586–87, 1593 (Creighton, 1894; Shrewsbury, 1970).

The plague spread from Lincolnshire to Yorkshire in 1631 where Redness and Armin were 'furiously infected, and 100 persons died'. It broke out at Acomb (1.5 miles west of York) and was brought into the suburbs of York 'by a lewd woman from Armin [clearly an infective] and in that street are since dead some four score persons. It has not yet got within the walls, except in two houses, forth which all the dwellers are removed to the pest-houses'. The disease was effectively controlled and confined almost entirely to one parish. The Council minutes for 29 August 1631 recorded that some persons in St Lawrence churchyard (an extra-mural parish) 'were visited with the infection of the plague'. Six watchmen were set to keep people away from the churchyard, all beggars and wanderers were removed from the city, a plague lodge was erected and the parish was effectively isolated from the rest of the city, but by 31 August there were '16 dead thereof in St Lawrence and one in St Margaret's parish'. Rigorous public health measures were enforced vigorously: there were restrictions on movement, strict isolation of the sick and all those suspected of being infected along with any contacts were immediately quarantined. Special attention was given to ensure that St Lawrence parish remained isolated and watch was 'sett at the end of St Nicholas forthwith to keepe out strangers and to keepe in the inhabitants without Walmgate Bar' but the orders were still not suspended and individuals and families continued to be isolated until the end of March 1632, well after any threat had disappeared (Galley, 1998). The epidemic at York in 1631 is an example of successful public health measures against an infectious disease spread person-to-person.

The plague also appeared in 1631 in west Yorkshire at Mirfield (180 died) and Heptonstall ('near forty houses infected') and then in west Lancashire at Dalton-in-Furness (360 deaths) and the adjacent Isle of Walney (120 deaths).

Although there were only a few, small outbreaks in England in the following year, 1632, the story of Slaidburn, Lancashire, an upland and isolated parish in the Pennines, is of interest because all the victims were

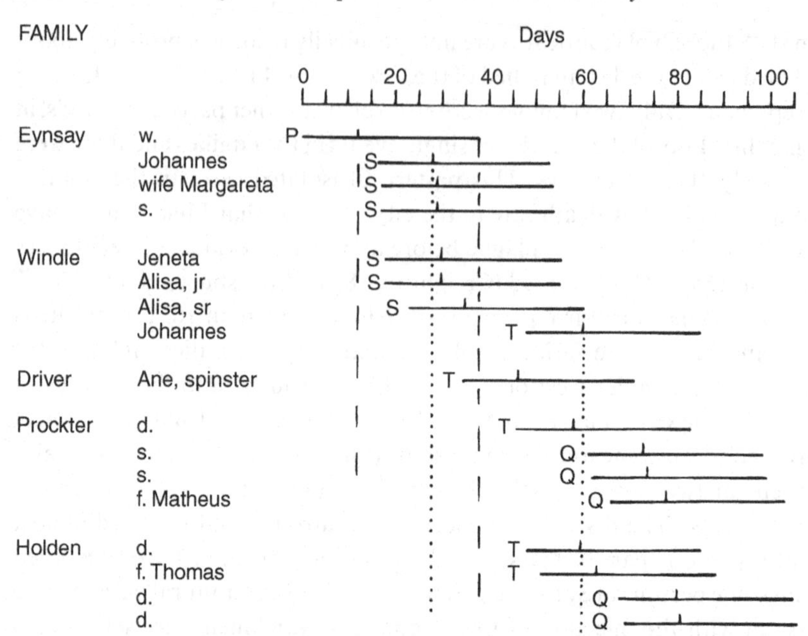

Fig. 9.8. Suggested sequence of infections in the small outbreak of plague at Slaidburn, Lancashire, in 1632. Scale: days after 30 June 1632. For further details and abbreviations, see Fig. 5.6. jr, junior; sr, senior.

carefully documented in the registers and this minor epidemic can be fully analysed; it conforms exactly with the pattern of this infectious disease that we have described previously. All the deaths were confined to the hamlet of Dalehead, which is now submerged under the Stocks Reservoir. The infection may have persisted over winter from the outbreak in the Furness peninsula; it was introduced about 30 June 1632 by widow Eynsay (primary), who died on 5 August (Fig. 9.8). There was a high household contact rate and she directly infected three co-secondaries in her own family and three further co-secondaries in the Windle family. Tertiary infections began with Ane Driver and followed in the Prockter and Holden families after 15 August 1632, with a total of 17 pest deaths confined to five families. Analysis of the records suggests that the epidemiological parameters were: latent period = 12 days, infectious period = 25 days. R_0 has been calculated as follows: primary to secondaries = 6; secondaries to tertiaries = 0.8 (i.e. <1); tertiaries to quaternaries = 1.

After a break of some 2 years when England was apparently free, plague was reintroduced in 1635, probably from abroad via the east coast ports of Hull and Yarmouth and also the port of London. The epidemic continued in Hull in 1636 after the winter, died down in the autumn and reappeared

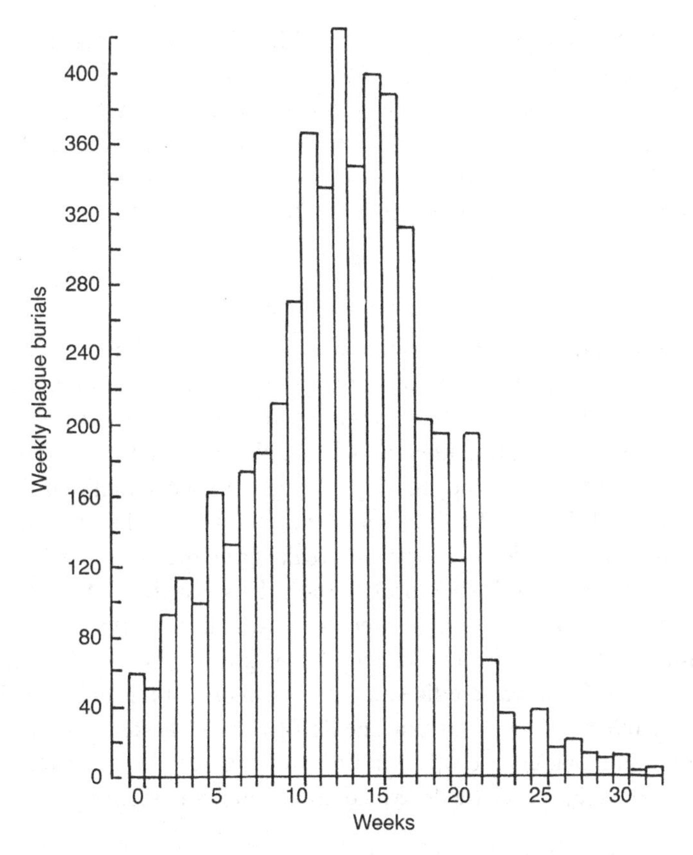

Fig. 9.9. Weekly plague burials at Newcastle-upon-Tyne in 1636. Abscissa: weeks after 14 May 1636. Data from Shrewsbury (1970).

again in 1638; it did not subside until late 1639 and the plague was very persistent but there may have been reinfections that restarted the outbreak. Shrews-bury (1970) quoted accounts that described the deaths of 2730 inhabitants of Hull during this 4-year period. Plague was also present in Yarmouth and Lowestoft on the east coast in 1635 and it continued in East Anglia in 1636–37, being recorded in Norwich, Yarmouth, King's Lynn, Bury St Edmonds and in Cambridgeshire and also extended to Leicester, Horton and Melton Mowbray in the East Midlands.

The northeast corridor also had a major outbreak of plague in 1636, probably emanating from the focus at Hull but possibly coming from a separate introduction from overseas via the port of Newcastle-upon-Tyne, which suffered a 'tremendous visitation', with a death toll of over 5000. The weekly toll of plague burials (Fig. 9.9) shows the characteristic Reed and

Frost dynamics over the 8-month period May–December 1636, peaking in July–August; a type (i) epidemic. The Scots Privy Council prohibited any colliers or persons coming from Newcastle to enter any Scottish town. Gilling (Yorkshire) and Barnard Castle and Gateshead (some 500 deaths reported) in County Durham were also afflicted in the outbreak in 1636, although the epidemic was very mild in Durham City.

9.5 Widespread plague in 1644–46

Plague was still widely distributed in England in 1637 and Shrewsbury (1970) has given detailed accounts of these outbreaks, most of which were probably minor but the pestilence continued to grumble on. For example, in Cheshire in 1641 the names of 300 people who died in the parish of Astbury are recorded in the registers, plague having originally broken out in the home of William Laplore whose entire family died in the following month, a characteristic start to an outbreak with a high household contact rate. In Burton (Wirral, Cheshire) in 1641 'This year from Midsomer till Michmas ye Plague was', a characteristic seasonality of a type (i) epidemic. In Shotwick (4 miles from Chester) the plague raged for more than 4 months in 1641 but was confined to the parish except for an isolated case 'when the infection was carried to the Red Lyon in Chester'.

Widespread plague re-erupted in England and persisted over the period 1644–47; the major foci are described in the following sections.

9.5.1 The northeast corridor

An epidemic in Newcastle-upon-Tyne in 1644 is regarded as a maritime importation via the port and was reported as 'very hot' in Tynemouth Castle by October. It continued through the winter and broke out again in the spring and summer of 1645 (a type (ii) epidemic). It spread to County Durham in autumn 1644 and was also prevalent in the following year; it broke out in Durham City in November 1644 where, quite unusually, it continued actively during December and January before finishing in February 1645 in the parish of St Oswald. This appears to be a type (ii) epidemic that did not flare up again in the spring. The burials are not marked with a 'P', but analysis of the registers shows clearly that the majority of the deaths were single events in individual families and that the infection did not spread through households, probably indicative (together with a relatively low mortality) of the fact that the population had been repeatedly exposed to the plague previously.

Plague was recorded in 1644 in Leeds (Yorkshire) and broke out again in the following year, 1645, appearing in June and following typical Reed and Frost dynamics over the next 6 months with a sharp rise to a clear peak in late July with 1300 deaths (a type (i) epidemic). Outbreaks were also recorded in 1645 in numerous towns in west Yorkshire, including Halifax, Wakefield, Bradford, York and Whitkirk.

9.5.2 The West Country

Outbreaks of plague were reported extensively in western and south-western England at this time. It appeared in 1644 in Oxfordshire (Banbury, where 161 died between April and November, and Oxford), Herefordshire (Yarkhill) and Devon (Tiverton). This outbreak spread markedly in the following year, 1645, and was reported in:

Dorsetshire (Dorchester, Poole, Sherborne)
Hampshire (Winchester)
Wiltshire (Wooton Bassett)
Gloucestershire (Bristol)
Somerset (East Coker, where 70 died; Rode, where 280 died; Wiveliscombe, where 440 died; Dunster; Yeovil)
Devonshire (Ottery St Mary, Totnes, Colyton)
Worcestershire (Worcester)

There are additional points of interest in three of these epidemics. The plague at Colyton began on 16 November 1645 and continued until January 1646, which was a cold month with much snowfall; it re-erupted in April 1646 and then continued until the late autumn when 459 had been buried, a typical type (ii) epidemic with perhaps 40% of the population dying.

The outbreak at Dunster, Somerset, in 1645 apparently began in the castle garrison and the governor related that the infected soldiers died 'suddenly with an Eruption of Spots'. A surgeon, who was serving as a common soldier there, persuaded the governor to let him treat the sick soldiers and

> he took away a vast quantity of Blood from every sick Person at first coming of the Disease, before there was any Sign of a Swelling; he bled them till they were like to drop down, for he bled them all standing, and in the open Air; nor had he any Porringer to measure the Blood; afterwards he order'd them to lie in their Tents; and tho he gave no Medicine at all after Bleeding, yet, which is very strange, of those very many whom he treated after this manner, not one died.
>
> *(Shrewsbury, 1970)*

Evidently, the appearance of the spots and not a swelling was the clinical feature.

The monthly burials at Ottery St Mary in 1645 were: October, 15; November, 18; December, 79; and in January 1646, 37. This plague epidemic was at its peak in winter and Shrewsbury (1970) consequently excluded bubonic plague as the cause. It appears to be an unusual type (ii) epidemic of haemorrhagic plague, with the typical start in autumn but an abnormally high winter mortality and the pestilence did not break out again in spring. It is comparable with the epidemic in 1644–45 in the parish of St Oswald, Durham (section 9.5.1).

9.5.3 East Anglia and the Midlands

Boston was an ancient seaport on the River Witham in Lincolnshire and the epidemic began in 1645 with the usher of the town school, who was seized with a violent fever of which he died within 'two or three days having some Red spots'. After three more persons had been seized with a similar fever and had 'all died spotted', the municipal authorities took fright, segregated the infected households and appointed day and night watchers, searchers, and buriers. 'Our searchers say the spots are like the spots which were in Plague time but none of them had any swelling or sores save only one that died on Friday last, the searchers affirmed he had a swelling on the outside of his thigh and had like spots also as the other, Our physicians seem doubtful whether it were the plague or not, but we for our parts are afraid and take it for granted to be the Plague' (Shrewsbury, 1970).

A minor outbreak of plague also occurred in King's Lynn, across the Wash, beginning on 20 May 1645 and in Kelvedon (Essex) in October.

Plague also erupted in Cheshire (in Chester again), Lancashire, Yorkshire (where many townships received plague relief) and County Durham, as well as in Derby. Axon (1894) described in detail the violent outbreak that began in May 1645 in Manchester (Lancashire) and followed the characteristic Reed and Frost dynamics (Fig. 5.2D), dying out in the late autumn (type (i) epidemic), during which time the town was isolated from the rest of England. Some 1100 persons died in this epidemic.

9.6 The mid-17th century

After the widespread epidemic of 1645, plague grumbled on in England until 1651. It reappeared in 1646 in some of the places that it had visited in the previous year and broke out anew in the southwest and in the Midlands

(Oxford, Northampton and Loughborough). These outbreaks continued in 1647 and St Ives (Cornwall) was grievously affected: 535 plague deaths occurred between Easter and October 1647, a high percentage mortality in a type (i) epidemic. Plague was probably introduced into Pembrokeshire from Ireland in 1652 (Howells, 1985).

9.6.1 Chester as a focus

Plague was rampant again in Chester in 1647, when there were 1875 burials in 16 weeks (Richards, 1947) and cabins for the infected were erected under the Water Tower and in the adjoining salt marshes. An eye-witness gave the following account: 'The plague takes them very strangely, strikes them black of one side and then they run mad, some drowne themselves, others would kill themselves; they dye within a few hours; some run up and down the street in their shirts to the great horror of those in the City' (Richards, 1947). This behaviour corresponds with the victims of the plague at Athens (section 1.2.1). The outbreak is said to have lasted from 22 June 1647 to 20 April 1648, when 2099 died (Axon, 1894).

We have found, tucked away in the parish registers of Great Budworth (Cheshire), the list of plague burials at the village of Barnton during a small epidemic from 11 April to 9 June 1647. There were only 17 victims from 9 families who were all buried at Barnton and not in the churchyard at Great Budworth. Only two children were listed and all the other victims were adult. The outbreak began with three co-primaries in three separate families who infected 8 secondaries ($R_0 = 2.6$) who produced 4 tertiaries ($R_0 = 0.5$). The epidemic was confined to this one small village and, apparently, had no impact on the rest of the parish. There must have been many other such small epidemics in little communities in England in the 16th and 17th centuries brought in by an apparently healthy infective. Where did he or she come from at Barnton?

The people of Liverpool were alarmed about the plagues reported at Chester and Warrington in 1647 and it was ordered 'that strict watch shall be kept by the townesmen'. The *Constables Accounts* show that Manchester also took precautions to prevent ingress to the town of those who were thought to be still infected (Axon, 1894). However, the pestilence erupted in Liverpool in the early spring in the following year, 1648, and cabins were built and families were removed to them so that the epidemic lasted only 2 months with 'the death of about eight or nine persons of mean quality' (Axon, 1894). Evidently this epidemic did not explode, perhaps because of the precautionary measures. The authorities in Liverpool were apprehensive of the importation of plague from Ireland and issued the following orders 2 years later on 2 April 1650.

Whereas it is certainly reported, that the sicknes in Dubline, wch by reason of the intercourse from thence may prove dangerous to this towne; it is therefore ord'red, that all owners and passengers comeing from thence shalbe restrained and debarred from comeing into this towne unless they cann make oath that they have not beene in anie infected place, nor brought over anie infected goods or passingrs from thence, and be allowed of by Mr. Maior; a Warrant to be drawn up for ye Guard to examine all passingrs comeing from thence, until they be sworne and examined, wch was donne accordingly.

However, in spite of these precautions, there are unconfirmed reports of 200 plague deaths in Liverpool in 1650 (Axon, 1894).

There was also an outbreak of plague in Shropshire (contiguous to Cheshire) in 1650: it erupted at Shrewsbury on 12 June and reached its peak in August 1650 but it is noteworthy that the epidemic was unevenly dispersed in the town, with 10 times as many burials in St Chad's parish than in St Mary's (corresponding to the pattern of the plague in 1631). Total plague burials exceeded 300.

Plague then erupted in Whitchurch, 18 miles to the north of Shrewsbury, on 2 August 1650; it died down in the winter but reappeared during April–June 1651 (Fig. 9.10). It began late in the summer for a type (i) epidemic and persisted through winter as in a type (ii) epidemic but did not explode in the following spring.

Plague struck Cheshire again in 1654, with an epidemic in Chester and four deaths in a single family in Tarvin on 10, 16 and 18 May.

Neston, a small port on the River Dee, 10 miles north of Chester, suffered from an outbreak of plague in 1665. The victims are not designated in the registers, but the epidemic probably began in May 1665 and followed the characteristic Reed and Frost dynamics with the typical slow build-up, followed by a peak in September–October before rapidly dying down in November but with deaths recorded until 4 February 1666: an extended type (i) epidemic. The total of plague deaths was estimated to be about 140. Analysis of the burials suggests that one-third of the families showed multiple deaths with a high household contact rate whereas two-thirds of the families recorded a single burial. At the height of the epidemic many of these single deaths were infected simultaneously because there were four to five burials on the same day. Examples of the spread of the infection within households are as follows:

(i) George Leene (son of William Leene) brought the infection into his

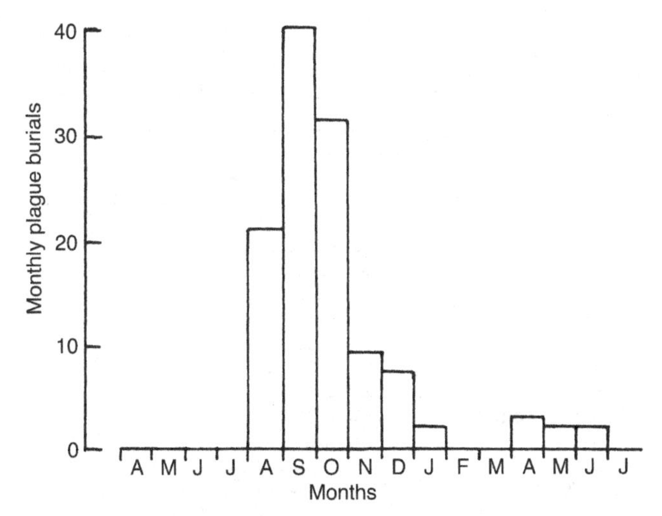

Fig. 9.10. Monthly plague burials at Whitchurch, Shropshire, from April 1650 to July 1651. Data from Shrewsbury (1970).

family and died on 13 September 1665. He infected other members immediately at the end of his latent period and they died on 24 (father), 25 (sister), 26 (brother), 26 ('base child which William Leene's wife nursed'), 28 (mother), 28 (sister) September.

(ii) Independent infections were introduced into the Tindale household with deaths on 27 (apprentice), 28 (son) August and 2 (Mr George Tindal, head of house), 4 (wife) September. A death on 15 September 1665 (daughter) was a secondary case and on 4 October ('A welshman. Lived at Mr. Tindale's') a secondary or tertiary case.

Analysis of the last victims of this plague from September 1665 to February 1666 reveals how the epidemic lingered because of the very long latent and infectious periods (Fig. 9.11). After the burial of Elizabeth daughter of Thomas Walley on 10 December 1665 there is an entry in the registers 'The Plague ceased. Thanks be unto God', but Elizabeth had infected her brother Thomas 2 days before she died, presumably when she was showing symptoms, and he 'dyed of ye sickness' on 14 January 1666, having first infected members of the Griffith and Barlow families (Fig. 9.11, Part 2) who died at the start of February 1666.

The burials of Elizabeth and Thomas Walley were separated by 35 days, confirming that this is the minimum length of time for the duration of latent and infectious periods of the disease. Another example of a long interval is Samuel Bennett, who died on 16 November 1665 and was almost

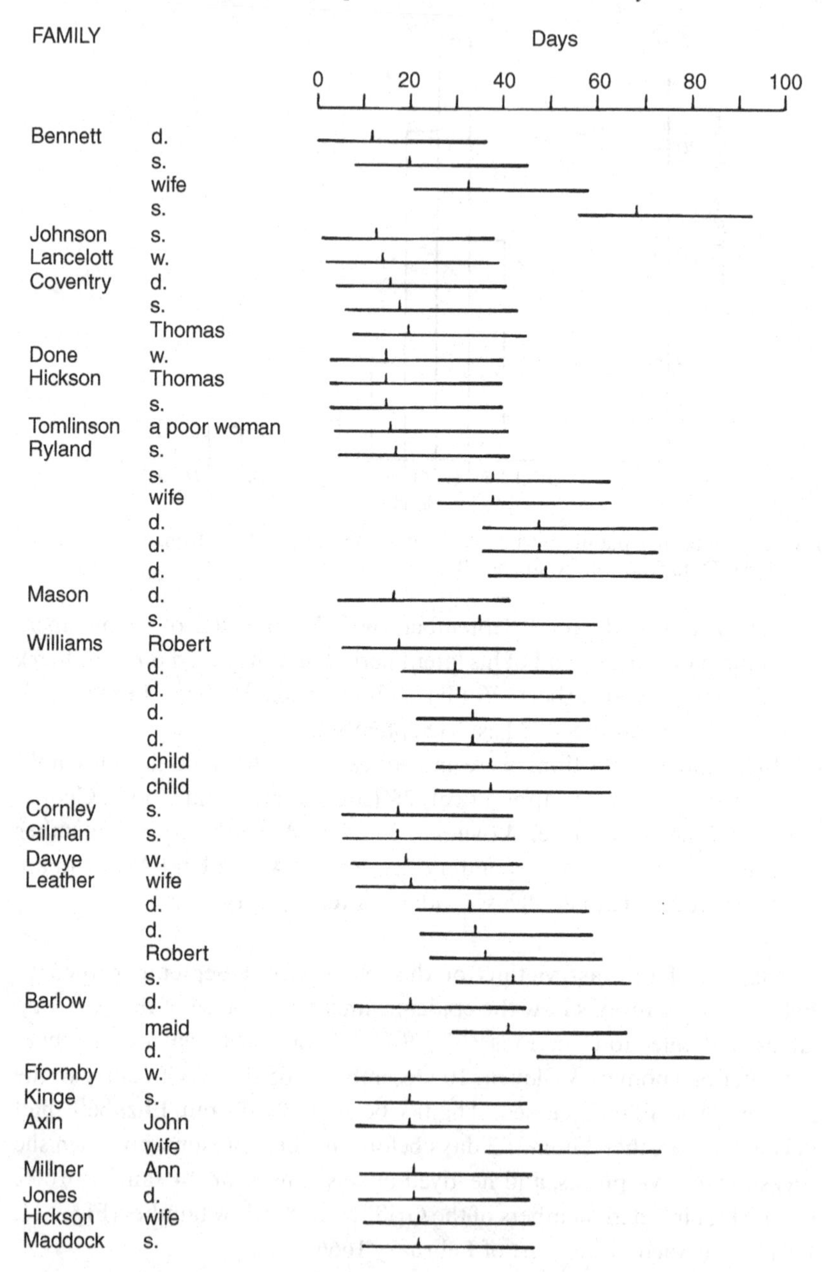

Fig. 9.11. The closing stages of the plague epidemic in the parish of Neston, Cheshire, May 1665 to February 1666. The epidemic lingered because of the long incubation period. Scales: days after 15 August 1665 (Part 1) and 28 August 1665 (Part 2). For further details and abbreviations, see Fig. 5.6.

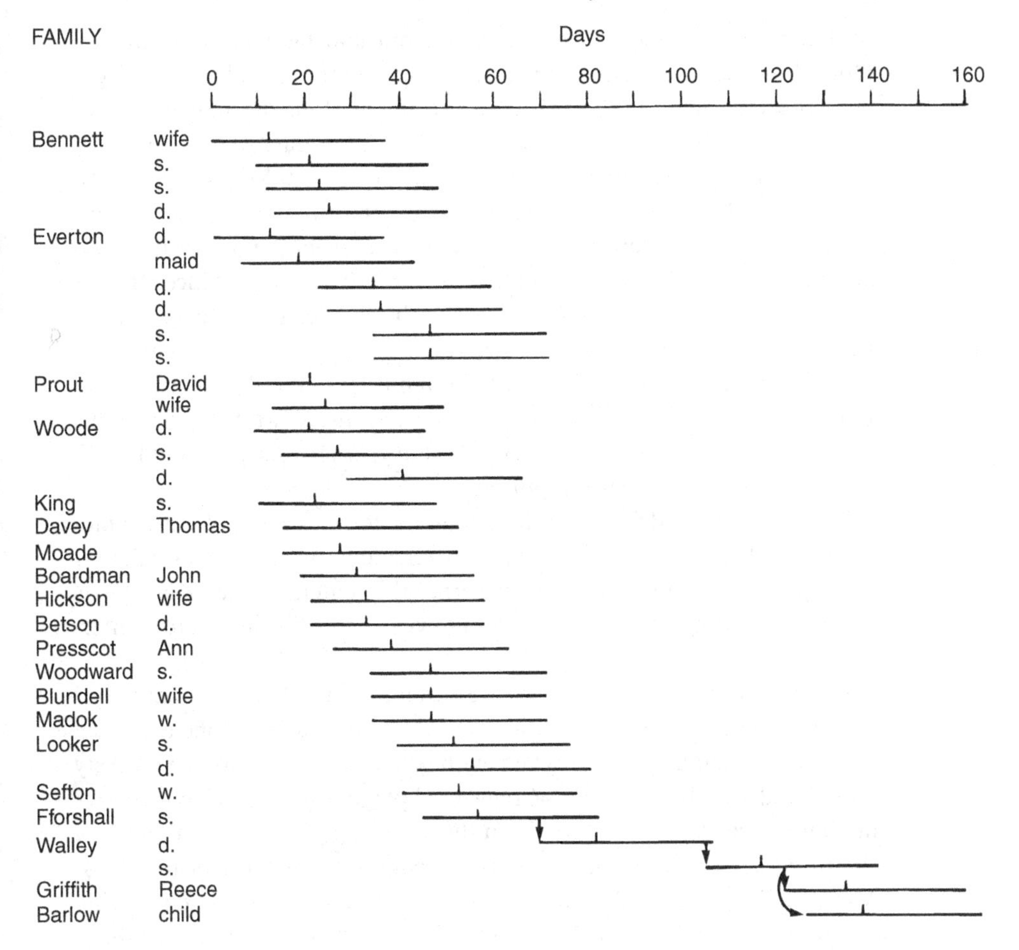

Fig. 9.11. (Part 2)

certainly infected by his mother, who had died on 12 October, when she must have been in the terminal stages of her illness and showing symptoms, again a 35-day interval (see top of Fig. 9.11, Part 1). The long latent and infectious periods account for the very slow elimination of the plague and the premature announcement of its cessation.

9.6.2 Spread of the plague of 1665–66 through the metapopulation

The plague in London in 1665–66 spread to the Provinces in several different ways. A solitary infective traveller brought the disease to an isolated community, as in the major epidemic at Eyam (see Chapter 10) and probably in the minor outbreaks at Epsom and Godalming in Surrey. Doubtless there were other unrecorded and minor epidemics. The plague also spread radially northwards to the Midlands along the main highways, whereas it exhibited a complex wave-like spread in East Anglia and Kent. It moved along the northeast corridor, appearing from Yorkshire to Newcastle-upon-Tyne (again), but without further information it is not possible to determine the pattern of spread nor the origins of the outbreak in this apparently separate metapopulation of the Northern Province. It may have come via the port at Newcastle with an infective on a boat from London. In contrast, Cumbria in the northwest again escaped completely, probably because of its isolation. Although it was described as the Great Plague of London, the epidemics in the provinces, with some notable exceptions (e.g. Eyam, Colchester, Salisbury), were generally mild, with relatively few plague deaths reported.

The plague spread from London along the lower reaches of the Thames on the Kent coast during 1665 and 1666, reaching Chatham and Rochester with quite severe mortality. Plague probably entered other Kent ports directly, including Deal, Dover and Sandwich, spreading inland to Canterbury.

On the south coast in Hampshire, Southampton suffered another major epidemic in 1665, presumably entering via the port, and in 1666 the naval base at Portsmouth was stricken with what was probably a relatively minor epidemic. Basingstoke (46 registered plague burials) and Winchester in Hampshire also had outbreaks in the following year, 1666. Shrewsbury (1970) makes the following noteworthy comment on the registers of St Maurice, Winchester '12 of the entries . . . are marked with an O, denoting death "from spots" as plague was called'. Evidently, even Shrewsbury admitted that the key identifying features of the plague were the tokens or red spots.

The plague arrived in Wiltshire by mid-summer 1665, probably by an infective travelling westwards from London rather than from Hampshire. It was severe in Salisbury in the summer of 1665, continuing through the winter and breaking out with renewed violence in 1666 and not dying out until November 1666. The epidemic radiated outwards and minor outbreaks occurred in other Wiltshire populations.

Shrewsbury (1970) believed that East Anglia was again invaded by plague from London in 1665 almost synchronously through its ports and thence by spread overland. Yarmouth on the Norfolk coast was attacked 28 May, Harwich (a port in Essex) probably at about the same time but was not officially recognised until 2 September, Colchester (Essex) in early

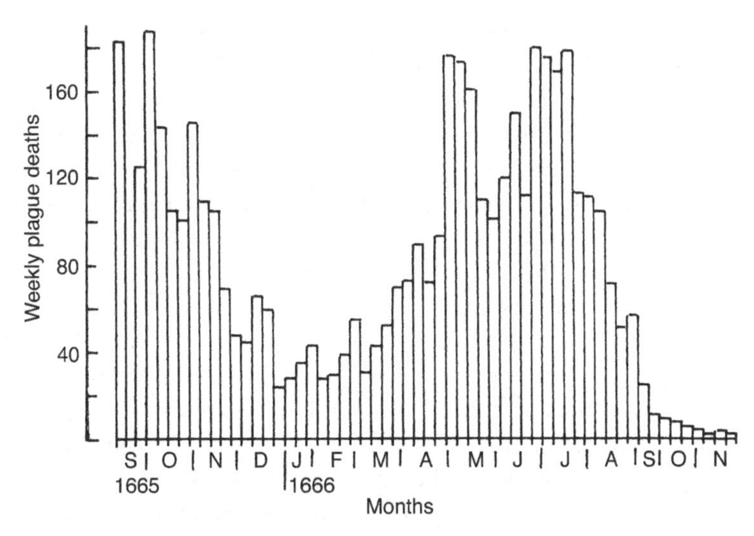

Fig. 9.12. Weekly plague deaths at Colchester, Essex, from September 1665 to November 1666. Data from Creighton (1894).

August, Ipswich (Suffolk) on 13 September and Norwich (Norfolk) in the first week of October 1665. It spread to a number of localities in Essex and continued into 1666, when Braintree was badly affected with 665 plague deaths (probably some 50% of its population) and with a case mortality rate of 97% because only 22 patients recovered (Shrewsbury, 1970). Evidently, plague was not invariably fatal. This spatial and temporal pattern suggests that plague was exchanged through the east coast ports via local sea-traffic and then spread inland, probably via the river system in many cases (see Fig. 13.1).

Among the towns in East Anglia that were affected at the start of the 1665 epidemic, Colchester suffered badly and Shrewsbury (1970) estimated that it may have lost between one-quarter and one-third of its inhabitants. The time-course of this outbreak is shown in Fig. 9.12; the plague deaths have a clear double peak and the outbreak clearly persisted over an 18-month period. From its reported start in August 1665, the deaths peaked in September–October, suggesting that an infective probably initiated the epidemic in June but plague deaths were not recorded at this time. Plague deaths continued at an average of about 30 per week through the winter, a high value for this season of the year, particularly as there was a severe frost in December and January. The outbreak then gained momentum in March and April 1666 and showed peaks of about 175 deaths per week in May and July 1666. It disappeared by the end of November

1666 and appears to be an unusual type (ii) epidemic with a very early start that led to a greatly extended outbreak.

This persistence of the epidemic through into 1666 was also seen at the East Anglian ports of Ipswich, Harwich and Yarmouth. There were relatively few plague deaths in Norwich in autumn 1665, but 13 were recorded in January 1666 and the epidemic regathered momentum in the spring and rose to a peak of 200 deaths per week in August 1666 (type (ii) epidemic). The second phase of the outbreak lasted from March to December 1666.

The plague moved outwards in East Anglia in 1665 and, in addition to the outbreaks in Essex, it was reported in Cambridge (413 burials), Royston, Oundle, Ely, Peterborough and King's Lynn and was said to be active in 20 parishes in Buckinghamshire. Several of these outbreaks continued into 1666: the plague broke out again in Cambridge in June 1666 and Sturbridge fair was cancelled. The Bishop of Ely also cancelled all fairs in his diocese because Peterborough (where it continued from summer 1665 to spring 1667) and Ely were also suffering from this outbreak. Pest-houses were erected outside Cambridge and the population apparently suffered a severe plague mortality so that the harvest could hardly be gathered. It was not until January 1667 that the Vice Chancellor notified the scholars that they might return to the colleges.

Ramsey in Huntingdonshire was attacked in 1666 and Shrewsbury (1970) quoted an account that plague was imported in winter (February) in a parcel of cloth from London. The tailor who made up the cloth and all his family died of the disease and the subsequent epidemic destroyed more than 400 of the townsfolk, although the first plague burial noted in the parish register was that of a woman who was buried in her own garden on 16 July 1666. Presumably the epidemic was started by an infective traveller who brought the cloth from London and the high household contact rate ensured that the infection ran through the family. The infection must have spread to another family and persisted in the population, even with the low contact rate in winter, but the epidemic slowly gained momentum during the following spring.

9.7 Overview of plagues in England

Biraben (1972) has collated for each year, 1347–1670, the numbers of localities in the British Isles in which plague epidemics were reported. Obviously, the list is not complete and some mortality crises that were not plagues are included, but the data-series, which is plotted in Fig. 9.13, provides a good picture of the changing pattern of the spread of the

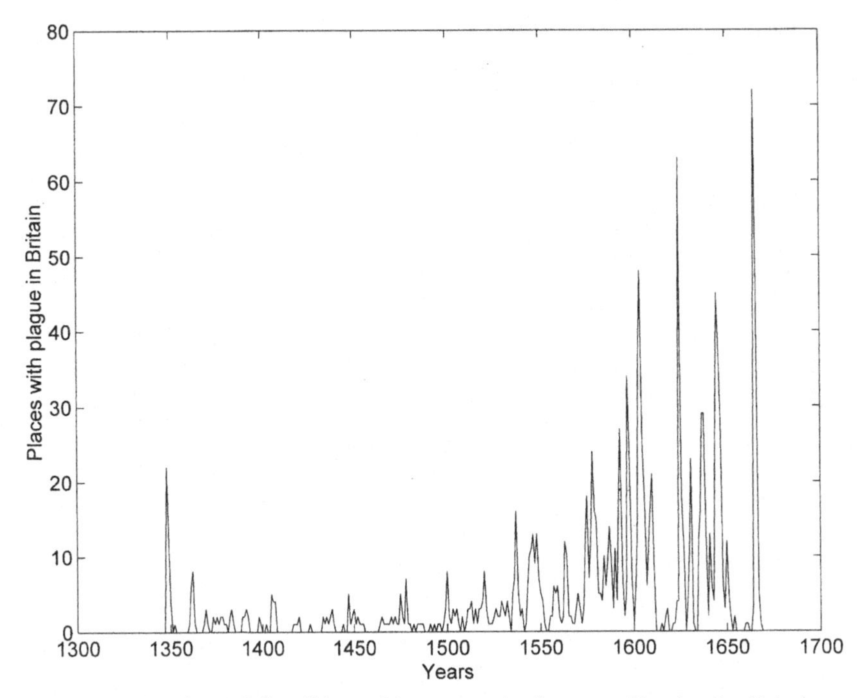

Fig. 9.13. Number of localities with reported plague epidemics in Britain, 1347–1670. Data from Biraben (1975).

outbreaks. It is not a series of the total annual mortality. After the Black Death, the epidemics grumbled along sporadically at a low level for 150 years but after 1500 the spread gradually increased in terms of the number of places attacked. Spectral analysis of the data-series (see section 2.8) reveals a non-stationary short-wavelength oscillation ($P = 0.05$), the period of which was 5–7 years and a medium-wavelength oscillation that is discussed in section 11.7.

We have previously described a persistent, short-wavelength oscillation in English wheat prices that caused sharply changing cycles of malnutrition; these had complex and serious effects on infant mortality, the outcome of pregnancy and the epidemics of lethal infectious diseases (Scott *et al.*, 1995; Duncan *et al.*, 1996a,b, 1997, 1998, 1999; Scott & Duncan, 1998, 1999a, 2000). Did this oscillation in wheat prices also drive the outbreaks of plague?

The English wheat prices series, 1450–1649 (data from Bowden, 1967, 1985), is shown in Fig. 9.14. It falls into two periods: for the first 100 years the level was low and the trend was steady with low amplitude oscillations superimposed thereon, but after 1550 the trend rose markedly with large-

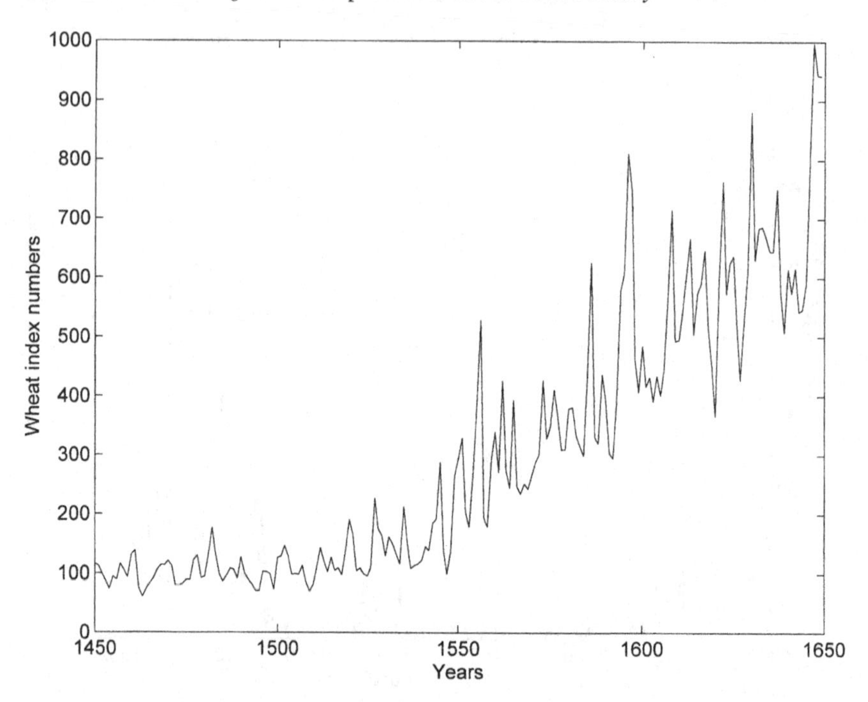

Fig. 9.14. English annual national wheat price index, 1450–1649. Data from Bowden (1967).

amplitude oscillations. Spectral analysis reveals a short-wavelength oscillation: period = 8 years (1450–1550) and 6.7 years (1550–1649). But filtering the two series shown in Figs. 9.13 and 9.14 revealed no cross-correlation between them. Equally, filtering for the medium-wavelength oscillation revealed no cross-correlation between the two series.

We conclude that fluctuations in the wheat prices did not drive the plague epidemics in England. The pestilence was not truly endemic in this separate island metapopulation (unlike France, see section 11.2) and the outbreaks outside London were governed largely by the unpredictable arrival of infective(s) from overseas or the metropolis. But England was part of the enormous metapopulation of continental Europe from whence came the infectives, so that the epidemiology of the plagues in England was governed by the spatial components of the pestilence on the Continent, factors that are described in Chapter 11.

10

Plague at Eyam in 1665–66: a case study

10.1 The traditional story of bubonic plague at Eyam

The story of the plague in the Derbyshire village of Eyam is well known: the causative agent was brought in a box of cloths from London and when this was unpacked a rat and/or fleas jumped out and so brought bubonic plague. The inhabitants, led by their vicar, nobly drew a cordon sanitaire around their village, preventing all egress and ingress and so confined the epidemic to their little community. The events of this outbreak have been described in detail many times (Creighton, 1894; Batho, 1964; Shrewsbury, 1970; Bradley, 1977; Clifford, 1989; Race, 1995) but all assume without question that this was an outbreak of bubonic plague, although, as we shall see, the details of this epidemic demonstrate clearly that again, as at Penrith in 1597–98, it is a biological impossibility that *Yersinia pestis* was the causative agent.

The upland parish of Eyam is situated on the eastern edge of the Peak District towards the southern end of the Pennines, lying on the slopes above the Derwent valley. It was a remote and isolated spot 300 years ago and the approaches to the village were by rough and narrow tracks (Clifford, 1989). Bakewell, the local market town, lay 7 miles to the south. The main settlement in the parish was the village of Eyam with hamlets at Foolow (2 miles away), Grindleford Bridge (1.5 miles away) and Woodlands (2 miles away). It is suggested that the majority of cottages were stone-built, with stone-slab roofs and frequently stone floors (Bradley, 1977); the cottages where the plague began in 1665 can be seen today. These conditions in northern central England, 70 miles from the nearest port, with the sparing use of fire and badly fitting windows would be completely unsuitable for the establishment of a colony of black rats, which would much prefer thatched roofs and warmth.

Weather conditions in winter in the Peak District are severe (Clifford, 1989) and in 1665 'in December a great snow is said to have fallen, with a hard and severe frost . . . The weather at the commencement of 1666 was exceedingly cold and severe' (Wood, 1865; Bradley, 1977). The plague at Eyam began in autumn 1665, continued over the hard winter and then began again in spring 1666 (see below; standard type (ii) epidemic). It is impossible that a colony of warmth-loving black rats and their infected fleas could have survived through such winter weather conditions and perpetuated bubonic plague.

The traditional story is that Alexander Hadfeild, a tailor, received a box of cloth or materials relating to his trade that was brought from London and arrived in the late summer of 1665 (about the end of August). The first person to die from the plague was George Viccars who was buried on 7 September 1665 (Eyam parish registers; Clifford & Clifford, 1993). Traditionally, he has been described as a journeyman, working for the tailor (Clifford, 1989), who unpacked the box and, seeing that the goods were damp, hung them before the fire to dry. This story was told to one of the authors when he was a boy at school in York 60 years ago. Viccars was suddenly seized with violent sickness and grew worse on the second day, becoming delirious and showing large swellings on his neck and groin. The plague spot (the fatal token) appeared on his breast on the third day and he died on 7 September (Wood, 1865; Bradley, 1977). This story is traditionally interpreted as follows: the cloth harboured (presumably blocked) fleas infected with *Yersinia pestis* from London where the plague was raging and that these bit George Viccars and infected him with bubonic plague (Clifford, 1989). Even the fervent proponents of bubonic plague see difficulties in this interpretation, particularly the rapidity of the onset of the disease, since only a few hours is insufficient time for the incubation of bubonic plague (Bradley, 1977). Batho (1964) suggested an alternative explanation that has nothing to do with boxes of cloth: a number of visitors came to the village to participate in the Eyam wakes on 20 August 1665 and some may have brought an infected flea from Derby where he believed that plague was raging. This latter view is suspect because Shrewsbury (1970) and Bradley (1977) found evidence that the plague did not come to Derby until 1666. Batho (1964) surprisingly concluded, however, that the traditional story of a blocked flea coming from London was the more probable explanation. Others have suggested that this was an outbreak of anthrax or measles (for a summary, see Clifford, 1989) or that a human flea might have acted as the carrier (see Bradley, 1977). In any event, these stories of a blocked flea carrying bubonic plague over very long distances through the countryside and starting an epidemic are quite impossible.

The next person to die of plague, Edward Cooper, was buried 15 days later on 22 September and if an infected flea *had* bitten George Viccars no epidemic of bubonic plague could have been established among the hypothetical rodents of the village in so short a space of time; it would have initially involved one or two rats becoming infected with *Yersinia pestis*, developing the disease, infecting their fleas, dying and the transference of the fleas to other rats. This cycle would need to have been repeated many times before the epidemic spread through the supposed population of roof rats; there would have been no possibility of establishing an epizootic and the domestic rats would start dying in numbers. Only then would the fleas transfer to the humans.

The role in the village of George Viccars, the first person to be infected and die, is not clear in these traditional anecdotal accounts, but Clifford (1989) implied that he was not a native and did not have a family with him, presumably because there is no other mention of the surname Viccars in the parish registers. However, it is known that he was survived by his children, although Clifford (1989), most disappointingly, does not say where they were living.

10.2 Origins of the plague at Eyam

The only real evidence of the details of the plague at Eyam is contained in the register of burials, which was meticulously kept (Clifford & Clifford, 1993). Nevertheless, analysis and objective interpretation of these data can, as we shall show, be fleshed out to tell a human story. A total of 260 died in the outbreak and the monthly plague burials are shown in Fig. 10.1; it was a type (ii) epidemic, beginning in September 1665, with a small peak in October, and continuing at a low level through the winter. The outbreak then followed characteristic Reed and Frost dynamics; it picked up very slowly in April–May 1666 before exploding with a peak mortality in August. Thereafter, monthly burials declined slowly and the epidemic ended in November 1666. These details correspond exactly with other type (ii) epidemics in other isolated communities that arrived in late summer (e.g. Penrith in 1597–98); it was clearly the same infectious disease with a long incubation period that we have described in all other outbreaks of haemorrhagic plague.

The story of the origins of the Eyam plague stem from the oral traditions that were collected by Wood (1865) and written before the etiology of

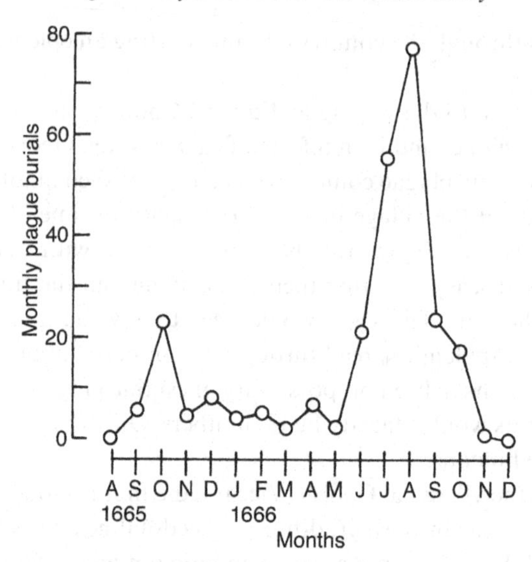

Fig. 10.1. Monthly plague burials at Eyam, August 1665 to December 1666.

bubonic plague was understood, but his account has subsequently been greatly modified to fit with accepted opinion. He clearly assumed that it was an infectious disease and did not invoke the rats and fleas of bubonic plague of which he knew nothing. It is on his original account and the invaluable details contained in the parish registers that we base our analysis of the origins of the plague and the events of 1665–66.

Wood says that Mary Cooper, the widow of a lead miner (they had married in 1652) had, according to the registers, remarried Alexander Hatfeild (trade not stated) on 27 March 1665, some 5 months before the plague erupted. She was living with her two sons in the middle of three cottages to the west of the churchyard and had taken in as a lodger George Viccars who was a travelling tailor and the first victim. Clifford (1989) stated that Wood did not record whether Viccars was a regular visitor, calling on Eyam as part of an established round, or whether he was a stranger. Nor does he record from whence he came.

During the 17th century, there were several Viccars families living in the parishes of Baslow, Dronfield and Edensor, all close to Eyam. One family living in Edensor, 5 miles south of Eyam, is of note: George Viccars married Ann Allin in 1638; he may have been born to Elizabeth Viccars in 1611 at Dronfield. George and Ann had seven children at Edensor and if this were *the* George Viccars, the travelling tailor who brought the plague to Eyam, he would have been aged about 54 years in 1665 and he would still have had to support some of his children. The Viccars family continued living at Edensor and there are records of baptisms and marriages in the register but, in any event, the problem remains: where did George Viccars become infected in autumn 1665 in the course of his travels? We have no idea of the area that he covered in his itinerant work. There is little evidence of plague in Derbyshire in 1665 and no rise in mortality at Edensor in 1665; the statements that Derby suffered from an epidemic in 1665 (Batho, 1964; Race, 1995) as well as in 1666 (Shrewsbury, 1970) were not confirmed by Bradley (1977), who found no excessive mortality recorded in the registers. Nevertheless, Bradley quoted an account by Hutton that in '1665 Derby was again visited by the plague . . . The town was forsaken: the farmers declined the market place'. Shrewsbury (1970) quoted evidence that Newark (37 miles from Eyam) lost one-third of its inhabitants from plague in either 1665 or 1666. Alternatively, Viccars may have been to London to collect cloth or he may have associated with an infectious carrier who came from there. Viccars was infected about 1 August and was lodging with Mary Cooper in Eyam when the first secondary infection took place on 16 August 1665 (see below), probably sufficient time to have made a journey from London. Viccars lodged with Mary Cooper for at least 3 weeks before he died.

10.3 The first phase of the epidemic

The simple account given by Wood, unencumbered by any bubonic plague baggage, corresponds exactly with the origins of the plague at Penrith (section 5.3) where a stranger, Andrew Hogson, took lodgings in the town in the autumn and then died of plague. Wood's version of events gains credence when the next victim to die at Eyam was Edward Cooper, son of Mary Cooper, aged 3 years, indicating a high household contact rate. However, his brother, Jonathan Cooper, aged 12 years, was not buried until 28 October 1666 and so was clearly a tertiary case, probably infected by his younger brother in the terminal stages of his illness (Fig. 10.2). Alexander Hadfeild (Mary's second husband) died much later in the epidemic, from a reinfection, on 3 August 1666. Mary Cooper apparently survived the outbreak, in spite of losing her entire family, and married, for the third time, Marshal Hole on 5 June 1672. Clifford (1989) says in contradiction to this that Mary was married for the third time to John Coe, although this is not recorded in the parish registers.

The epidemic spread through the village in a clearly defined and predictable way, quite unlike the erratic behaviour of bubonic plague. Viccars

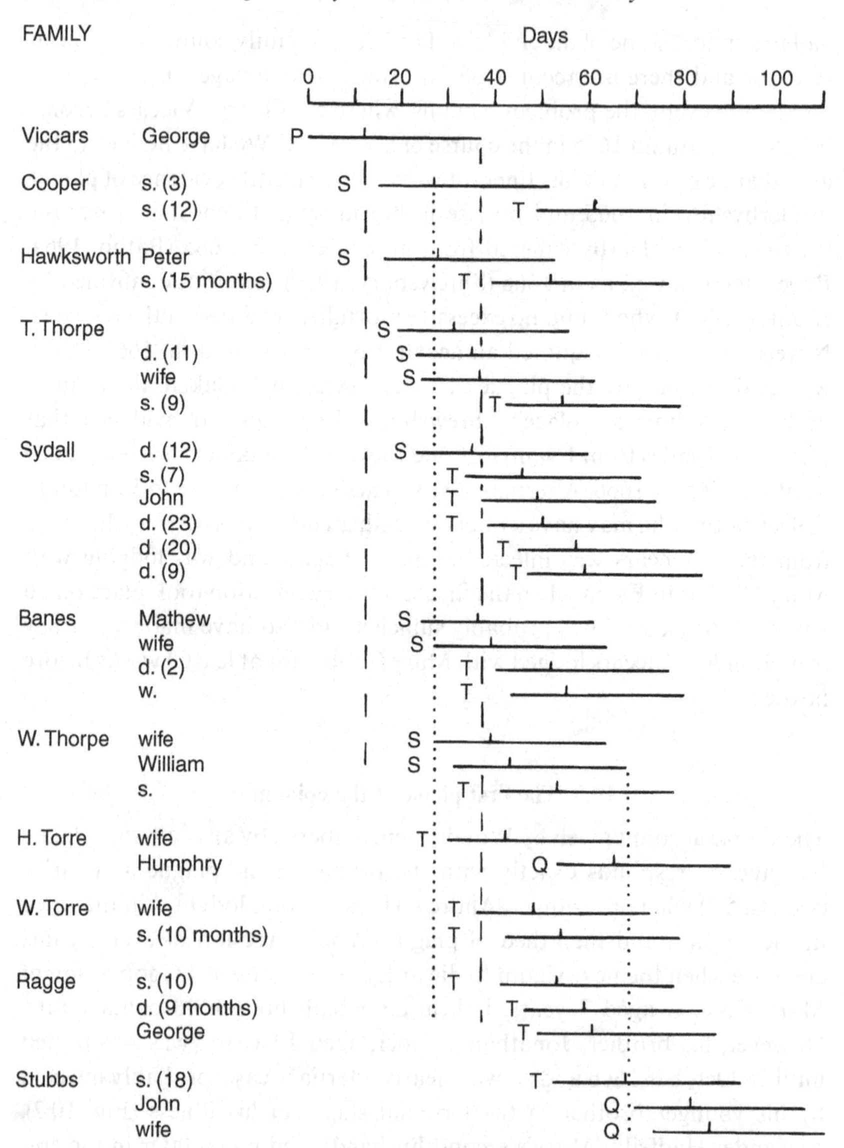

Fig. 10.2. Suggested sequence of the infections during the first phase of the epidemic at Eyam. Scale: days after 1 August 1665. For further details and abbreviations, see Fig. 5.6.

introduced the infection into five other families: Hawksworth, T. Thorpe, Sydall, Banes and W. Thorpe, see Fig. 10.2. These sequential family infections, as the first phase of the outbreak gathered momentum in the late summer of 1665, may be summarised as follows:

(i) The Hawksworth family lived near Mary Cooper (Clifford, 1989). Peter Hawksworth senior was a secondary case, the first to be infected outside Mary Cooper's house by Viccars, who, in turn, infected his 15-month-old only son Humphry (a tertiary) who died on 17 October. His wife Jane apparently survived.

(ii) Thomas Thorpe was another near-neighbour of Mary Cooper (Clifford, 1989); his daughter Mary (aged 11) and his wife Elizabeth were all co-secondaries infected by Viccars. The son, Thomas (aged 9 years), was clearly a tertiary infection within the family. However, three other children of Thomas Thorpe died of plague 6 months later during the second phase of the outbreak: Alice (aged 13 years) was buried on 15 April 1666 and she had probably infected her two brothers, Robert and William, who were both buried on 2 May 1666.

(iii) The Sydall family lived on the opposite side of the road to Mary Cooper and the house still stands today (Clifford, 1989). Sarah (aged 12 years), daughter of John Sydall, was infected by Viccars on about 24 August 1665 and she probably then sequentially infected the son Richard (aged 7 years), John Sydall (head of household), daughter Ellen (aged 23 years), daughter Elizabeth (aged 20 years) and daughter Alice (aged 9 years) as co-tertiaries (Fig. 10.2). John Sydall's wife Elizabeth, appears to have escaped the outbreak, in spite of nursing six members of her family, and she married John Danyell on 24 April 1666, just 5 days before the death of her daughter, Emmott (aged 22 years), who had survived the first phase of the outbreak.

(iv) Mathew Banes and his wife, Margaret, were co-secondaries infected by Viccars and they then infected their daughter Martha (aged 2 years) and the widow Mary Banes (possibly the mother of Mathew Banes) as tertiaries.

(v) William Thorpe may have been related to Thomas Thorpe, although we have found no evidence in the registers to support this. His wife Mary was infected about 28 August 1665 and William 3 days later. They were probably not infected by Thomas Thorpe (see Fig. 10.2), but were probably two further secondaries infected by Viccars. The burial of Humphry, son of William Thorpe, on 17 October is recorded only in the Bishop's Transcripts.

Presumably, either George Viccars visited these nearby houses in his work as a tailor or they came to visit their neighbour Mary Hadfeild, née Cooper.

The plague struck next at the families of Humphry Torre and William Torre; they may well have been father and son, respectively, and Peter Hawksworth (secondary, died 23 September, see above) was married to Jane Torre, the daughter of Humphry Torre. For this reason, we believe that the Torre families were not infected by Viccars but that the plague was brought by Peter Hawksworth immediately he became infective. Sythe, wife of Humphry Torre, was infected about 31 August 1665 and she then transmitted the disease to her husband Humphry (a quaternary case). Hawksworth also infected Amie, wife of William Torre (his sister-in-law) and Humphry (aged 10 months) their son; William Torre died on the same day as his wife. The Torre families were related to Rowland Mower (section 10.5) and may also have been connected with the Sydalls, because Rowland Torre was betrothed to Emmott Sydall.

Three members of the family of George Ragge were tertiary infections who could have been infected by any of the early secondary infectives (Hawksworth, Thomas Thorpe, Sydall and Banes families). Tradition suggests that the Ragges lived next door to the Sydall family so that they may have been infected from any of the little collection of cottages where the outbreak began. Son Jonathan (aged 10 years), daughter Alice (aged 9 months) and George Ragge himself died over a space of 9 days (see Fig. 10.2). His daughter Ellen and wife Elizabeth (who may have died in 1699) apparently survived.

Likewise, the Stubbs household was another tertiary infection that was introduced about 25 September 1665 from any of the families listed above, plus the families of William Thorpe and the Torres. Hugh Stubbs (aged 18 years), son of John and Ann Stubbs, was infected about 25 September and he probably infected both his parents (quaternaries).

The contact rates, R_0, for the first phase of the epidemic are: primary to secondaries = 10; secondaries to tertiaries = 1.8; tertiaries to quaternaries = 0.2.

There are striking similarities between the initial phases of the type (ii) epidemics at Penrith (section 5.3) and Eyam. Both were northern upland communities where the epidemic began in the autumn. Both were started by the arrival of a symptomless, infective stranger who took lodgings and *stayed* in the town or village. They contracted the disease some distance away and the long incubation period allowed the plague to be transmitted. These epidemics were not begun by, say, a drover passing through. Viccars was resident in Eyam for at least 22 days and up to 34 days; Hogson stayed in Penrith for at least 15 and probably up to 33 days. Their abodes can reputedly be identified today. It is impossible that pneumonic plague could be brought into the community in this way by terminally ill, infected persons on their death beds with 48 hours to live.

Both Hogson and Viccars first infected people in the house where they were lodging (high household contact rate), probably soon after they arrived, and so began the outbreaks but, whereas Hogson at Penrith infected only the Railton family (where he was probably lodging) with three secondaries, Viccars infected five other households (all living nearby) at Eyam with a total of 10 secondaries. For a plague to spread and be perpetuated it must be transmitted to other households; continuing infections within a household are a dead end. The difference between the two localities may be because the Eyam outbreak began 15 days earlier and the warmer weather of August promoted the spread of the disease, or it might be thought at first that it was because Viccars was entering neighbouring cottages in his capacity as a travelling tailor and so infecting these families. However, over a period of 3 weeks a travelling tailor might be expected to work more widely throughout the village. Possibly, as we suggest above, these families where the secondaries were established were infected when they visited their neighbour Mary Cooper in the cottage where Viccars was working. The difference between the numbers of secondaries meant that the outbreak was established more quickly at Eyam and so produced a larger autumn peak there than at Penrith.

Once blind acceptance of bubonic plague has been abandoned, the sequence of events faithfully catalogued in the parish registers fit readily into the pattern of a 'standard' infectious disease which was transmitted person-to-person.

10.4 The second phase: maintenance of the epidemic through the winter

As in the 1597–98 plague at Penrith, the epidemic was only just maintained at Eyam through the winter of 1665–66 (Fig. 10.3). Hugh Stubbs (aged 18 years; died 1 November 1665) probably infected both his parents (Fig. 10.2) but Hannah Rowland (aged 15 years), who died only 4 days later on 5 November, was not infected by Hugh Stubbs. It is impossible to suggest who may have infected her because there were some 26 victims who died between 29 September and 24 October who could, potentially, have been the source. These two teenagers were critically responsible for the continuation of the epidemic, not only within their families but also, more importantly, infecting other families (Fig. 10.3). Elizabeth Warrington (aged 18

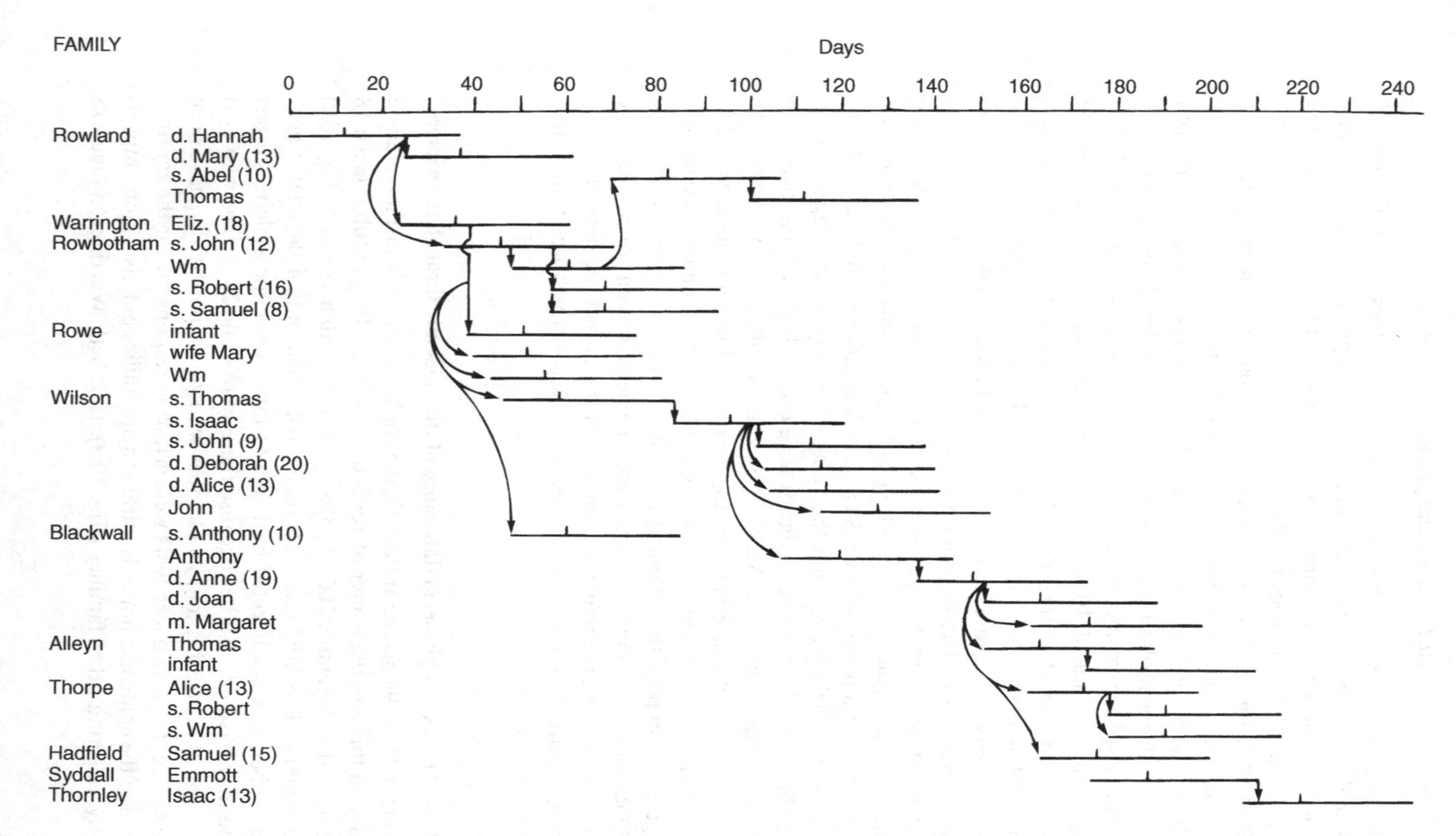

Fig. 10.3. Suggested sequence of infections during the winter in the second phase of the epidemic at Eyam. Scale: days after 30 September 1665. For further details and abbreviations, see Fig. 5.6.

years) was infected on 23 October 1665 by either Hannah Rowland (Fig. 10.3) or Hugh Stubbs (Fig. 10.2); she was the same age as Hugh and represents a crucial link in the chain as the epidemic limped along through November. Firstly, she infected sequentially the three members of the Rowe family probably by visiting frequently; William Rowe had married May Warrington (perhaps the elder sister of Elizabeth) on 16 June 1665 and their infant was the first to die on 14 December, followed by the two parents on 15 and 19 December. Secondly, Elizabeth Warrington probably infected Anthony Blackwall (aged 10 years), who died on 24 December 1665 (Fig. 10.3), although he may have been infected by Mary Rowland (aged 13 years; see Fig. 10.3). Anthony Blackwall may not have infected anyone else. Thirdly, Elizabeth Warrington probably infected Thomas, the son of John Wilson on about 15 November who died on 22 December. We see that because of the long incubation period the plague proceeded in a desultory fashion with few deaths in November and December.

Meanwhile, the infection had continued through the Rowland family; Mary (aged 13 years) was infected by her sister Hannah, but neither could have infected their brother Abel, aged 10 years (Fig. 10.3), who must have contracted the disease from another source. Abel infected his father.

John Rowbotham (aged 12 years) was another early infection (on 2 November) in the second phase of the epidemic. He was infected by either the parents of Hugh Stubbs (Fig. 10.2) or by Hannah Rowland. He died on 9 December having infected first his father William and then his two brothers Robert and Samuel, who both died on 1 January 1666 (Fig. 10.3).

The epidemic continued through January 1666 (when there were only two plague deaths) solely because of two key people: Anthony Blackwall (father of Anthony, above) and Isaac Wilson (brother of Thomas, above); see Fig. 10.3.

Isaac Wilson (age unknown) was infected about 22 December 1665, the day that his brother Thomas died, and he probably caught the disease from him but, among the few infectives around in December, the Rowbotham family or Abel Rowland might have been responsible (Fig. 10.3). Isaac infected his brother John, two sisters and his father John (who did not die until 1 March 1666). More importantly, he was the only person who could have infected Anthony Blackwall senior, thereby maintaining the link in the chain of events. Anthony Blackwall could not have been infected on 15 January 1666 by his son Anthony, who had already died on 24 December (see above, Fig. 10.3). The Blackwall family now succumbed. The daughter Anne (aged 19 years) was the next tenuous link in the chain; she was infected on 13 February and died on 22 March 1666 (note the long interval in plague mortality) having infected her mother and sister Joan. Critically, Anne Blackwall reinfected the family of Thomas Thorpe deceased (see above) and also introduced the plague into the Alleyn family (Fig. 10.3). Thomas Alleyn was infected on 1 March and died on 6 April having infected his infant child.

Alice Thorpe (aged 13 years), both of whose parents had died in the preceding September (section 10.3) was infected on 11 March and died on 15 April 1666. She infected her two brothers, Robert and William on 26 March and they both died on 2 May 1666. The disease was tenuously maintained, therefore, all through the period December 1665 to May 1666 when there were few deaths.

The next slender link in the chain of events was the death of Emmott Sydall (aged 22 years; Clifford 1989) on 29 April 1666. She had survived the first outbreak of plague in her family in the preceding October (see above) but now was infected on 24 March either by Alice Thorpe (it will be remembered that the Thorpes and Sydall families were originally near neighbours) or by Thomas Alleyn. Emmott Sydall's mother Elizabeth had also survived the October outbreak and she married a widower, John Danyell (who later died of plague on 5 July), on 24 April 1666, just 5 days before Emmott died and presumably before she showed symptoms. This wedding was probably a critical event, because the next link in the chain was Isaac Thornley (aged 13 years), who was infected on this day, either by Emmott Sydall or by Robert and William Thorpe. Did Isaac Thornley (who subsequently spread the epidemic to some 15–17 people and so caused the explosion in summer 1666; Fig. 10.4) contract the disease from Emmott Sydall when they both attended the church for the wedding of her mother?

Elizabeth Danyell (née Sydall) died later from the plague intestate on 17 October (one of the last victims) but about the day of her death she made a verbal declaration to her neighbour Rebecca Hawksworth, who testified to that effect and a Letter of Administration was drawn up which affirmed that her only surviving son Josiah (aged 3 years) was to be put in care of her trusted friend Robert Thorpe and that her estate was to be used for his upbringing (Clifford, 1989).

As at Penrith in 1597–98, we see that, when we have full details, the plague at Eyam persisted from November to the following May only by the skin of its teeth. Household contact rates remained high but the critical interfamily infectivity was very low over this period. Figure 10.3 shows how the plague, thanks to its long incubation period, persisted at Eyam over 6

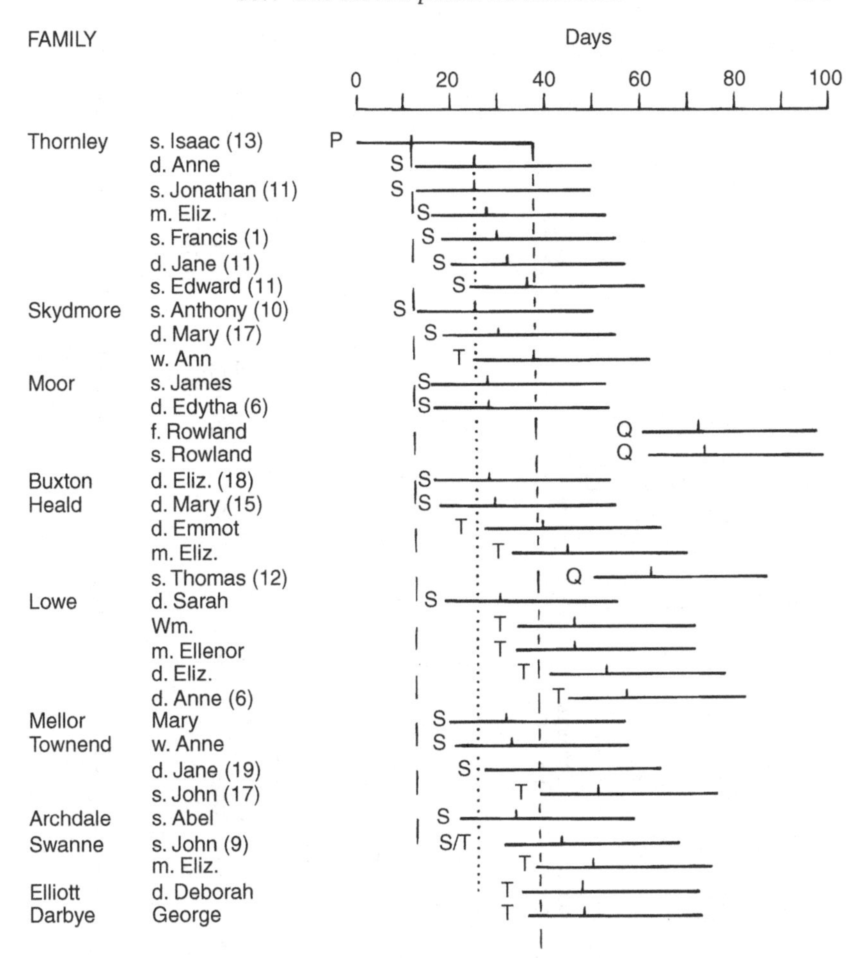

Fig. 10.4. Suggested sequence of infections during the start of the third phase of the epidemic at Eyam. Isaac Thornley is regarded as a new primary case who initiated the explosion in the late spring of 1665. Scale: days after 24 April 1665. For further details and abbreviations, see Fig. 5.6.

months via a few crucial transmissions to new families. There must have been many unrecorded autumnal outbreaks of plague in England in the 16th and 17th centuries that did not persist over winter.

It is interesting that so many of these interfamily infections at Eyam were via the teenage children. They were responsible for going to each other's houses during the winter and so maintaining and spreading the epidemic.

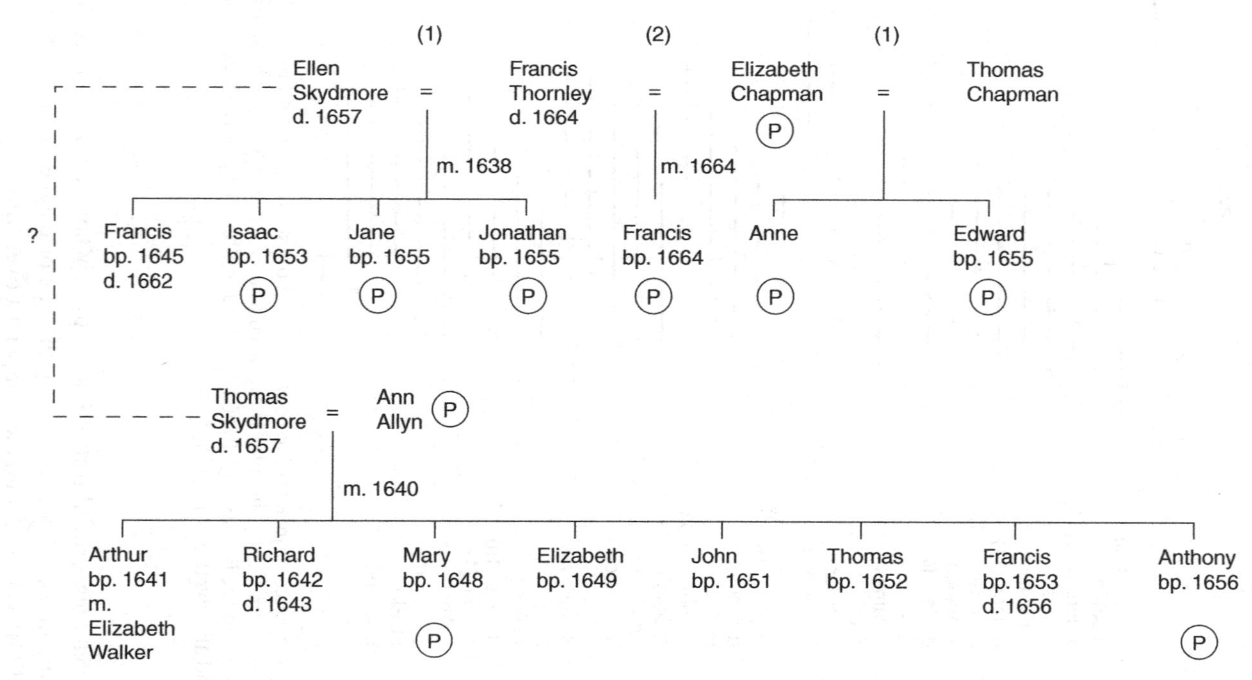

Fig. 10.5. Interrelationships of the Thornley and Skydmore families. Plague victims marked with P in a circle. m., married; d., died; bp., baptised.

10.5 The third phase: explosion of the epidemic in summer 1666

Isaac Thornley (aged 13 years), who became infectious on 6 May 1666, directly infected 16–18 individuals in eight families in addition to his own, a dramatic change in the epidemiology of the disease, as shown in Fig. 10.4. Since the epidemic re-started in phase three with the infection of Isaac Thornley, he is regarded as a new primary. The infections began simultaneously on the following day, 7 May (thereby confirming the length of the latent period) in the Thornley and Skydmore families who were interrelated, as shown in Fig. 10.5. Francis Thornley married Ellen Skydmore in 1638 and, when she died in 1657, he remarried in February 1664 Elizabeth Chapman, a widow who had two children by a previous marriage. They had a single son, Francis, who was baptised in November 1664 but by that time the father, Francis Thornley, had died in September 1664. Thus, at the outbreak of the plague, 12 months later, Elizabeth Thornley had six children in her care, two of her own who took the Thornley surname, one that she bore to Francis and three that were born to Francis Thornley and Ellen Skydmore. This entire family was to die of the plague. Anne and Jonathan (aged 11 years) were both infected on 7 May (one day after Isaac became infectious), the mother Elizabeth on the 10th, Francis (aged 1 year) on the 12th, Jane (aged 11 years, twin sister of Jonathan) on the 14th and Edward (aged 11 years) on the 18th.

Thomas Skydmore was probably the brother of Ellen Skydmore and he married Ann Allyn in 1640. She may have been related to Thomas Alleyn who died of the plague on 6 April (see section 10.3). Thomas Skydmore had died in 1657, leaving his widow with seven surviving children (Fig. 10.5). Ann Skydmore clearly retained close connections with the Thornley family because Anthony (aged 10 years) was also infected on 7 May and his sister Mary (aged 17 years) was infected on 18 May 1666. Again, therefore, the infection was carried between the families by the children. Ann Skydmore, the mother, was probably a tertiary infection from her son and the remaining five children apparently survived.

James and Edytha (aged 6 years), children of Rowland and Elizabeth Mower (= Moor), were both infected by Isaac Thornley on 10 May and were buried on 15 June 1666. Rowland Mower died later on 29 July with his son Rowland on the following day but they were not infected by the other children. The will made by Rowland Mower on 26 June 1666, 11 days after his two children died, throws further light on how the infection was spread at this time and also on the interrelationships in the village. He was a cooper and made the following bequests:

To John Torre, brother-in-law, 10 shillings [Mower's wife Elizabeth was a member of the Torre families, who died in the first phase of the plague.]

To Robt. Marsland, my brother, 12 pence [Not traced at Eyam.]

To Elizabeth, wife of Henry Clarke, my sister, 10 shillings [Not traced at Eyam.]

To Thomas, Robert and Edyth Bockinge, the children of Francis Bockinge, 5 shillings each [Evidently Rowland Mower was friendly with the Bockinge family; their daughter Edyth was infected on 20 June and died on 17 July; son Thomas died on 30 July and son Robert died on 3 August 1666. Edyth Bockinge could have infected Rowland Mower and his son and so reintroduced the infection into his family, thereby bridging the gap in the sequence of deaths.]

To the children of James Mower [family not affected by the plague] Thomas Ragge and William Abell [both families infected] 12 pence each

That Jane French, my tenant, 'shall have and enjoy the house wherein she now dwelleth' at a yearly rent of two pence [She survived the plague and died in 1675.]

To his wife Elizabeth [who survived] and his son Rowland [who subsequently died] all houses, lands, real estate and worldly goods. But if they should both die he re-distributed his bequests, including four pounds to his true and lawful apprentice George Cowper [who survived the plague and married in 1672].

Elizabeth Buxton (aged 18 years) daughter of John and Katharine Buxton was also infected by Isaac Thornley on 10 May (Fig. 10.4). Two daughters of Robert and Elizabeth Heald were next infected by Isaac Thornley, Mary on the 11th and Emmott on the 21st May. There were two subsequent tertiary infections and one quaternary infection in this family.

Sarah, the daughter of William and Ellenor Lowe, was then infected by Isaac Thornley, on 12 May and she probably then infected her parents and both her two sisters, all of whom died in early July.

Mary Mellor was next infected by Isaac Thornley on 13 May and died on 18 June.

Anne Townend, a widow, and her daughter Jane (aged 19 years) were next infected by Isaac Thornley on the 14 and 20 May. Her son John was a tertiary infection.

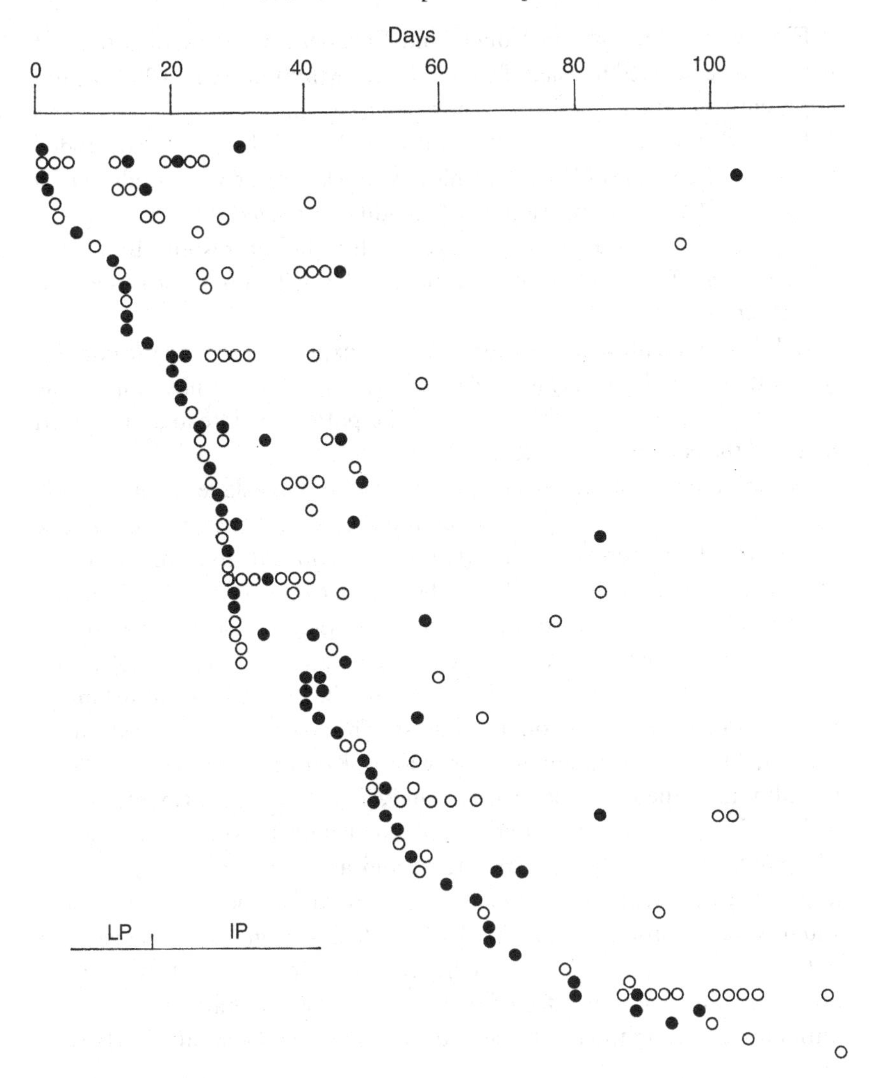

Fig. 10.6. Diagram to show the deaths in the families in the final stages of the epidemic at Eyam, following the infections shown in Fig. 10.4. Each horizontal line represents one family, arranged sequentially in order of the first victim. Closed circles, adults; open circles, children. Scales: days after 5 July 1666; LP, latent period; IP, infectious period.

Finally, Abel, the son of Robert and Elizabeth Archdale, was the last person to be infected by Isaac Thornley on 15 May and died on 20 June but apparently no other members of the family were affected.

It is evident that the plague at Eyam entered its third phase and exploded in June 1666 because of Isaac Thornley, whose key role corresponds to that of George Viccars, who originally brought the plague into the village 9 months before. Once again, it was a youngster who was responsible for the transmission of the disease between the families with a total of 17 secondary infections.

In Fig. 10.6 we illustrate how the epidemic exploded through the population, with 56 deaths in July and 78 in August 1666, but it is now impossible to determine the lines of infection and the pattern of spread as the third phase of the epidemic exploded.

The inhabitants of the extreme western end of the village, an area called Shepherd's Flatt, were few in number and they shut themselves up in their houses and did not cross a small stream that divided them from the village at Fiddlers Bridge. The story of how the plague came to two families on the Flatt has been embroidered, but the registers suggest that again a youngster introduced the disease. Robert Kempe, aged 10 years, son of Lydia Kempe, widow, died first on 31 July 1666 and he infected the entire family: Elizabeth (aged 19 years) on 11 August, Thomas (aged 23 years) on 12 August, Michael on 15 August and widow Kempe on 22 August. Their neighbours on the Flatt, the Morten family, may have been also infected by Robert Kempe; Sarah (aged under 4 years) died on 18 August, her mother Margaret (née Bagshaw) on 20 August and an infant on 24 August. The father, Mathew Morten, was the only survivor at Shepherd's Flatt (Wood, 1865). Rebecca Morten of Shepherd's Flatt died of plague on 4 August 1666 but we have not been able to trace her relationship with Mathew Morten. She could not have been infected by Robert Kempe, so that there were two initial infections on the Flatt; she may have infected the family of Mathew Morten.

Two features that have been recorded in our analyses of other plagues are also evident at Eyam. Firstly, during the third phase of summer 1666 the families differed sharply; some showed the typical high household contact rate, with secondary infections through most of the family (e.g. the Talbotts, Bockinges and the Kempes); many families, in contrast, had only the single initial infection; a few families, in contrast again, had multiple *initial* infections in which the entire family may have died (e.g. the Hancockes and Naylours).

Secondly, when we examine the plague deaths in a family over the

12-month period of the epidemic, we see that, when the first waves of the infections in some families were completed, there were some members who survived but were then infected from another source and died, sometimes many months later. Some of these may have been orphans who were taken into other people's houses. The majority of these late reinfections must have been exposed when the plague first struck the family; were they resistant or did they contract the disease and recover? If so, why did they succumb later in the epidemic? One explanation might be that the initial family infection was in the first phase in autumn, when infectivity was low, whereas the later victims succumbed in the summer of 1666 when the infective agent was more virulent. Examples are Alexander Hadfeild (Mary Cooper's second husband) who died on 3 August, and Alice, Robert and William Thorpe (see Fig. 10.2), who died in April and May 1666. However, this explanation will not suffice when the first infection in the family was in July or August (the Darbye, Taylor, Talbott and Glover families).

We see, again, that the basic unit of the epidemic was the household; the disease usually spread readily within this unit because of a high contact rate with a resultant high mortality to add to the plague statistics. But for the plague to be perpetuated it had to be transmitted to other households where, except in high summer, the effective contact rate appeared to be lower. This difficult transmission during the autumn and winter was most readily achieved into the households of relatives. Otherwise, the infective agent was most frequently transmitted between households in the winter and spring at Eyam by the youngsters. Unbeknownst to everybody they would have been moving around, apparently completely healthy, in the infectious state for about 20 days.

10.6 Percentage mortality of the population during the epidemic

The number of those dying of plague is recorded as 260 in the Eyam parish register but it is difficult to estimate what proportion of the population this represented. Wood (1865) concluded that over 80% of the population had died and that there were only 83 survivors but this has been queried by several authors. Bradley (1977) has shown that at least 48 burials were of families living in Foolow and Woodlands and so it is not clear whether Wood was referring to the township of Eyam or the whole parish.

Clifford (1989) has summarised the statistical evidence as follows:

(i) The Eyam Hearth Tax return for 1664 records 59 taxed and 101 poorer, untaxed households, which, he suggested, indicated a population of about 800, i.e. plague mortality = 33%. It was possible to reconstruct the listed families and then to trace the subsequent history of the survivors of the plague which he estimated as 430, i.e. plague mortality = 38%.

(ii) Batho (1964), analysing the Comptom Ecclesiastical return of 1676, estimated that the population of Eyam 10 years after the plague was about 750 adults and concluded that considerably more than 300 survived the plague.

(iii) The Rev. John Green, rector of Eyam and a contemporary of Wood, concluded from a study of the annual burial records that the population size was considerably in excess of 350 in 1665. Analysis of the registers for the 3 years preceding the plague shows that the mean annual burials and baptisms were 27 and 47, respectively, in broad agreement with Clifford's estimate (above) of a population of about 750. With such a marked excess of baptisms over burials during 1662–64, Eyam was a rapidly expanding community.

In conclusion, plague mortality at Eyam probably lay between 33% and 38%, a devastating blow to what was probably a naive community, but of the same order of magnitude as at Penrith in 1597–98.

10.7 Public health measures during the plague at Eyam

It is well known that in late May or early June 1666 the Rector of Eyam, William Mompesson (aged 28 years) persuaded his parishioners to draw a cordon sanitaire and to confine themselves within a circle of about half a mile around the village (Wood, 1865). The Earl of Devonshire arranged for food to be left at his own expense at the southern boundary of the village. Other supplies were left at a well beside the top road to Grindleford, about a mile outside the village and at an ancient stone circle at Wet Withens. When payment was required, the money was placed if possible in running water; alternatively, holes were drilled in the Eyam boundary stone and the money was covered with vinegar to act as a disinfectant. Pest-houses were erected.

These quarantine measures were effective and there were no plague deaths outside the parish, although there were at least 48 burials in the outlying hamlets of Foolow and Woodlands (Bradley, 1977). None of the proponents of the bubonic plague hypothesis seems to realise that these measures would be effective only against an infectious disease spread person-to-person and that a cordon sanitaire would have no effect on a hypothetical rat population which would be free to move to adjacent villages and so continue the spread of the epidemic. These quarantine measures were not an innovation, unique to Eyam, but had been practised routinely since the previous century, as we have seen at Penrith, Carlisle and York. Many villages and towns in England today still have their plague stones; it is just that Eyam has had better public relations agents to promote its story.

In reality, these quarantine measures were not enforced in the early days of the epidemic; Wood (1865) recorded that some people did leave the village in the spring including the Bradshaws of Bradshaw Hall, and a number of children were sent away, including those of the rector William Mompesson. A few others fled to the neighbouring hills, erected huts and dwelled therein until the approach of winter (Bradley, 1977).

The second decision taken in June 1666 by the villagers was that there were to be no further organised funerals and burials in the churchyard. People were advised to bury their own dead in gardens, orchards or in the fields (Clifford, 1989). Marshall Howe, a lead miner, living in Townhead, is said to have contracted the disease but recovered (although he subsequently lost his wife and son) and, since he believed himself to be immune, he volunteered to dispose of the bodies where the families were unable to perform this task, and he then claimed a burial fee and frequently appropriated their chattels. Wood (1865) recorded that when he was dragging out the body of a man called Unwin, whom he believed to be dead, his victim regained consciousness, called out for a drink and survived the plague.

The third decision taken by the Rector and villagers was to close the church and to hold the services in the open air, thereby avoiding crowding together indoors. The services were held in a natural amphitheatre called Cucklett Delph (now known as the Delph) where the people did not need to crowd together but kept in their family groups, which were separated by some distance from each other and Mompesson preached from a rock (Wood, 1865).

When the plague was over, Mompesson ordered that all woollen clothing and bedding should be burnt and he set an example by burning his own effects so that, as he said in a letter to his uncle, he had scarcely enough to clothe himself. Presumably they believed that the infection might be contained in the clothing because the symptoms of the pestilence could reappear 3 weeks after a person had died.

10.8 The nature of the infectious agent

It is clear from the foregoing that at the time of the epidemic the people of Eyam believed that it was caused by person-to-person infection or perhaps by contagion. In contact with their neighbours and certainly when dealing with strangers it was considered that the minimum safe distance was about 12 feet (Clifford, 1989), presumably out-of-doors.

The symptoms of the plague at Eyam are described by Wood (1865) as shivering, nausea, headache and delirium. In some, these affections were so mild as to be taken for slight indisposition

> until a sudden faintness came on when the maculae, or plague-spot, the fatal token, would soon appear on his breast, indicative of immediate death. But in most cases the pain and delirium left no room for doubt: on the second or third day, buboes, or carbuncles, arose about the groin and elsewhere; and if they could be made to suppurate, recovery was probable, but if they resisted the efforts of nature, and the skill of the physician, death was inevitable.

One of the symptoms of the plague was a sickly, sweet cloying sensation in the nostrils. One evening, when Mompesson and his wife were returning to the rectory, she is said to have exclaimed 'How sweet the air smells', which filled the rector with alarm because he realised its import. She died a few days later. Wood (1865) recorded that a tradition in the hamlet of Curbar, 2 miles southeast of Eyam, during an isolated outbreak of plague in 1632, was that a woman on leaving a house where some person was suffering from plague said to her husband 'Oh! my dear how sweet the air smells'. She took the distemper and died. Mompesson wrote after the plague to his uncle 'My nose never smelt such noisome smells'. It appears that this sweet smell appeared just before the terminal stage when the classic symptoms began. Could it be early evidence of necrosis of the internal organs?

Undoubtedly some contracted the disease and recovered, as recorded by Mompesson in his letters. Marshall Howe believed that he was immune when he had recovered and Wood (1865) recorded that there was a general belief that a person was never attacked twice. Others, apparently were resistant; they were in close contact with infectives but did not contract the disease. Mompesson is an example: throughout the epidemic he went among his parishioners and administered to the dying; he helped them with the writing of their wills; his wife died in his arms. He did not succumb, although for much of the time he had a painful leg infection that his wife feared was a symptom of the plague. Afterwards, he wrote,

> the pest houses have been long empty . . . During this dreadful visitation, I have not

had the least symptom of disease, nor had I ever better health. My man had the distemper, and upon the appearance of a tumour I gave him some chemical antidotes, which operated, and after the rising broke he was very well. My maid continued in good health, which was a blessing; for had she quailed, I should have been ill set to have washed and gotten my provisions.

The following story of Margaret Blackwall (whose house still stands) is told. Apart from her brother, the other members of the family had died earlier in the plague, but she eventually contracted the disease and appeared to be in the terminal stages. Her brother cooked his breakfast and poured the excess fat into a jug that he left in the kitchen and when he left the house he was certain that she would not be alive when he returned. Shortly after his departure, Margaret, who was delirious, was overcome with a great thirst; she left her bed and, finding the warm fat which she took to be milk, drank it greedily, probably causing her to vomit. When her brother returned, not only was Margaret still alive but clearly much stronger; she recovered and remained convinced that the bacon fat had cured her (Clifford, 1989).

It is noteworthy that Mompesson wrote on 20 November 1666 'all our fears are over, for none have died of plague since the eleventh of October . . .'. Evidently, a 40-day interval was considered to be sufficiently long for all risks of infection to be passed; this value may be compared with the estimated 37 days between infection and death in the plague and with the 40 days of official quarantine in London and Carlisle.

11

Continental Europe during the third age of plagues: a study of large-scale metapopulation dynamics

Hitherto, after the pandemic of the Black Death in 1347–50, we have described the plague epidemics in successive centuries in England and have seen that it was an isolated island metapopulation that was completely dependent, if plagues were to be maintained, on regular injections of fresh infectives from overseas, which came, in the main, from the ports of continental Europe. We explore in this chapter the dynamics of the plague in the enormous metapopulation of Europe and show that it can be considered as an aggregation of subpopulations.

Consideration of plagues in Europe is complicated because, certainly by the 16th century, in addition to the 'standard' (haemorrhagic) plague that we have described in England, there were a minority of outbreaks of genuine bubonic plague. A nucleic acid-based confirmation of the existence of *Yersinia pestis* has been obtained in the dental pulp of bodies buried in the Provence region in the 14th, 16th and 18th centuries (Drancourt *et al.*, 1998; Raoult *et al.*, 2000). In addition, Twigg (1984) concluded that bubonic plague was present in some of the epidemics in the 6th century in the coastlands around the Mediterranean. Black rats had not reached northern Europe by the time of the plague of Justinian (Shrewsbury, 1970), but Twigg presumed that they were present in the ports and major cities in the warmer climate around the Mediterranean and would have been able to sustain small outbreaks of bubonic plague.

11.1 Frequency of epidemics in the metapopulation of Europe

Biraben (1975) has assembled and collated an enormous amount of information on the history of the pestilence in Europe after 1347. He has determined the number of localities in which plague epidemics were reported in each year. Obviously this list is not complete, particularly for the

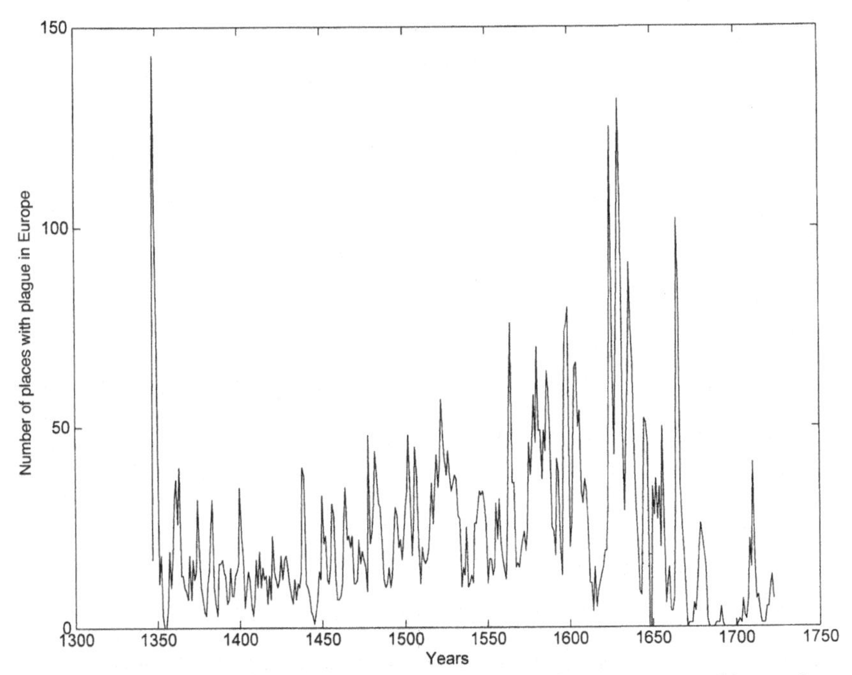

Fig. 11.1. Number of localities in which plague epidemics were reported in northwest Europe, 1347–1722. Area covered: Spain, Portugal, Italy, France, Britain, Ireland, Benelux countries, Germany, Austria, Bohemia, Switzerland, Poland, the Baltic, Scandinavia, Hungary, Croatia, Slovenia, Dalmatia and northwest of the former USSR. Data from Biraben (1975).

14th and 15th centuries, and it includes mortality crises that were not the result of infectious diseases. Nor are the data-series as valuable as the lists of annual plague deaths in a city, as for London shown in Figs. 8.1 and 8.2. Nevertheless the series, shown in Fig. 11.1, illustrates firstly the fluctuating spread of the plague and it can be seen that plague was present somewhere in almost every year in Europe. The infection might be considered to be pseudo-endemic, by which we mean that it was present in a handful of widely scattered places every year in the metapopulation from which infectives travelled out to start epidemics in fresh localities in the following year. Secondly, the basal, endemic level rose steadily over the 300 years after the Black Death, as the plague gradually established itself. Finally, the plague ceased abruptly after 1670, as in England, with only erratic epidemics thereafter and these were not necessarily haemorrhagic plague. The picture is complex, as would be expected, since it represents the sum of the records from several constituent subpopulations that were widely separated and enjoyed different climatic conditions. Nevertheless, spectral

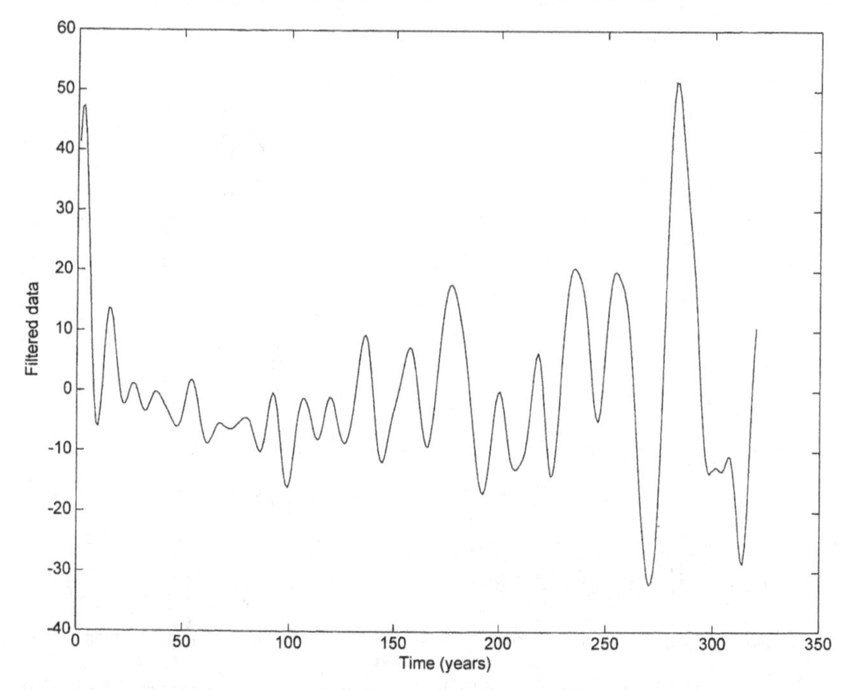

Fig. 11.2. Number of localities with plague in Europe, 1347–1666 (see Fig. 11.1) filtered to reveal a medium-wavelength oscillation. Filter window = 20 to 30 years.

analysis (see section 2.8) of this data-series reveals underlying oscillations:

(i) A short-wavelength cycle, with a period of about 6 years, that persisted from the Black Death until 1646 ($P = 0.05$).
(ii) A medium-wavelength oscillation, with a period of about 12 years, that was significant ($P = 0.001$) during 1347–1447 but was not significant thereafter and disappeared after 1550.
(iii) A non-stationary oscillation in the 20–25 year waveband that appeared after 1475 ($P = 0.05$; see Fig. 11.2) effectively replacing the 12-year oscillation.

11.2 Plagues in France: the endemic situation

11.2.1 Oscillations in the frequency of the occurrence of epidemics

The number of places with the plague in France in each year, taken from the data given by Biraben (1975), is shown in Fig. 11.3, wherein it can be seen that France was the focus and epicentre for the plague in Europe from the time of the Black Death to 1670. Since there must have been many

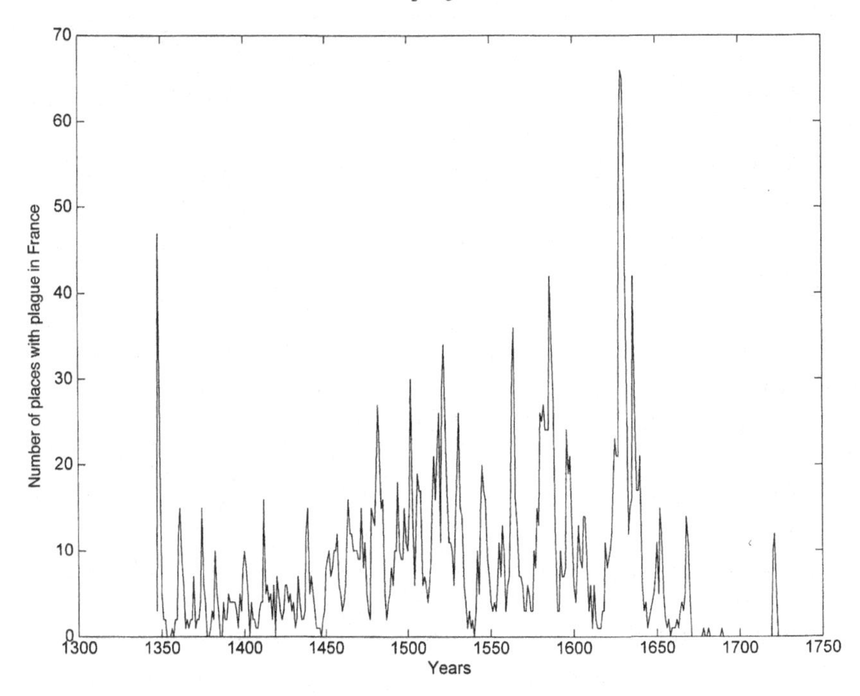

Fig. 11.3. Number of localities in France in which plague epidemics were reported, 1347–1722. Data from Biraben (1975).

localities where outbreaks went unreported, it is probable that plague was pseudo-endemic and cycled round the towns of France throughout this period. Infectives also spread out from France, travelling to adjacent metapopulations by trade routes, across mountain ranges or by boat and so caused sporadic epidemics there.

Inspection of the data-series for France in Fig. 11.3 suggests that there was a change in the character of the outbreaks in 1436 after which the oscillations become more pronounced and this is confirmed by spectral analysis, which shows that the most significant oscillations were:

(i) A strong 22–25-year cycle ($P = 0.005$), which emerged clearly after 1436 and persisted thereafter; it is shown after filtering in Fig. 11.4.
(ii) A 13-year oscillation ($P = 0.005$) detectable throughout the whole period.
(iii) A non-stationary short-wavelength oscillation, the wavelength of which changed as follows: 1346–1436 = 6.4 years; 1437–1506 = 9 years; 1507–1621 = 8.2 years ($P = 0.05$); see Fig. 11.5.

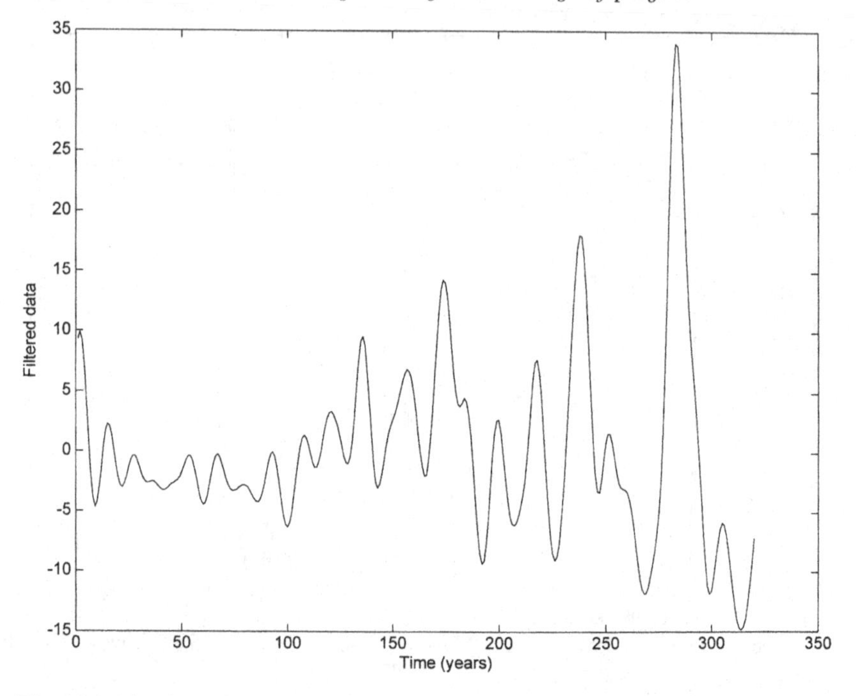

Fig. 11.4. Number of localities with plague in France, 1347–1667 (see Fig. 11.3) filtered to reveal a medium-wavelength oscillation. Filter window = 20 to 30 years.

It can be seen that these oscillations correspond broadly with those detected in the data-series for the whole of Europe and, indeed, we shall suggest that the dynamics of the epidemics in France drove the oscillations in the rest of the vast supermetapopulation of Europe.

There are few firm meteorological data for Europe over much of this period of study but the date of the grape harvest in the vineyards of northern and central France, Switzerland, Alsace and the Rhineland has been taken as proxy for summer temperatures in central continental Europe: late harvest dates are indicative of mostly cold average temperatures during the vine-growth period, April to October (Ladurie & Baulant, 1980). Figure 11.6 shows the number of days after 1 September before the start of the grape harvest plotted annually, 1484–1668. Spectral analysis reveals a 7-year oscillation therein ($P = 0.05$) that was strongly developed after 1573 ($P < 0.005$). This cross-correlates with the non-stationary, short-wavelength oscillation shown in Fig. 11.5 (ccf = -0.3) but with a 1-year lag, i.e. warm summers were followed by widespread plague in France in the following year. This rather weak correspondence does not necessarily

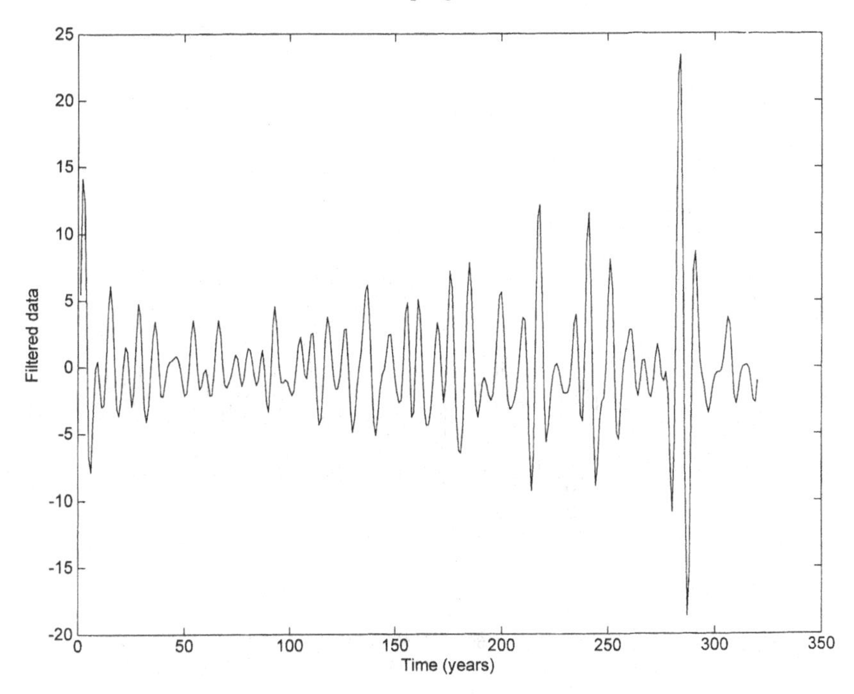

Fig. 11.5. Number of localities with plague in France, 1347–1667 (see Fig. 11.3) filtered to reveal a short-wavelength oscillation. Filter window = 5 to 12 years.

represent a causal relationship between the two oscillations but may be indicative of a mild autumn (and/or winter) during which the disease more readily overwintered.

A data-series for the Paris corn prices, 1431–1688, has been determined by Baulant (1968) and is shown in Fig. 11.7: oscillations were limited between 1450 and 1525 with a low, steady level, but after 1520 there was a marked rise in the trend, with clear oscillations that progressively increased in amplitude. Spectral analysis reveals a short-wavelength oscillation in these corn prices that was initially of low amplitude and non-stationary (wavelength = 4 to 7 years) for the period 1431–1551. Thereafter, the cycles became more regular, with a wavelength of 5.8 years and with oscillations of a much greater amplitude (Fig. 11.8). This short-wavelength cycle in corn prices cross-correlates with the series of the number of places affected by plague in France (see Fig. 11.3) for the period 1431–1531 (ccf = +0.39; see Fig. 11.9), i.e. high corn prices (and hence poor nutrition) cross-correlated with widespread plague in the same year (i.e. at zero lag) but after 1531 this association becomes progressively weaker.

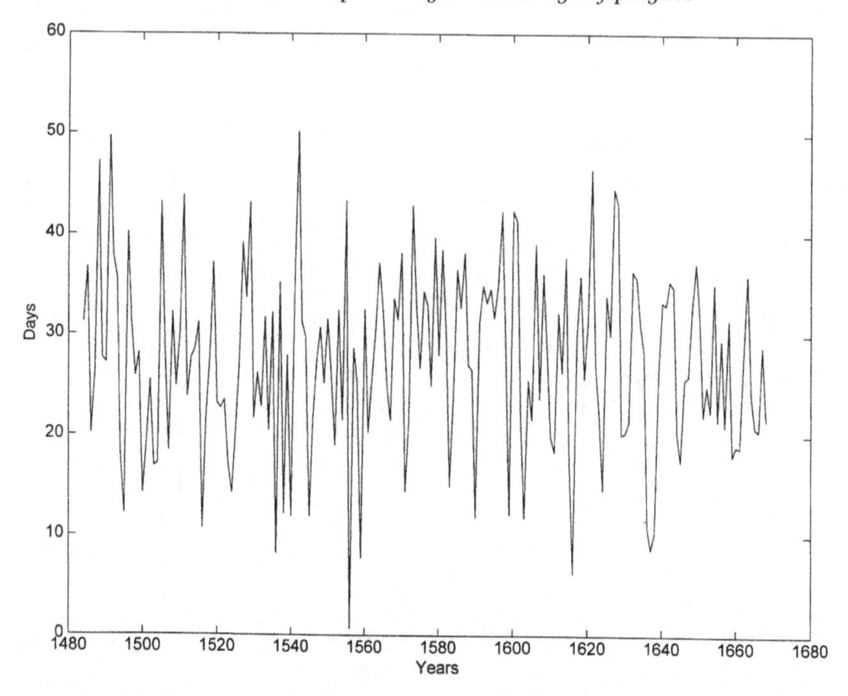

Fig. 11.6. Annual date of the grape harvest in the vineyards of northern and central France, Switzerland, Alsace and the Rhineland, 1484–1668. Ordinate: number of days after 1 September before the start of the grape harvest. Data from Ladurie & Baulant (1980).

Again, this correlation does not necessarily imply a causal relationship but there appears to be some form of weak and possibly interacting association in France between summer temperatures, high corn prices and widespread plague. On the other hand, there is no immediate explanation for the medium-wavelength oscillations in the 22–25-year and 13-year wavebands.

It is generally agreed that the outbreaks of plague in France appeared less frequently and were of milder intensity during the period 1450–1520, which marked the end of the Hundred Years War, the start of a demographic rise and the general reconstruction of areas that had been ravaged by battles, food shortages and natural disasters. The worst epidemics were in 1464, 1478–84, 1494, 1502 and 1514–19 (see Fig. 11.3). An estimated 40 000 people died in Paris in 1466, although it has been suggested (Kohn, 1995) that the death tolls from plague during this period were exaggerated. France was much slower than Italy to introduce preventive measures: Brignoles in Provence was the first town, in 1451, to prohibit the entry of

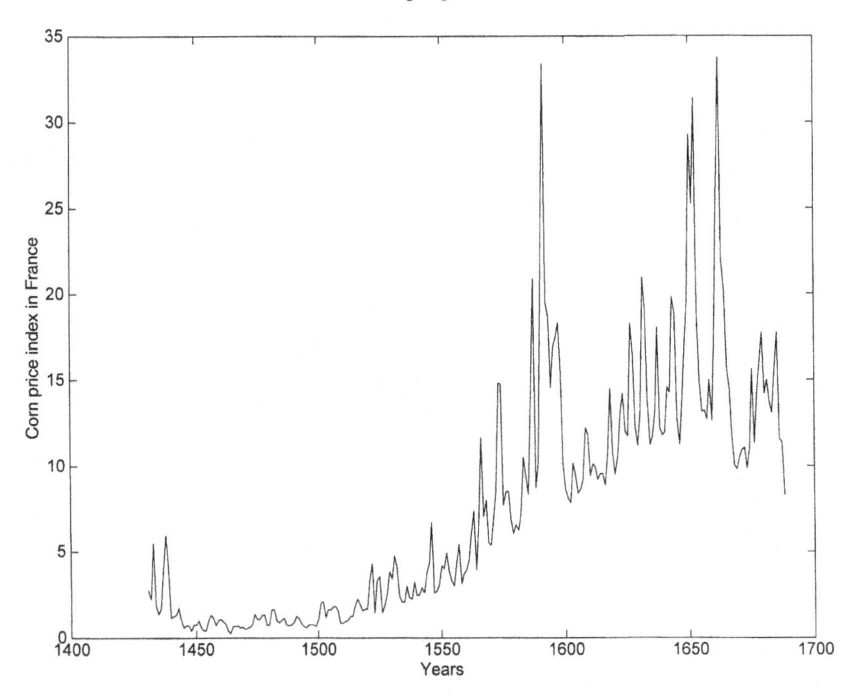

Fig. 11.7. French corn price index, 1431–1688. Data from Baulant (1968).

travellers if they came from a town with plague; subsequently they also expelled sick persons and required a bill of health from travellers. By 1520, many towns had appointed plague bureaux charged with ensuring that public health measures were enforced.

There were frequent and virulent outbreaks in France during 1520–1600 that were accompanied by food shortages, famines, flooding, peasant uprisings and religious wars. Epidemics occurred with greater frequency (see Fig. 11.3) and, during this time, the health bureaux often hired armed men to enforce the plague regulations and to maintain civic order. The epidemic in 1564 in Lyons was particularly deadly and the town was almost paralysed after 2 months, with one-third of the houses closed; some of the victims recovered (showing that not everyone died), only to die of hunger. This plague spread widely through Provence and Languedoc in summer 1564, reaching Nimes in mid-July. Autumn brought a lull, although local outbreaks were reported, most interestingly as late as mid-December until a *severe winter finally ended the epidemic* (Kohn, 1995), showing that the plague was sensitive to winter temperatures even in southern France.

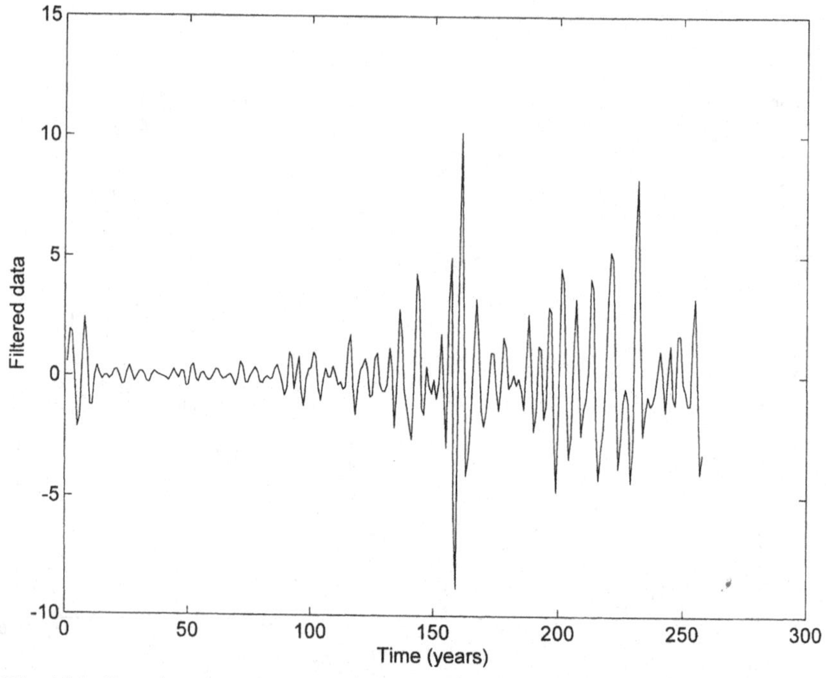

Fig. 11.8. French corn price index, 1431–1688 (see Fig. 11.7) filtered to reveal a short-wavelength oscillation. Filter window = 3 to 10 years.

11.2.2 The 17th century in France

The period 1622–46 saw the most grievous and widespread outbreaks of plague in the history of France (see Fig. 11.3) in contrast to the most serious epidemics that occurred in 1665–66 in England. The miseries were exacerbated by peasant revolts, pillaging by soldiers and virulent outbreaks of other diseases. The epidemics spread throughout France; Biraben (1975) listed almost 400 localities in France that reported outbreaks over the 300 years of the age of plagues and, of these, nearly 300 places suffered at least once between 1622 and 1646. Anti-plague regulations and measures were by now properly established and the poor were more generously treated. Bourg-en-Bresse paid young boys and girls in 1636 to be shut up in newly fumigated houses for 40 days (the standard quarantine period, section 13.3) to test the efficacy of the disinfections.

A serious epidemic struck Lyons in summer 1628; passing soldiers were accused of having carried the plague with them 'as their baggage', perhaps because one of the first reported deaths was that of *a soldier lodging in a*

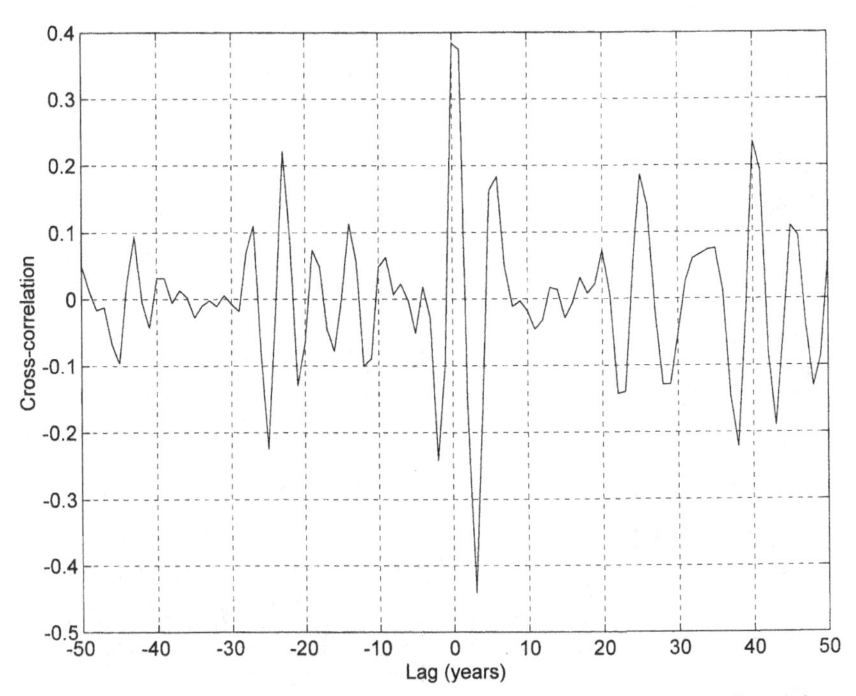

Fig. 11.9. Cross-correlation between the French corn price index (Fig. 11.7) and the number of localities with plague in France (Fig. 11.3), 1431–1531. Filter window = 3 to 10 years.

nearby village. This is an important observation and can be compared with the strangers lodging at Eyam (section 10.2) and Penrith (section 5.1) who initiated the epidemic. The usual health regulations were brought into force: guards were posted at the city gates, health certificates were required and, once again, a 40-day quarantine was imposed. The city became a vast hospital; the streets and houses were strewn with corpses, which sometimes were buried hastily in gardens and cellars. The epidemic abated somewhat at the end of December, but broke out again with great force in early 1629, gradually diminishing from March through the summer months. An estimated 35 000 died in Lyons during the 12-month epidemic (Kohn, 1995).

11.2.3 Regional differences within the metapopulation of France

France represents a very wide area to act as a plague focus, even though infectives would be able to travel freely, particularly in the later years, and there are regional differences in climate within the metapopulation. We have divided France into three areas: a southern zone with a warmer

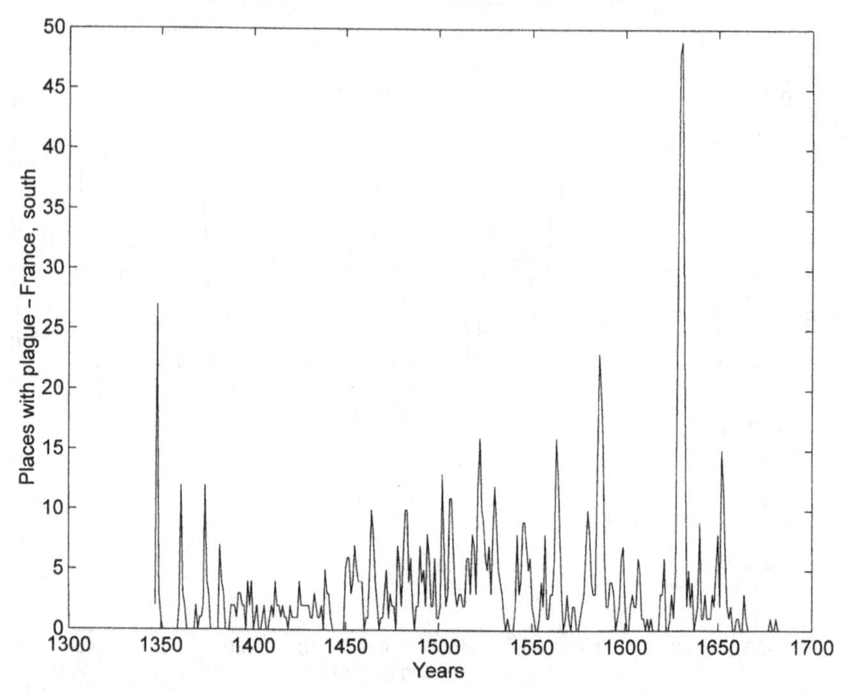

Fig. 11.10. Number of localities with plague in southern France (south of latitude 40° N), 1347–1688.

climate (south of latitude 46° N), a northern central zone and the coastal zone of the north and northwest. The number of localities showing plague summarised in Fig. 11.3 have been subdivided accordingly into these three areas and the respective data-series are plotted in Figs. 11.10 to 11.12 – note the different scales on the ordinates.

Inspection of these plots shows that

(i) The south and central regions suffered most severely in 1628–31 with widespread plague, a period that was preceded and followed by times of fewer attacks.
(ii) The south and central regions exhibited similar patterns of spread and both showed a clear increase in frequency after 1450.
(iii) The coastal region is a smaller area and plagues there were relatively less frequent than in the other regions of France; epidemics were sporadic during 1350–1450 but became more widespread thereafter.

Spectral analysis reveals 21- and 13-year oscillations in the south, central and coastal regions, 1347–1666, with the following statistical significances:

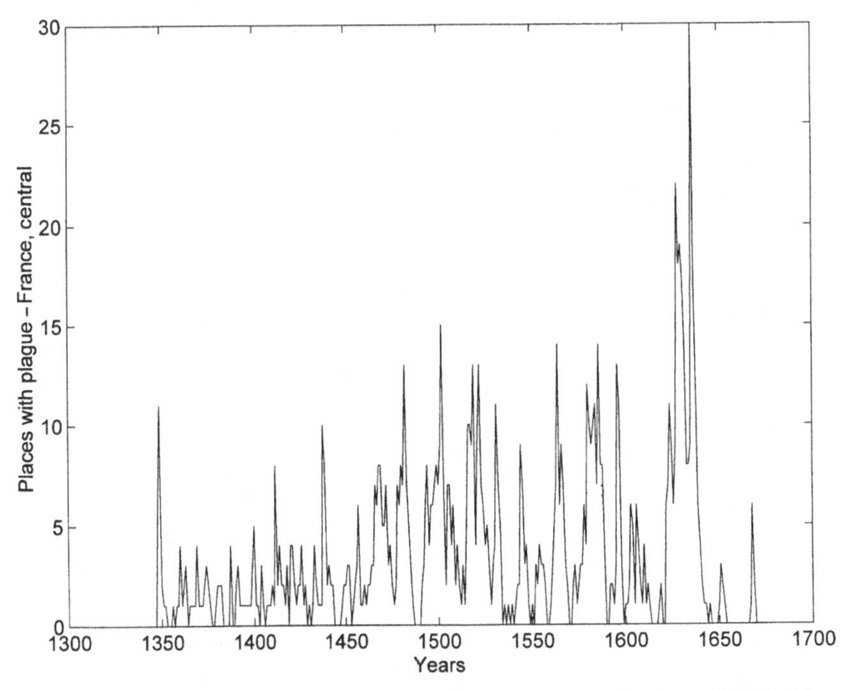

Fig. 11.11. Number of localities with plague in northern/central France, 1347–1688.

Cycle	South	Central	Coastal
21-year	$P = 0.005$	$P = 0.005$	$P = 0.05$
13-year	$P = 0.01$	$P = 0.05$	NS

NS, not significant.

The 21-year oscillation was the stronger and both oscillations were of less importance in the coastal regions of the west and northwest. The filtered medium-wavelength oscillations in southern and central France are shown in Figs. 11.13 and 11.14 respectively; they cross-correlate significantly and both also correlate with the coastal region (ccf = +0.7 at zero lag), suggesting that these cycles were present throughout France.

11.2.4 Plague centroids in France

Difficult aspects of the geographical spread of the plague to measure and present are the direction and velocity of propagation of the epidemics. Centroid movements are one approach to this problem (Cliff, 1995). A

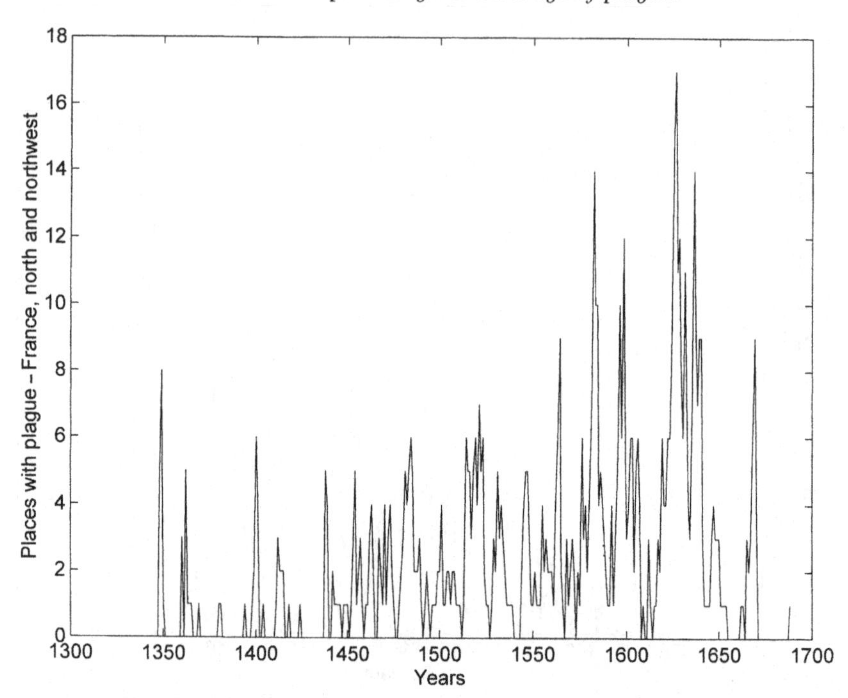

Fig. 11.12. Number of localities with plague in the coastal region of north and northwest France, 1347–1688.

centroid is defined as the mean centre of reported disease incidence and by plotting centroids for successive outbreaks and linking them in sequence the general direction of movement can be displayed (section 2.12). Unfortunately, we do not have a record of the number of cases of plague for each town but only its location, so that the results cannot be properly weighted. Nevertheless, the centroids plotted in Fig. 11.15 for each 50-year period show clearly how the plague moved around the metapopulation over a period of 250 years. For this study, we have included within the metapopulation the towns over the Belgian border, Flanders and Luxemburg because the plague moved readily and frequently to this region.

In the 50 years after the Black Death (Fig. 11.15A), the epidemics were concentrated in a band from north to south of the metapopulation (between longitudes 3° E and 5° E). During 1400–49 and 1450–99 the plague became more centrally localised (Fig. 11.15B and C), a process that continued during 1500–49; plague was reported only in Rouen (on the coast) in 1538 (Fig. 11.15D) and only in Luxemburg in 1540. This central localisa-

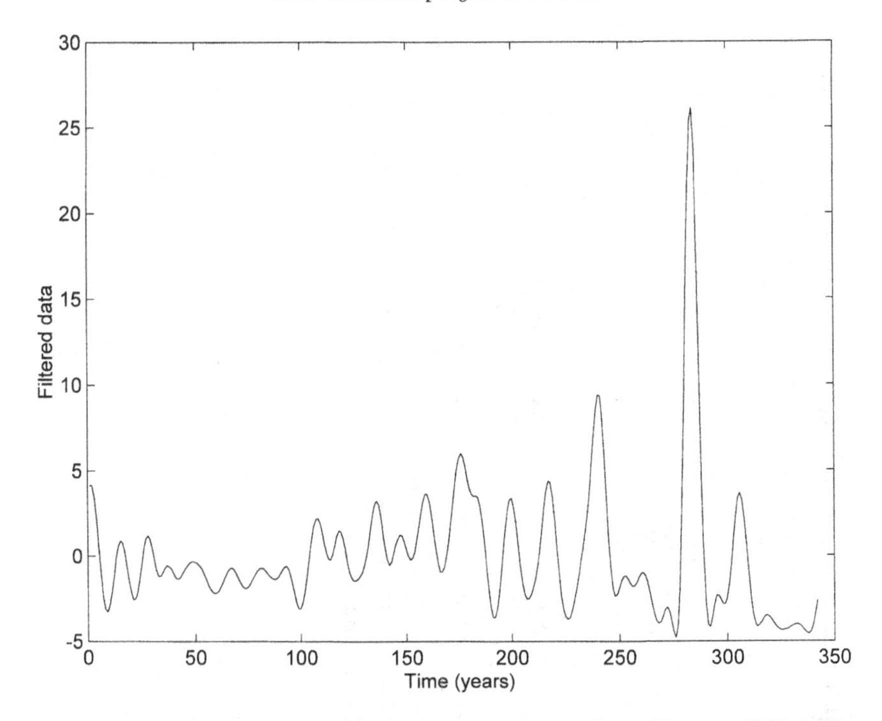

Fig. 11.13. Number of localities with plague in southern France, 1347–1688 (Fig. 11.10), filtered to reveal a medium-wavelength oscillation. Filter window = 16 to 32 years.

tion of the outbreaks continued for the next 100 years, moving slightly to the north and west (Fig. 11.15E), but as the 17th century progressed the centroids became more erratic and widespread.

It can be seen that, although epidemics continued to break out in southern France, the centroids were never located there after 1400; the activity was concentrated around the centre of the metapopulation and there was a considerable exchange of the infection with the Holy Roman Empire to the northeast.

11.2.5 Localities with a high frequency of epidemics

The localities in France where plague was most commonly reported in each cohort are summarised in Table 11.1, which, to some extent, modifies the picture presented by the centroids. Epidemics broke out repeatedly in towns right across the metapopulation (Fig. 11.15) and the disease moved over long distances in each year. The inhabitants of those towns that were

Table 11.1. *Localities in France that recorded the largest number of outbreaks of plague in successive cohorts, 1347–1649*

Cohort					
1347–99	1400–99	1450–99	1500–49	1550–99	1600–49
Strasburg (20)	Bourg-en-Bresse (14)	Bourg-en-Bresse (26)	Toulouse (26)	Amiens (28)	Luxemburg (20)
Paris (16)	Haute-Auvergne (13)	Nantes (24)	Rouen (23)	Troyes (24)	Paris (20)
Avignon (12)	Paris (12)	Toulouse (24)	Tours (21)	Luxemburg (20)	Bordeaux (19)
Burgundy (12)	Strasburg (12)	Paris (14)	Dijon (21)	Orléans (18)	Amiens (18)
Montpellier (10)	Limoges (10)	Villefranche (14)	Nantes (18)	Paris (18)	Angers (18)
Toulouse (10)	Saint-Flour (9)	Châlon-sur-Marne (14)	Nîmes (17)	Bourg-en-Bresse (17)	Lille (17)
Marseilles (7)	Angers (8)	Poitiers (13)	Bordeaux (17)	Cambrai (15)	Nantes (15)
Provence (7)	Amiens (7)	Amiens (12)	Bergerac (15)	Toulouse (15)	Rennes (15)
Limoges (6)	Arras (7)	Limoges (11)	Cambrai (15)	Dijon (12)	Troyes (15)
Luxemburg (6)	Nantes (7)	Strasburg (11)	Amiens (14)	Rouen (12)	Lure (14)
Apt (5)	Poitiers (7)	Troyes (11)	Bourg-en-Bresse (14)	Lyons (12)	Rouen (13)

The number of epidemics suffered by each locality in each cohort given in parentheses.
Data from Biraben (1975).

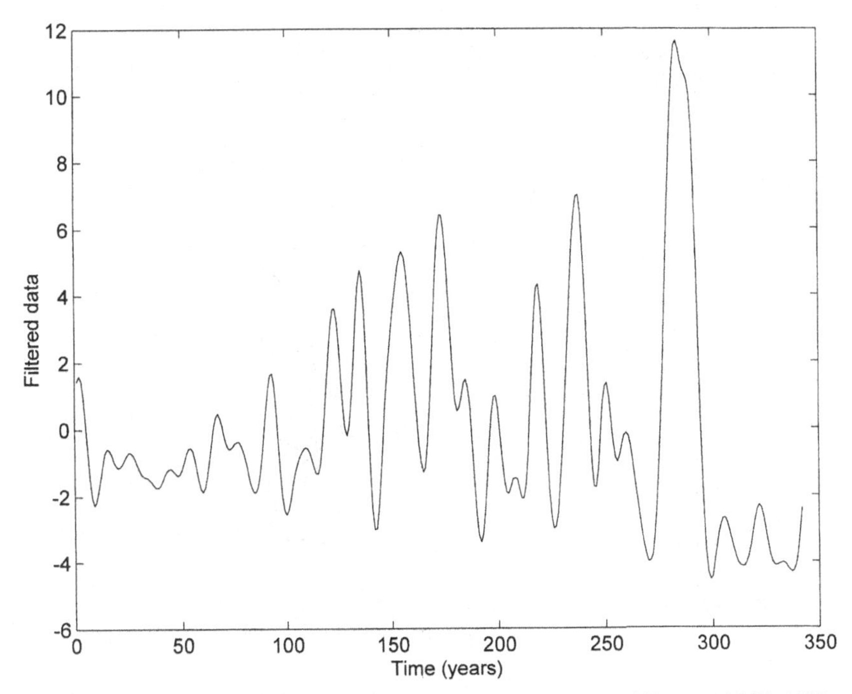

Fig. 11.14. Number of localities with plague in northern/central France, 1347–1688 (Fig. 11.11), filtered to reveal a medium-wavelength oscillation. Filter window = 16 to 32 years.

most frequently affected must have lived almost continuously for over 100 years in fear of the pestilence.

The Black Death arrived at Strasburg in 1349 but thereafter the city was free until 1358, after which it was repeatedly attacked for the next 150 years (Table 11.1). The Middle Ages were Strasburg's golden period; its wealth stemmed from the activity of its merchant class and also from its location at the centre of numerous waterway and road communications. The Customs House was built in 1358 to store and collect taxes on goods going through the city: exports were textiles and grain, imports were glass, hides, fur, silk and spices. Its location at the centre of trade routes was the reason that it received regular visits from travelling infectives.

Bourg-en-Bresse was heavily and regularly hit by plague after 1400 and epidemics continued to erupt there frequently for the next 200 years (Table 11.1). Probably the most important feature of the topography of Bourg-en-Bresse (where the first plague hospitals were established in France in 1472) was that it was the gateway to Geneva and to one of the passes over the Alps. It lies on the western edge of the Jura mountains, some 37 miles

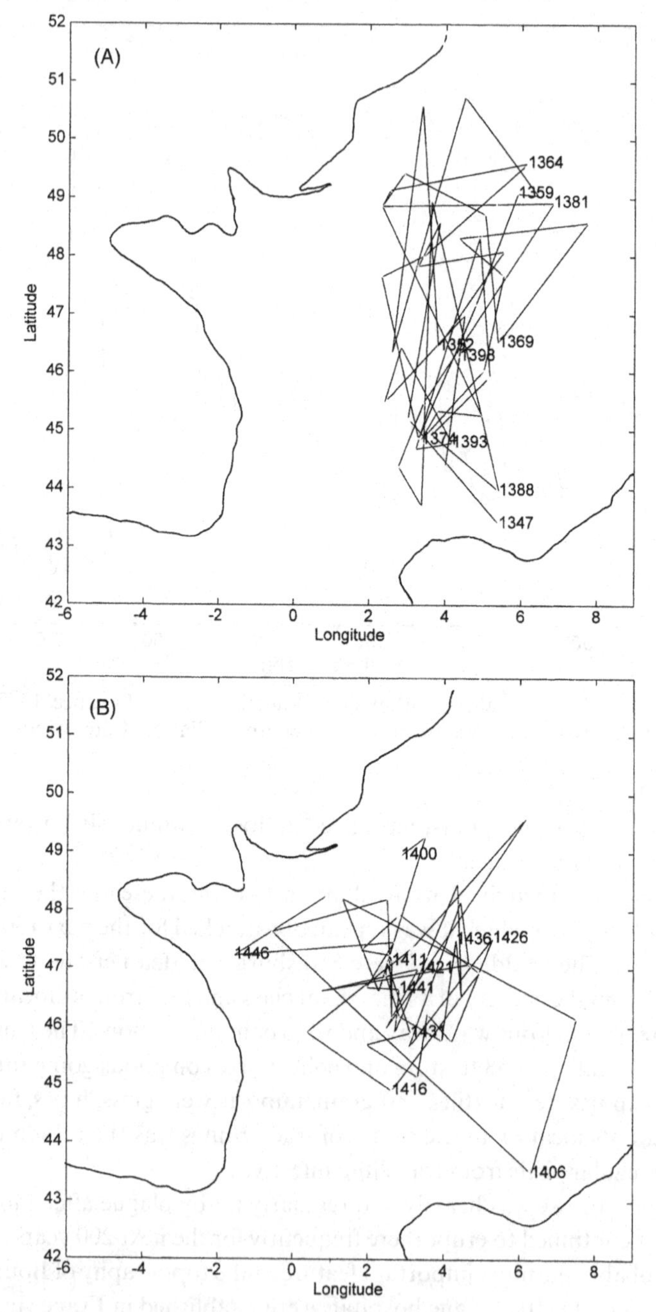

Fig. 11.15. Plague centroids in France. (A) 1347–99; (B) 1400–49; (C) 1450–99; (D) 1500–49; (E) 1550–99. The numbers on the figure are dates of years given at set intervals.

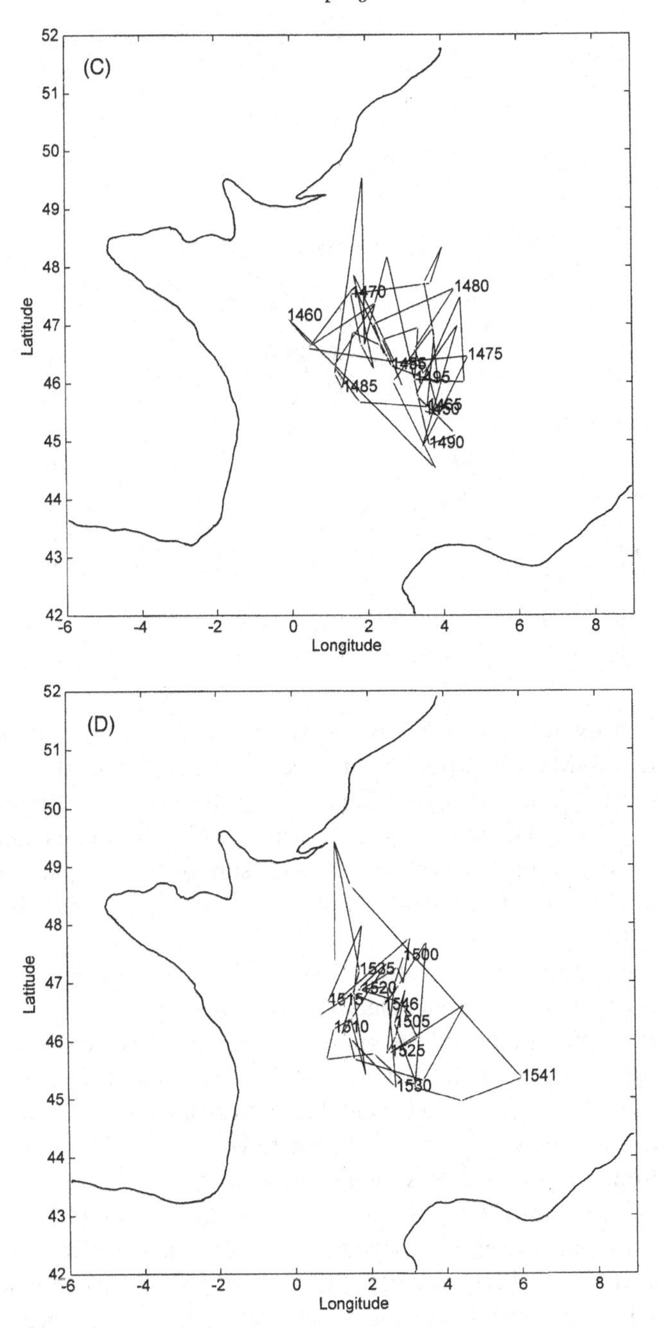
(C)
Latitude
Longitude
52
51
50
49
48
47
46
45
44
43
42
-6
-4
-2
0
2
4
6
8
1460
1470
1480
1475
1485
1490
(D)
Latitude
Longitude
1500
1535
1520
1515
1546
1505
1510
1525
1530
1541

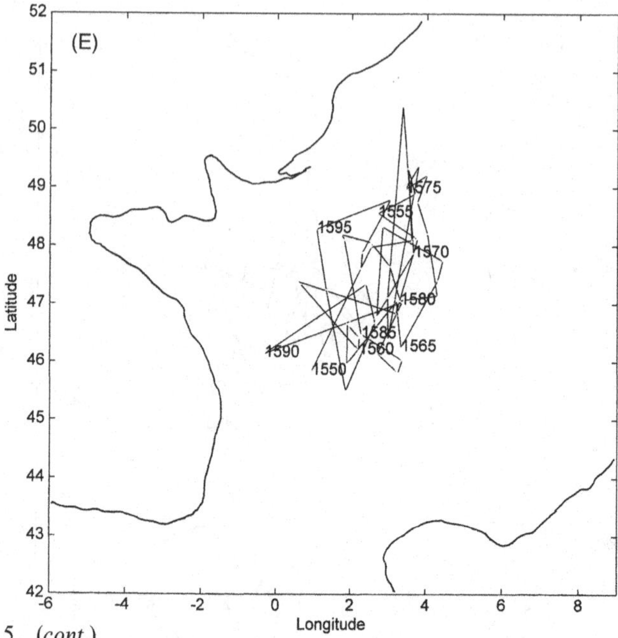

Fig 11.15 (*cont.*)

northeast of Lyons, close to the rivers Ain, Saône and Rhône and to the Troyes–Lyons–Marseilles trade route (see Fig. 11.22); all of these towns also appear in Table 11.1. Lyons lay at the confluence of the Rhône and Saône and attracted many foreign merchants with its four annual fairs. Its silk weaving, printing and other industries supported a population estimated at 60 000 in the mid-16th century and it was frequently subject to plague epidemics.

Visitations of the pestilence to Luxemburg are also included in Table 11.1. That city was strategically placed on what are now the boundaries of modern Belgium, France and Germany and originated as a Roman crossroads. Siegfried, Count of the Ardennes, built a castle on the site in the year 963 that later developed into a formidable fortress, known as the Gibraltar of the North, that was the origin of the town. After 1550, Luxemburg suffered from repeated and persistent epidemics.

To conclude: Table 11.1 and Fig. 11.15 show the important features of the plague in France after the Black Death. After the progressive, wave-like spread of the Great Pestilence, the pattern changed and the plague was established as endemic; it was not present everywhere, rather grumbling along, with an epidemic in a few widely scattered but important places every year, spreading from which there may have been a clustering of local epidemics. The number of places affected fluctuated regularly, with a medium-wavelength cycle and superimposed thereon were the major epidemics when the infection was widespread. The plague became progressively more widespread, reaching its peak in the 17th century. The towns listed in Table 11.1 were large and important and an epidemic frequently persisted there for more than 1 year. Some were sited on the major international trading routes (see below and Fig. 11.22), others were on the internal lines of communication, particularly at crossroads. We conclude that the plague was maintained and spread in the metapopulation by the long-distance movement of infectives.

11.3 Italy

Although Italy possessed a common language, local patriotism was too strong to permit the growth of a national unity during the age of plagues and it consisted of three main groupings of states:

(i) The City States of the North: Venice, Milan (Lombardy), Genoa and Florence (Tuscany), immensely wealthy and jealous of one another.
(ii) The Papal States.
(iii) The Two Sicilies to the south, the most backward and poorest of states, consisting of two very different regions, Naples and Sicily, and ruled by the King of Aragon in Spain (Taylor & Morris, 1939).

It is evident from the work of Carmichael (1986) that, at least in northern Italy, the authorities took public health measures during plague epidemics very seriously, with physicians examining suspect cases and this point is discussed in detail by Carmichael. They distinguished between major and minor pests and there is the possibility that Italy suffered outbreaks of bubonic plague from an early date in addition to epidemics of what were clearly haemorrhagic plague. The southern Mediterranean coastal climate, proximity to seaports along the length of the land and a trading centre with the Levant and North Africa all produce conditions that may have allowed the development of bubonic plague epidemics that could have spread inland but would not persist. Bubonic plague in Asia today is regarded as an endemic disease of rural areas where an epizootic is established but during the period after the Black Death minor *epidemics* of bubonic plague in Italy may have begun in a port and then possibly have spread inland. It is interesting that Carmichael (1986) recorded: 'In contrast to the devastating effects of plague on rural areas in the fourteenth century, the countryside was usually spared this scourge during the fifteenth century. Thus the

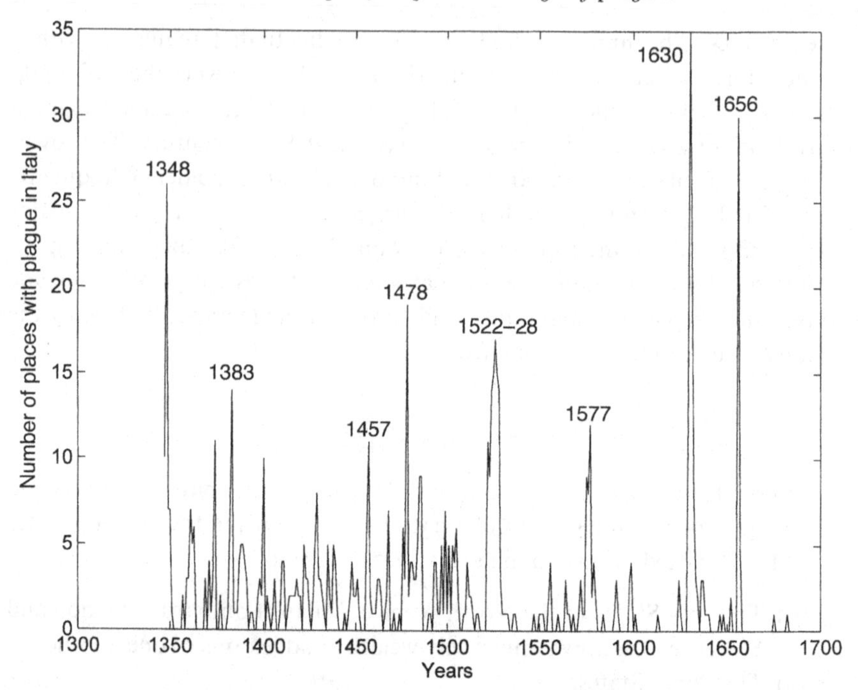

Fig. 11.16. Number of localities in Italy in which plague epidemics were reported, 1347–1688. Major epidemics indicated. Data from Biraben (1975).

basic experience confronting fourteenth- and fifteenth-century observers on the frequency and extension of plague led them to different conclusions about how plague spread.'

11.3.1 Occurrence and frequency of the epidemics in Italy

The number of places in which plague was recorded in Italy (Biraben, 1975) is shown in Fig. 11.16, although, as we say above, this may be a somewhat mixed bag of lethal infectious diseases. The picture presented for Italy is clearly different from that for France: the major, widespread epidemics stand out, as in the Black Death, in 1630 (when almost half the population of Milan was wiped out; Livi Bacci, 1997) and in 1656 (when half the population of Genoa and Naples died; Livi Bacci, 1997), but there were many years with no or only a single outbreak and after 1525 the plagues declined and the epidemics became more sporadic. Each burst of epidemics, whether of minor (perhaps bubonic plague) or major haemorrhagic plague must have been restarted by the arrival of fresh infective(s) from overseas.

Spectral analysis revealed no short-wavelength oscillation, unlike the situation in France (see Fig. 11.5), and we conclude that the epidemics were not related to environmental factors but were simply the result of the random arrival of infectives. In contrast, a significant 25-year oscillation is detectable by spectral analysis in the Italy data-series after 1447 ($P < 0.005$) and this cross-correlates significantly in the medium waveband with southern France (ccf = +0.65 at zero lag). We conclude that, because plague was not endemic in Italy, the medium-wavelength oscillation in the recorded spread of plague in southern France reflected the probability of spread of the infection to Italy, either by boat or across its northern frontier.

Regular fluctuations in plague intensity in Italy have been noted previously. Carmichael (1986) wrote:

> One of the most important features of epidemics in the early Renaissance is that many of them returned regularly . . . During the fifteenth century plagues assaulted cities in northern Italy more frequently than during the late fourteenth century. In the earlier century, pestilences that were described as bubonic plague visited major cities once in every fifteen to twenty years. In the 1400s plagues hit some cities, for example, Venice and Florence, at least once a decade.

Livi-Bacci (1997) also recognised the importance of plague cycles in Italy:

> For several parts of Tuscany between 1340 and 1400 I have calculated that on average a serious mortality crisis – defined as an increase in deaths at least three times the normal – occurred every 11 years; the average increase in deaths was at least sevenfold. In the period 1400–50 these crises occurred on average every 13 years and deaths increased fivefold. In the following half century (1450–1500) the average frequency declined to 37 years and the average increase to fourfold. With the passage of time, both the frequency and the intensity of the crises declined, as did the geographic synchronisation of their occurrence. Keep in mind that Tuscany is an exceptional case only for the abundance of historical sources to be found there.

11.3.2 Signs and symptoms

The city states in northern Italy led the rest of Europe in adopting measures for dealing with plague: quarantine, the pest-house and health boards. The physicians took great pains when examining cases to determine whether a true plague had broken out and one of the most consistent indicators of the pestilence that was feared by community leaders was the clustering of plague cases in a limited number of households (Carmichael, 1986), indicative of a high household contact rate and repeatedly found in our studies of plagues in England. The symptoms were violet and black spots and blotches ('certain' signs for which the Mantuan investigators searched; Carmichael, 1986) or small red spots, indicative of a haemorrhagic disease, the bubo and pustules. Onset was marked with a high temperature, vomiting, diarrhoea, cloudy urine and burning thirst (Cipolla, 1981). The acute fever in some was accompanied by madness and delirium 'so much so that many hurled themselves out of windows' (Carmichael, 1986).

The following reports of two autopsies of victims of the plague in 1656–57 in Rome and Naples respectively are of particular interest:

> The exterior part of the body was found to be covered by black petechiae with a black spot as large as a bean in the medial part of the right knee . . . The muscles of the abdominal wall were of bad colour, the fat tissue very dry, the omentum rotten, the guts all black, the peritoneum cyanotic, the stomach very thin, the spleen rotten, the liver doubled in size but of bad colour and consistency, the gallbladder full of black bile. Regarding the thorax, the pleurae were rotten, the pericardium very hard, the mediastinum and the sagittal septum livid, the heart livid with its tip black, both ventricles full of very dark blood. The lungs, of bad consistency and colour, were all covered with black petechiae.
>
> It was noticed that all the organs – namely, the heart, lungs, liver, stomach, and guts – were covered with black spots. Moreover, the gallbladder was found full of black bile which was very thick and fattish to the point that it adhered stickily to the inner part of the gallbladder. Especially, however, the major vessels of the heart were full of blood which was clotted and black.
>
> *(Cipolla, 1981)*

The similarities between the foregoing and the case reports, description of symptoms and autopsy reports of plagues in England, particularly in London, are striking.

11.3.3 Analysis of the spread of epidemics

Carmichael (1986, 1991) has described the clustering of plague deaths within families in three different epidemics. Obviously the data are incomplete but, nevertheless, it is possible to suggest how, in each case, the epidemic spread. There was an apparent clustering of plague deaths (believed by Carmichael (1986) to be a minor plague) in one street in the 1430 epidemic in Florence. Four households along this narrow short street suffered more than one death and a suggested analysis of events is shown in Fig. 11.17 in which the major gaps between the primary and secondary infectives (23–30 days) are evident. These are the fingerprints of haemorrhagic plague.

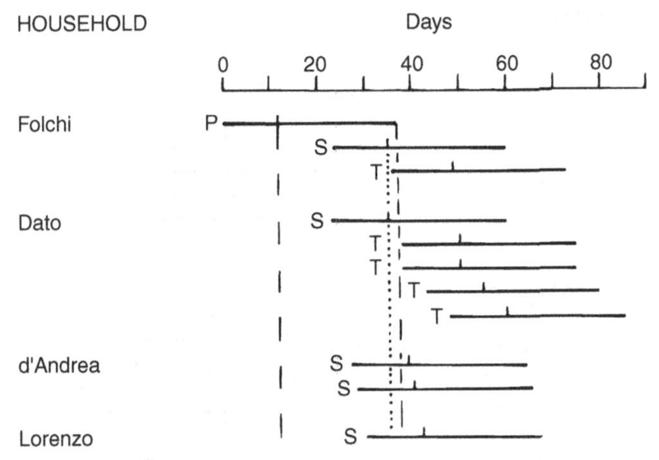

Fig. 11.17. Suggested sequence of infections in four households in Borgo Tegolaia, Florence in the summer of 1430. Scale: days after 4 June 1430. For further details and abbreviations, see Fig. 5.6. Data from Carmichael (1986).

An example of deaths in two families in the plague of Florence in 1456 shows the same household clustering, another characteristic of haemorrhagic plague (Carmichael, 1986) and a suggested analysis of the sequence of events is shown in Fig. 11.18.

Carmichael (1991) described the early stages of an outbreak of plague in Milan in early March 1468 and the events may be summarised (see Fig. 11.19) as follows:

(a) Cases in the parish of St Pietro della Vinaor:
- (i) The first cases died in early March, all in one large household [A].
- (ii) 25 March, a 13-year-old boy died after an illness of 11 days. Blood-tinged urine [Household B].
- (iii) 16 April, an 18-year-old girl died after an illness of 4 days with a bubo [Household C]. She lived next door to household A.
- (iv) 17 April, another man in Household C fell ill and was exiled with three siblings.
- (v) 23 April, a 6-year-old girl died with an enlarged inguinal gland [Household D].

(b) Cases in the neighbourhood of Cinque Vie:
- (vi) 27 April, Maestro Legutero died [Household E].
- (vii) 3 May, a 25-year-old female servant to Lord Scaramuzia Visconti died plus another death after a 7-day illness [Household F].
- (viii) 4 May, deaths of a 17-year-old boy [Household G, probably

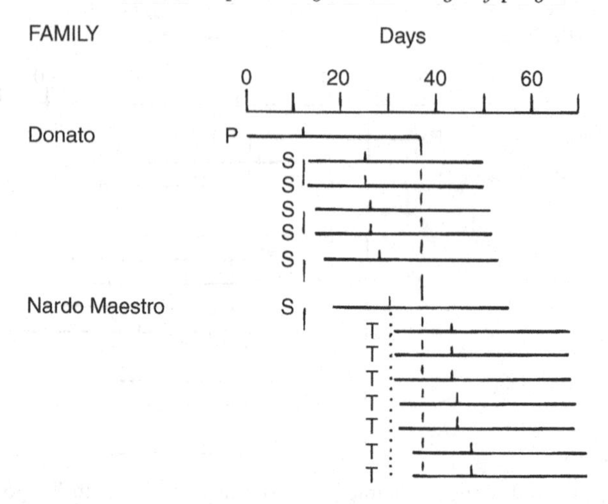

Fig. 11.18. Suggested sequence of infections in two families in Florence in the summer of 1456. Scale: days after 16 August 1456. For further details and abbreviations, see Fig. 5.6. Data from Carmichael (1986).

associated with Household E] and the wife of Guido de la Croce, builder [Household H].

(ix) 5 May, a 16-year-old cleric died with fever and vomiting [Household I].

(x) 7 May, a 19-year-old, Antonio da Robion working in a barber's shop, fell ill [Household K].

(xi) 7 May, a 14-year-old girl died [Household J]. People to whom she spoke in the Cinque Vie area subsequently died of plague.

(xii) 8 May, two daughters of Cristoforo da Cazeniga fell ill [Household L]; they lived in the neighbourhood of St Pietro dell Vigna and were probably infected from Households C or D.

The probable sequence of events shown in Fig. 11.19 has clear similarities with the epidemiology of the opening stages of the epidemics of haemorrhagic plague in England. One primary probably infected only one secondary ($R_o = 1$), who, in turn, infected four tertiaries ($R_o = 4$) and nine quaternaries ($R_o \approx 2$).

11.3.4 Plague epidemics in Italy

Figure 11.16 reveals that, after the terrible mortality of the Black Death, which worked its way progressively through the country, there were only a limited number of epidemics in Italy during the following 300 years.

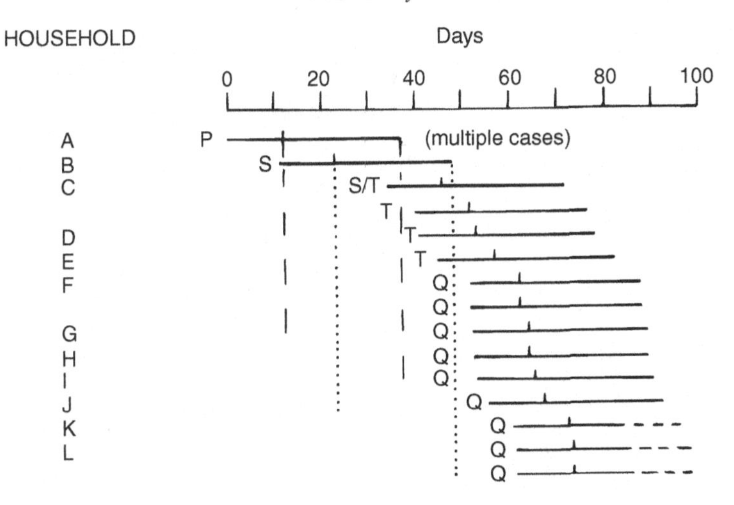

Fig. 11.19. Suggested sequence of infections in the epidemic in Milan in 1468. Households indicated A to L; for further details see text. Scale: days after 5 February 1468. For further details and abbreviations, see Fig. 5.6. Data from Carmichael (1986).

Perhaps 11 such outbreaks can be identified, of which five stand out as being the most widespread and severe: 1383, 1478, 1522–28, 1630 and 1656. Plague was reported in fewer than five localities in other years and after 1500 there were merely sporadic outbreaks in single towns. The authorities seem to have largely disregarded these smaller epidemics, often designating them as 'minor plagues'. Possibly some of these were outbreaks of bubonic plague in the warmer climate that did not spread and did not persist for more than a few months, particularly if the epidemic erupted in a seaport.

Time-series analysis of the data shown in Fig. 11.16 has suggested that the outbreaks had a periodicity of 25 years and this oscillation became more significant after 1447 (see above).

In this section we examine the sequence of events in six epidemics that erupted after Italy had been plague free for at least 2 years, so that these outbreaks were the result of fresh introductions.

1456–58 Plague was reported only in Milan in 1452 and thereafter there was a 4-year break before it erupted in Sicily, central Italy (two localities) and northern Italy (Fig. 11.20A). Were these separate introductions by sea via the ports at Palermo, Naples and Venice/Ravenna (as seems most likely) or did the plague spread in a saltatory fashion along the length of the country? It was not reported from Sicily in the following year, but appeared in the

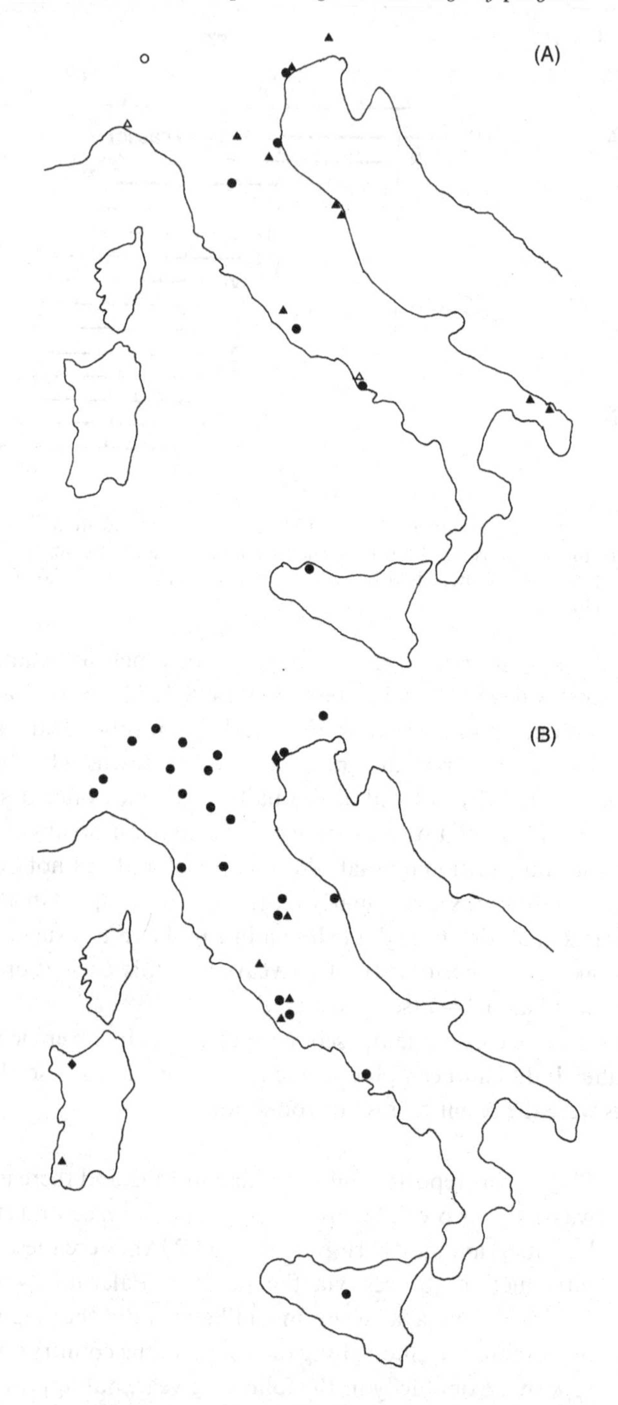

Fig. 11.20. Geographical distribution of the major plague epidemics in Italy. (A) 1456–58: open circle, Milan in 1452; closed circle, 1456; closed triangle, 1457; open triangle, 1458. (B) 1476–78: closed triangle, 1476; closed diamond, 1477; closed circle, 1478. (C) 1522–29: open circle, 1522; open triangle, 1523; closed triangle,

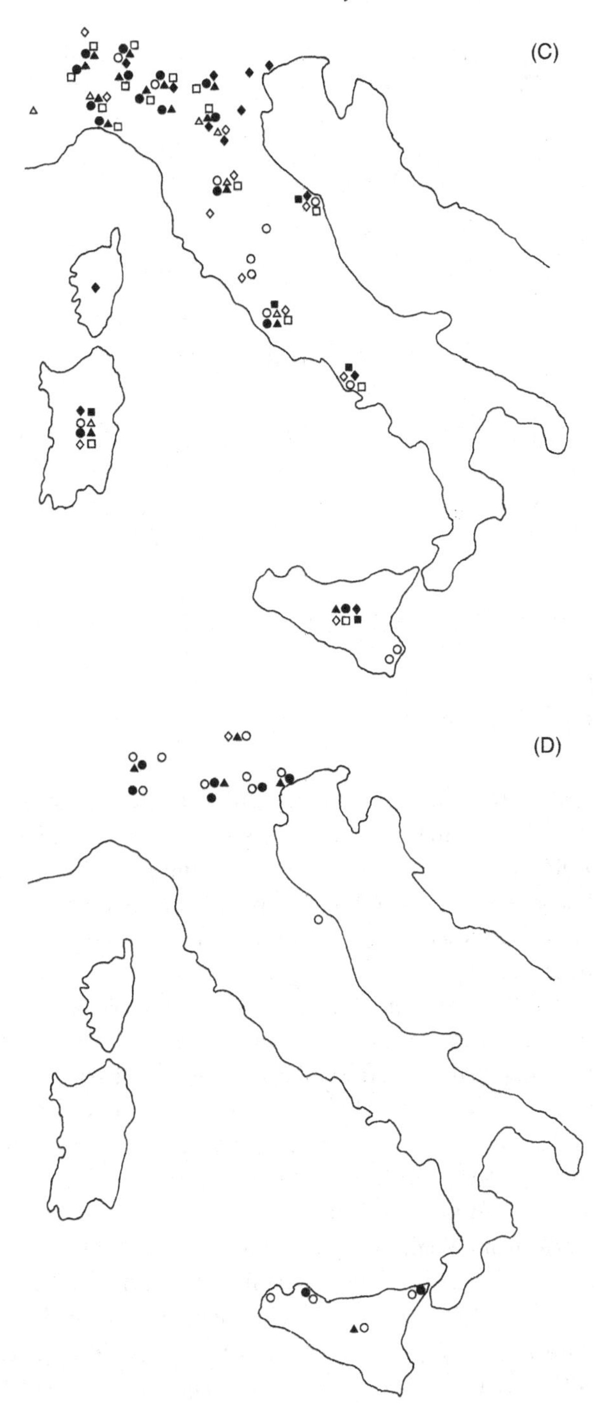

1524; closed circle, 1525; open square, 1526; open diamond, 1527; closed diamond, 1528; closed square, 1529. (D) 1574–77: open diamond, 1574; closed triangle, 1575; closed circle, 1576; open circle, 1577. (E) (*overleaf*) 1629–31: closed triangle, 1629; closed circle, 1630; open circle, 1631. Data from Biraben (1975).

Fig 11.20 (*cont*).

heel of Italy; it moved from Velletri to Rome but disappeared from Naples in the centre; it persisted in Venice and appeared in Bologna in the north. In the following year, 1458, plague was reported only in Naples (again) and appeared in Genoa. Thereafter, Italy was plague free until Venice had a prolonged 3-year outbreak, 1460–62.

1476–78 Forty thousand died in northern Italy. Italy had been largely plague free for 8 years before an epidemic began in 1476, when it was reported in southern Sardinia (it was absent from Corsica) and in central Italy (Fig. 11.20B). Plague persisted in Sardinia but appeared in Venice in 1477 and exploded in 1478, when a total of 30 000 died. It appeared in Sicily in 1478 and was now widespread in the north, probably having spread radially outwards from Venice to the west and south; 22 000 died in Milan and 200 died each day for the first 4 months in Brescia, with a total of 34 000 and a mortality rate of 90%. It died down in the following year, persisting only in Venice and Florence.

1522–29 After a virtual absence for 8 years, plague erupted widely in 1522 in Italy; it was reported in Sicily and Sardinia, where it came by sea, and in central and northern parts of the country. Apart from Naples and Ancona, most of the sites were inland and it is possible that the plague may have come to Milan and Lombardy from Lyons (where there was an epidemic) via the Alps. Figure 11.20C shows that this broad geographical spread of the plague persisted for the next 6 years: it remained in Sicily, Sardinia (but was not reported in Corsica until 1528) and in central Italy in Rome, but it expanded rapidly and moved around the northern states.

1575–77 Plague was reported only in Messina (Sicily) in 1573 and only in Trento (in northern Italy in the Dolomites) in 1574. In the following year a major epidemic erupted; it persisted in Trento and spread widely in the northern city states (Fig. 11.20D). It seems most probable, therefore, that this epidemic began in Trento and was not introduced from overseas but came over the Alps from Geneva where plague was reported from 1568 to 1574. This epidemic continued through 1576 and 1577 and was confined to northern Italy; an estimated 50% of the 180 000 inhabitants of Venice contracted the disease in this epidemic.

1629–31 Italy was free from plague from 1626 to 1628 but an outbreak erupted in the northern states in 1629. It may have entered the country via Venice (where 46 000 of the 140 000 inhabitants died) but it is more probable that it came to Milan or Turin over the Alps (Fig. 11.20E); plague was widespread over the borders in France, Germany and Switzerland and was recorded in Geneva and Basel at this time. Marseilles also had an epidemic in 1628 but the distribution of the outbreaks in northern Italy suggests that plague did not come along the coastal plain via Nice and Genoa. It has also been suggested that German and French troops carried the plague to Mantua in eastern Lombardy, where France was waging war against Austria and Spain in 1629. Eventually, some 280 000 died in Lombardy and other territories in northern Italy and the average mortality rate in the cities attacked has been established as 46% (Kohn, 1995). Plague was much more widespread in northern Italy in the following year (Fig. 11.20E) and Cipolla (1981) wrote:

> 'Coming from the north, the plague arrived in the Grand Duchy of Tuscany in the summer of 1630 and spread in the course of the autumn. The sequence of events is fairly clear: in the month of July a few persons

died, allegedly of plague, in Trespiano, a small village a few kilometers north of Florence. According to the Health Deputies of the time, the plague had been brought to Trespiano by a man who, violating the sanitary cordon, had been on a business trip to the infected city of Bologna. In August suspicious deaths were recorded in the neighboring village of Tavola and in Florence itself. On September 1, the plague made its appearance in Monte Lupo, 28 kilometers west of Florence, on the busy road to Pisa. By September 19, the plague was recognised in Prato, about 20 kilometers north-west of Florence, on the road to Pistoia. Before the end of the month the plague had reached Pisa.'

Strict, but unpopular, preventive measures were instituted when the plague reached Milan in October 1629, including the quarantining of all persons who had come into contact with those who had been infected and it was believed that the outbreak had been contained by these measures. But these regulations were relaxed in March 1630 during a carnival in Milan (which must have established a high effective contact rate) and plague broke out through the city and 3500 inhabitants were reportedly dying every day (Kohn, 1995). The epidemic continued through 1630 and 1631 and, once again, was confined to 15 cities in the north of the country (Fig. 11.20E) where it spread in two major waves in the autumn and winter of 1630 and the spring and summer of 1631. An estimated 10% of the population of Florence died in this epidemic. Victims suffered a sudden and high fever and developed foul-smelling boils; they were sometimes delirious and a terrible headache was the usual prelude to death. A large percentage of those who died were artisans and Capuchin friars or those who performed much of the custodial work during the epidemic. Nobody broke the standard 40-day quarantine and those appointed by the Office of Public Health dedicated themselves to their tasks of fumigation, burning mattresses and clothes, scrubbing floors and carrying away the dead for burial (Kohn, 1995).

1656–57 The health officers of the northern city states had a firmly established joint policy of excluding trade with any locality known to have a plague outbreak. However, Naples, then under Spanish rule, refused to suspend trade with Spain and the catastrophic plague of 1656–57 in Naples, Rome and Genoa may have entered Italy via this route. In these three cities, 218 000 out of a population of 498 000 died, although much of the rest of Italy remained plague free (Kohn, 1995).

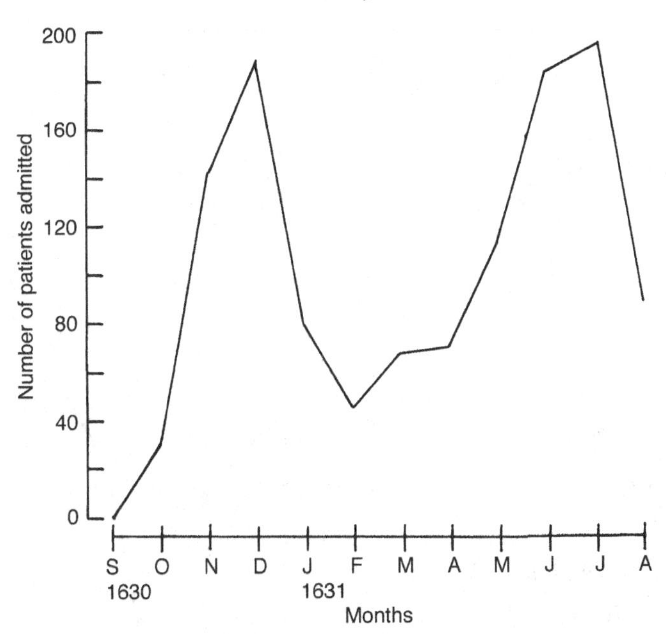

Fig. 11.21. Number of patients admitted to the pest-houses of Pistoia, Florence from October 1630 to August 1631. After Cipolla (1981).

11.3.5 Conclusions

The biology of the plague in Italy is quite different from the situation in France where haemorrhagic plague was virtually endemic somewhere throughout the 300-year period and France acted as the source for the majority of the epidemics in the remainder of Europe. Plague was never endemic in Italy, which we can consider as a separate metapopulation. It probably suffered from only about a dozen major epidemics in which the mortality rates were often very high, probably on average 40% but sometimes reaching over 60%, as in Verona in 1630–31. Naples, with about 300 000 inhabitants, suffered 150 000 deaths in 1656. The morbidity rate in the countryside has been estimated at 66% (Benedictow, 1987). These epidemics often lasted for 2–3 years in the big cities before burning out, a clear difference from the pattern in England, perhaps because of the warmer climate in winter and early spring. Although the plague persisted for longer in these cities in Italy, there was still a fall in the number of cases over the winter, as shown by the number of patients admitted to the pest-houses of Pistoia, Fig. 11.21 (Cipolla, 1981).

Each of these major epidemics was introduced from outside the

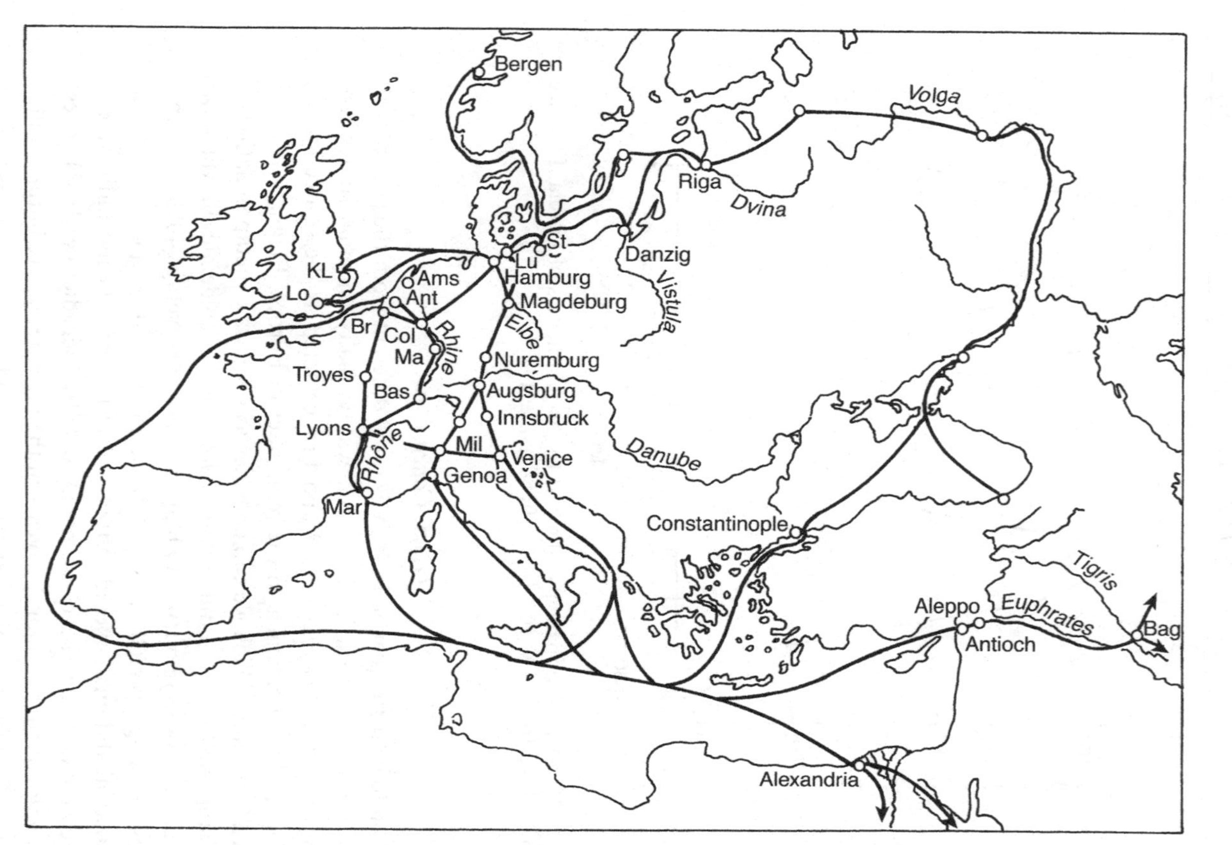

Fig. 11.22. Major trade routes of the Middle Ages in Europe. Ams, Amsterdam; Ant, Antwerp; Bag, Bagdad; Bas, Basel; Br, Bruges; Col, Cologne; KL, King's Lynn; Lo, London; Lu, Lübeck; Ma, Mainz; Mar, Marseilles; Mil, Milan; St, Straslund.

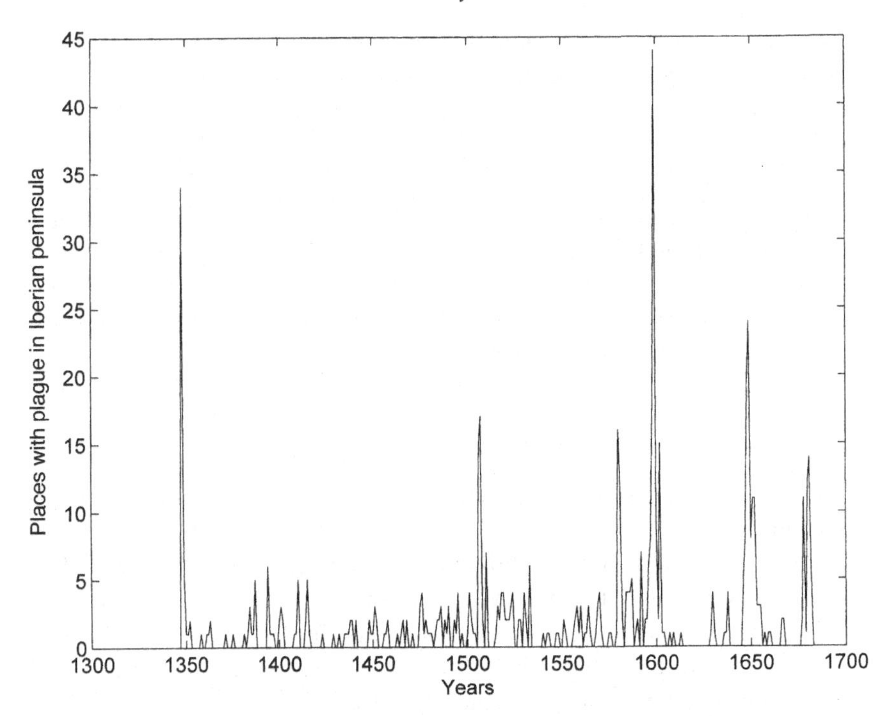

Fig. 11.23. Number of localities in the Iberian peninsula in which plague epidemics were reported, 1347–1688. Data from Biraben (1975).

meta-population, probably mostly from overseas in the 14th and 15th centuries and probably more commonly overland via the alpine passes to the north of the country in the 16th and 17th centuries. Those epidemics that came from overseas probably originated from the Levant and North Africa and entered via the ports at Venice and Sicily, or from France via Genoa and Leghorn. The epidemics tended to spread along the length of the country but in later centuries the outbreaks were confined largely to the northern city states where the pestilence was widespread. Saltatory transmission by apparently healthy infectives carrying a disease with a long incubation period travelling over the major trade routes over the alpine passes (Fig. 11.22) is consistent with the epidemiology, whereas a regular movement of rats and fleas infected with *Yersinia pestis* over a mountain range is an impossibility.

The health authorities of the northern states were the first to recognise the importance of strict quarantine measures, a cordon sanitaire (see quotation from Cipolla, 1981, above) and efficient diagnostic procedures, all applicable to a contagious disease, and their actions were probably successful to some extent in limiting the importations of the plague via the northern ports on both the east and west coasts.

Apart from the major, devastating epidemics there were many years when only a single, isolated outbreak was reported (see Fig. 11.16). Their cause remains unknown; they may have been 'minor plagues' that the health authorities identified, or some may have been bubonic plague brought in from the Levant or the ports of North Africa which infected the local rats under favourable Mediterranean climatic conditions. Apparently, the epidemic did not spread far and soon died out because it was not possible to form an epizootic with local resistant rodents.

11.4 The Iberian peninsula

Like Italy, plague was never endemic in the huge area of the Iberian peninsula, which was separated from France by the Pyrenees and can be considered as a separate metapopulation. The number of localities where plague was reported in each year is shown in Fig. 11.23: after the Black Death, although there were several outbreaks in Portugal and Spain in the 15th century, there were only five major epidemics and the last outbreak (1680–82) occurred after haemorrhagic plague had apparently disappeared in 1670. Time-series analysis does not reveal any significant oscillations in the data-series.

11.4.1 The major epidemics

The foci and spread of the infection in these major epidemics are shown in Fig. 11.24 and may be summarised as follows.

1506–7 Plague was reported only in Seville in 1504, was apparently absent in Spain in 1505 and then broke out again in Seville in the following year, which was probably its entry point. It was accompanied by a severe drought and food shortage. It had two foci (Fig. 11.24A): to the north and east of Seville and in a sector to the north and west of Madrid. Presumably, the epidemic was spread from Cordoba by apparently healthy infectives to Madrid, a distance of some 200 miles, and then spread radially therefrom. The epidemic persisted for 2 years in each locality and then disappeared, except in Seville and its environs where it continued in 1508. About 100 000 are said to have died in Andalusia, in southern Spain, alone.

1580–81 Lisbon was the only locality where plague was reported in 1579 and this was probably the original point of entry; it erupted into a full scale epidemic in the following year when the outbreak spread radially throughout southern and central Portugal (Fig. 11.24B). It did not persist there and, except for Oporto, had disappeared from Portugal by 1581 and moved to the extreme southwest of Spain, where it was reported in a cluster of localities, including Seville, where, alone, it persisted in 1582; the cause of the latter outbreak has been attributed to soldiers and slaves who disembarked in the port.

1596–1602 A prolonged outbreak. The only localities in the Iberian peninsula where plague was reported in 1594 and 1595 were Seville and Andalusia in southern Spain. Presumably, this area acted as the primary focus for the remarkable epidemic that broke out in 1596, when it was reported in three widely separated centres: it persisted in Seville (in the southwest) but spread to Madrid (in the centre) and appeared in Santander on the north coast, where the total number of deaths was about 2500 in a population estimated at 3000 to 3300; the lethality rate has been determined at between 85% and 90% and, since part of the social elite fled from the town, Benedictow (1987) suggested that the net morbidity rate must have been around 95%. This was not an outbreak of bubonic plague. The epidemic at Santander was probably the result of a separate entry of the pestilence from overseas when the ship *Rodamundo* docked, carrying cloth from the French port of Dunkirk. Plague persisted in these foci in 1597 and a limited further spread was reported (Fig. 11.24C). It was probably brought to Santiago de Compostela by the pilgrims. This slow spread continued in 1598; the transmission along the north Spanish coast and to Lisbon was probably by coastal shipping. The epidemic exploded in 1599 and spread widely throughout the peninsula (Fig. 11.24C). The small city of Segovia lost 12 000 inhabitants over 6 months and responded positively to the plague: temporary hospitals were established, the city gates were guarded to prevent the arrival of infectives, victims were rapidly buried and bedclothes were burnt. Thereafter, although plague persisted in a few localities, the epidemic was largely confined in 1600 to 1602 to the

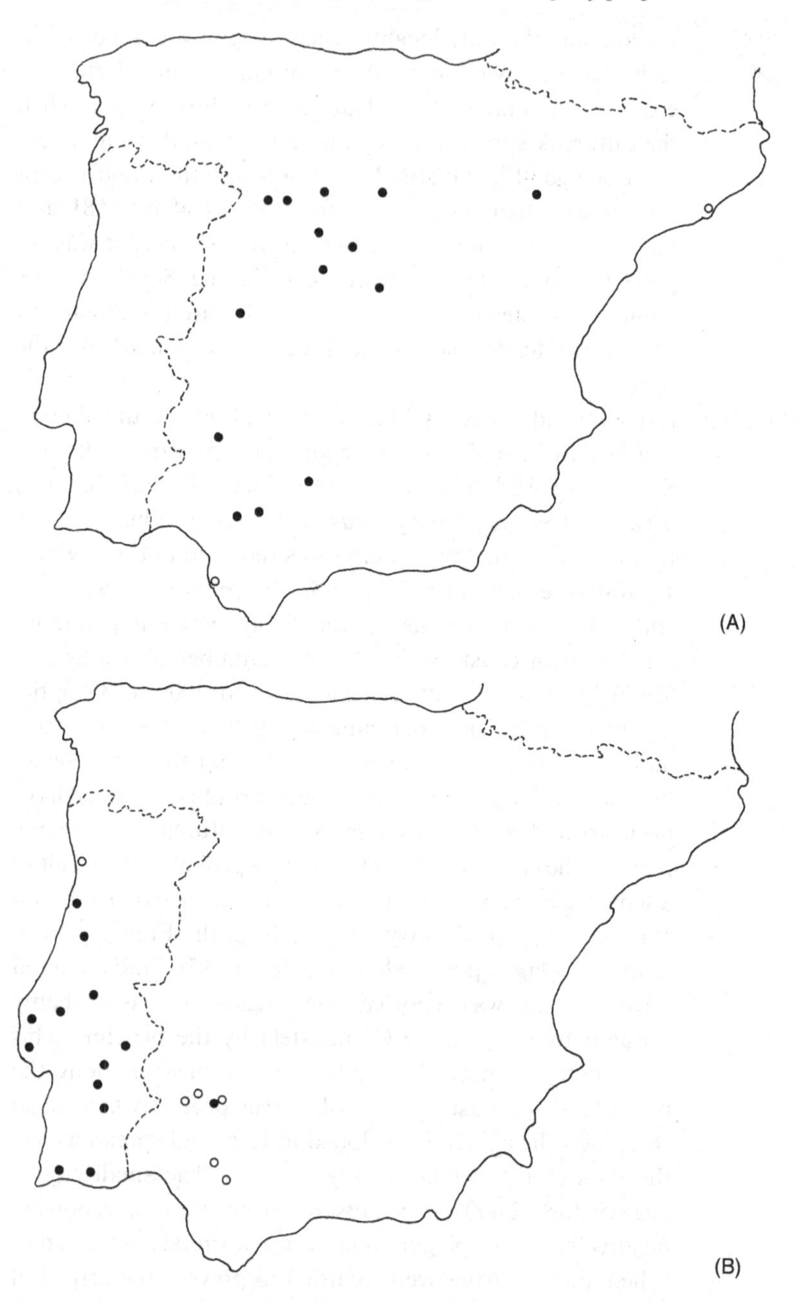

Fig. 11.24. Geographical distribution of the major plague epidemics in Spain and Portugal. (A) 1506–7: open circle, 1506 and 1507; closed circle, 1507. (B) 1580–81: closed circle, 1580; open circle, 1581. (C) 1596–1602: open square, 1596; closed diamond, 1597; closed triangle, 1598; open circle, 1599; closed circle, 1600; open

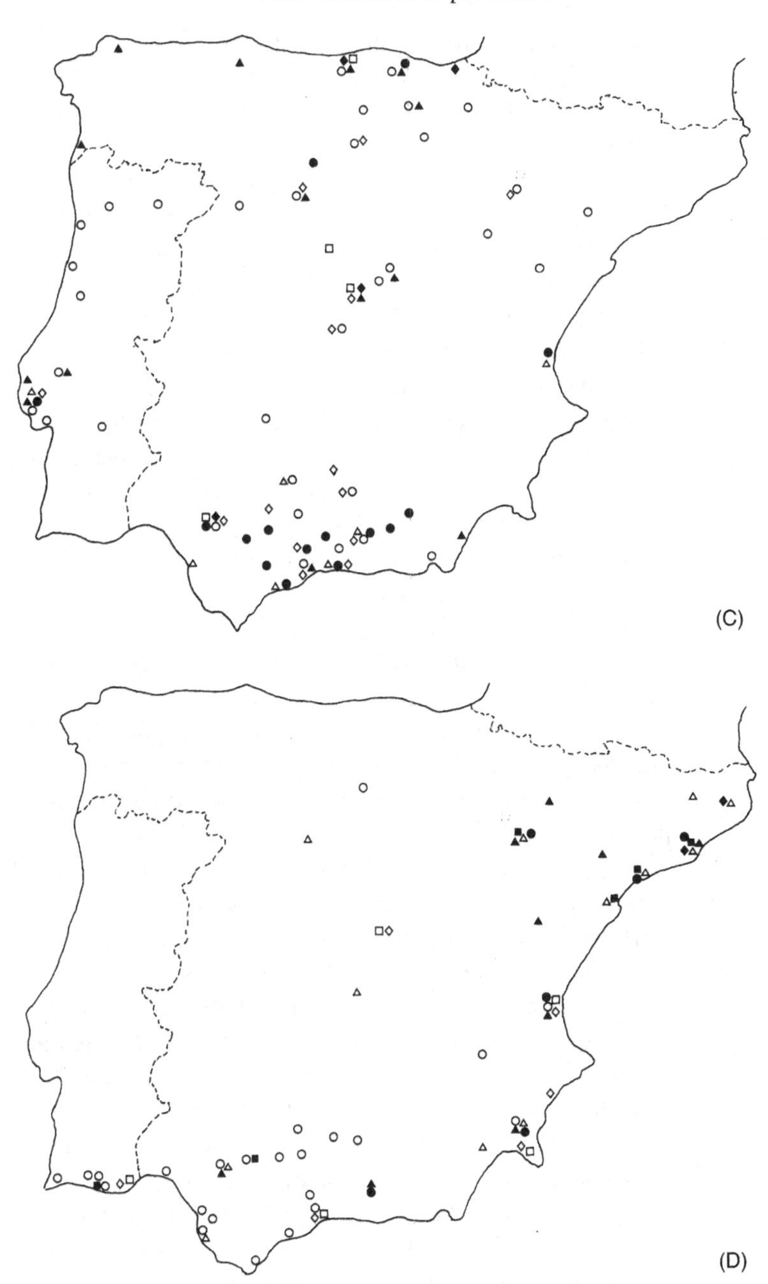

triangle, 1601; open diamond, 1602. (D) 1646–53: open square, 1646; open diamond, 1647; open triangle, 1648; open circle, 1649; closed square, 1650; closed triangle, 1651; closed circle, 1652; closed diamond, 1653.

southern coastal area to the east of Seville. The total death toll in Spain in this epidemic is estimated to have been 500 000 to 600 000 (6% to 8% of the population) (Kohn, 1995).

1646–52 After an absence of 8 years, plague appeared, surprisingly, in four widely separated sites in 1646: (i) Tavira (southern Portugal); (ii) Málaga, Seville and Andalusia where 200 000 died and the disease was traced to a shipment of silk (presumably infectives were in the crew); (iii) Valencia, where 30 000 died; and (iv) Alcalá de Henares (near Madrid). It persisted at these sites but spread little in the following year, 1647 (Fig. 11.24D). Plague was also reported in Catalonia (northeast Spain) and spread southwards along the coast from Valencia (perhaps by coastal sea traffic) in 1648. In 1649, plague was, as usual, widespread in the southern coastal strip of the peninsula to the east and west of Seville, but this area was largely free by 1650, although the outbreak continued in a cluster of localities in northeast Spain during 1650–52 (Fig. 11.24D) and also spread to Majorca and Ibiza. Nearly 500 000 died in total in this epidemic.

1678–82 The last major plague epidemic in Spain was reported in 1678 and its spread is illustrated in Fig. 11.25. Was this a late outbreak of haemorrhagic plague after it had largely disappeared in Europe or was it a different disease with high mortality? It was mostly confined to the southern coastal area of Spain around Andalusia, although it also appeared at Cartagena and Murcia, some 100 miles away, where it might have had a separate origin. We tentatively suggest that this might have been a major outbreak of bubonic plague. *Yersinia pestis* was causing outbreaks along the warmer Mediterranean coasts, being imported by sea into the ports and then spreading inland by infecting the local rodents. These outbreaks did not last long and bubonic plague never became truly endemic in the western Mediterranean coastal region probably because there were no resistant resident rodent species in which a buffer epizootic could form. The data contained in Fig. 11.25 are suggestive of bubonic plague with two coastal points of entry, Cartagena (where the spread was limited) and Málaga (where the epidemic was quite wide-

Fig. 11.25. The last major epidemic in Spain, 1677–82. Open diamond, entry port in 1677; closed triangle, 1678; plague reported in Andalusia in 1679 but no localities known; open circle, 1680; closed circle, 1681; open triangle, 1682.

spread in the hinterland). The pattern at Málaga is comparable to the spread of the plague at Marseilles in 1720 (Chapter 12).

In summary, the Iberian peninsula acted as a separate metapopulation and probably experienced only four major epidemics of haemorrhagic plague after the Black Death. Since it was never endemic, the pattern is completely different from that of France. The data suggest that each epidemic began by importation from overseas and did not spread from France over the Pyrenees, which formed an effective isolation barrier. Each of the four epidemics had its characteristic pattern, but Seville was implicated in all, twice acting as the point of entry. Seville shows little temperature variation through the year, with long, dry and hot summers and warm winters. It has acted for 2000 years as a port at the navigable head of the Guadalquivir River, as the great market place of the Guadalquivir Valley and as the crossroads between the northeast and west of the Iberian peninsula. The year 1492 is important in the history of Spain: it marked the final expulsion of the Moors, the unification of Spain under a single crown and the discovery of America. For 200 years thereafter, Seville became the gateway to the New World, the Mecca of European commerce and the principal city of Spain. Plague epidemics were particularly common in Andalusia and the area around Seville. Lisbon was the port of entry for the plague that began in 1580 and Portugal was again hit in 1598. In most of these epidemics, a radial or linear spread can be detected from the foci, but a saltatory transmission, over distances of 200 miles can also be seen, particularly from coastal regions to Madrid.

11.4.2 Epidemics at Barcelona

The data contained in Fig. 11.23 show that, apart from the major epidemics, Spain suffered frequently from isolated epidemics in single localities. These did not spread and did not persist. Were these outbreaks of haemorrhagic plague that did not establish themselves or was another infectious, lethal disease responsible? Where these outbreaks occurred in ports they may have been limited epidemics of bubonic plague. Barcelona, on the northeast coast of Spain, repeatedly, was the only locality to report a plague outbreak. Biraben (1975) has analysed the data for 35 outbreaks at Barcelona after the Black Death; he could only estimate the total number of deaths for the period 1350–1452, but thereafter he provided detailed information, often the daily number of deaths. The estimated number of deaths as a proportion of the population was below 1.5% in 11 of the epidemics and was presumably of little demographic significance. The epidemic of 1497 (0.25% mortality) showed single deaths daily from July to September, with no peak period.

The details of the other epidemics are summarised in Table 11.2 and mixed patterns emerge from the analysis. There was only one epidemic after 1589, namely that of 1651–54. It was present during 1648–50 and was part of the national epidemic during those years; the heavy mortality then (45%) is typical of haemorrhagic plague, although the long duration in one locality is not. The epidemic in 1507 also occurred during a national outbreak, although the 3-month break from August to October, which is much longer than the usual quarantine period for haemorrhagic plague, suggests either that plague continued undetected at a low level during this period, or that the population was reinfected, or that another disease was responsible.

Apart from these two epidemics, the outbreaks were confined to Barcelona and did not spread (see Table 11.2) and each infection was presumably introduced from overseas. The mortality in these isolated epidemics was relatively low, with an average of 10%. The greatest mortalities were in 1589 (28.8%), 1465 (20.0%) and 1530 (18.9%); if these years are excluded the average is reduced to 8%, suggesting that these lesser outbreaks may not have been typical haemorrhagic plague, or that the population had developed resistance in the face of continuous attacks, although the average mortality during the first 100 years was estimated at only 8%.

Where information is available (after 1452) the average duration of the outbreaks was 8 months (range = 4 to 14 months), excluding the national epidemics. The outbreak began in the autumn, was reduced in the winter and began again in the following spring in 1457, 1466, 1490 and 1558, reminiscent of the type (ii) epidemics recorded in England. The epidemic that lasted from August 1475 to September 1476 was unusual: a small number of deaths was recorded steadily over a 12-month period, including January–February, with only a small peak. If the longer-lasting outbreaks are excluded, the average duration of the seven remaining epidemics was 6 months, usually lasting from April to September, with a peak mortality around June (see Table 11.2).

Are a mixed bag of lethal epidemics summarised in Table 11.2? Why did they apparently not spread inland? Why did these one-off epidemics apparently cease after 1590? It will probably not be possible to answer these questions with any degree of certainty, but some tentative suggestions can be made. Barcelona in the 15th century was a city of merchants, navigators, traders and professionals and it acted as a major port that had traffic with the whole of the Mediterranean and so could have received a regular supply of infectives. The epidemics of 1507 and 1651–54 were probably part of the national outbreaks of haemorrhagic plague, albeit with the slightly different biology from those described in England because of the coastal Mediterranean climate where, today, the monthly average temperature, December–February, is 13–14 °C. These outbreaks at Barcelona may have been initiated separately from those in other parts of the Iberian peninsula and in 1648 there was limited spread into the hinterland (Fig. 11.24D).

The other epidemics in which there was a mortality of over about 4000 may also have been haemorrhagic plague; some of these started in the autumn, continued with a reduced mortality during the mild winter and then broke out again in the following spring, perhaps a modified type (ii) epidemic. However, the outbreak from October 1465 to November 1466, with a mortality of 5000 (20%) stands out as being completely atypical because Biraben (1975) has identified a steady build-up in mortality through the autumn and winter that peaked in February and thereafter declined steadily.

Table 11.2. *Plague epidemics at Barcelona, 1362–1654*

Start	Finish	Duration (months)	Records	Peak	Estimated total deaths	% age mortality	Number of other localities in Spain where plague reported
Jan. 1362	1363	—	—	—	(5000)	14.2	0
1371		—	—	—	(3000)	8.3	0
1375		—	—	—	(1000)	2.7	0
1394	1396	—	—	—	(4000)	10.8	3
1408		—	—	—	(2500)	7.0	0
1429		—	—	—	(3000)	8.8	0
May 1439	1441	—	—	—	(3500)	10.5	2
May 1448		—	—	—	(2000)	6.2	1
1452		—	—	—	(1500)	4.8	1
May 1457[a]	Nov. 1457	7	Daily	July–Aug.	3630	11.4	0
Oct. 1465	Nov. 1466	14	Monthly	Feb.	5000	20.0	0
Aug. 1475	Sept. 1476	14	Daily	May–June[b]	2100	7.8	1
Mar. 1483	Sept. 1483	7	Daily	June	1400	3.7	0
Mar. 1490[c]	Sept. 1490	7	Daily	May–June	3770	15.1	1
June 1494	Oct. 1494	4	Daily	July	600	2.3	0
Apr. 1501	Nov. 1501	8	Daily	June–July	2650	9.8	3
Feb. 1507	Nov. 1507[d]	10	Monthly	May	3500	11.9	16
May 1515	Nov. 1515	6	Daily	No clear peak	1160	3.8	0
May 1520	Sept. 1520	4	Daily	June–July	1600	5.0	1
Feb. 1530	July 1530	6	Daily	April	6400	18.9	2
Sept. 1557	Aug. 1558[e]	12	Daily	May–June	4500	13.6	1

May 1589	Dec. 1589	7	Daily	Aug.–Sept.	12 400	28.8	0
Jan. 1651	Apr. 1654[f]	40	—	—	20 000	45.5	16

Notes:
[a]Probably began in preceding year, but died down October to May. Type (ii) epidemic.
[b]Only a small peak; deaths recorded steadily over the 12-month period including the period January to February.
[c]Very small number of deaths recorded in the months November 1489 to February 1490. Type (ii) epidemic (?).
[d]This was part of the great national epidemic of 1506–7, although there was a clear 3-month break, August to October 1507.
[e]Daily deaths through the winter. Type (ii) epidemic?
[f]Breaks: August to September 1652; September 1653. Part of the national epidemic.
Data from Biraben (1975).

The other epidemics with low mortalities may have been repeated, but isolated, outbreaks of bubonic plague brought into the port from the Levant or North Africa. *Yersinia pestis* has been identified in corpses from the French Mediterranean coast from the 14th, 16th and 18th centuries and, although it could apparently survive in the warmer climate, the epidemics seem to have been short lived, persisting for up to 3 years, probably because there were no local resistant rodent species in which a permanent epizootic could be established.

Epidemics occurring every 8–10 years with a 10% mortality must have had a severe impact on the demography of the city unless there was an accompanying immigration to fill the ecological niches available. Biraben (1975) has estimated that the population before the Black Death was 42 000 which had fallen progressively to 31 000 by the end of the 1465–66 epidemic. It fell further to a low of 25 000 by the end of the 15th century before recovering steadily to 43 000 by 1589.

11.5 Germany, Austria, Bohemia and Switzerland

The geographical area described in this section formed the major part of the Holy Roman Empire (which also included the present-day Netherlands) during the age of plagues. The Black Death arrived in this area of central Europe in 1348 and the epidemic raged through 1349 to 1350 and had disappeared by 1352. Thereafter, plague was reported sporadically, usually with six or fewer localities affected in any one year, until about 1520, except for the period 1462–65 when terrible outbreaks that killed thousands struck in *widely distant cities* in Germany. In Regensburg, 6300 died in 1462 and 2500 in the following year; this outbreak is believed to have originated in the Rhineland region (Kohn, 1995).

Plague was absent in many years during 1352–1520 (Fig. 11.26), suggesting that the epidemics were fresh introductions, but became almost endemic after about 1520, with epidemics that were increasingly widespread. Spectral analysis of the data shown in Fig. 11.26 (1347–1688) reveals highly significant oscillations in the pattern of the epidemics with wavelengths of 14.25 years ($P < 0.005$) and 11.4 years ($P < 0.005$). A secondary oscillation with a wavelength of 22 years appeared after 1450 ($P = 0.05$).

The list of localities where plague was most reported frequently is shown in Table 11.3. For 100 years after the Black Death, a cluster of towns in the north heads the list: Hamburg was a large port on the River Elbe, 64 miles from its mouth, and was already a thriving trading town by the 13th century; the port of Lübeck was the major outlet for the Baltic trade. After

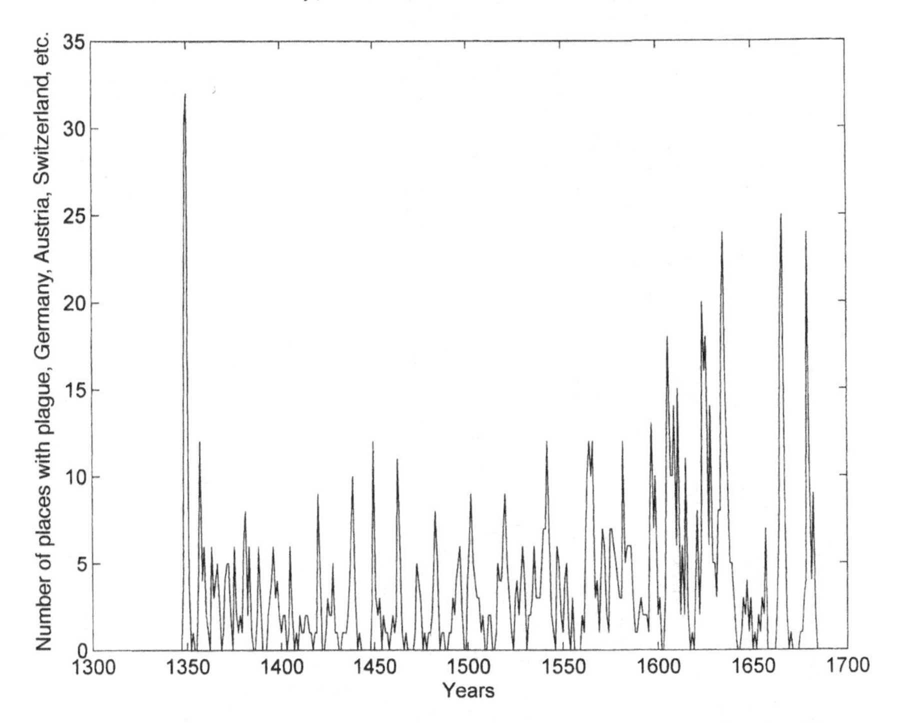

Fig. 11.26. Number of localities in Germany, Austria, Bohemia and Switzerland in which plague epidemics were reported, 1347–1688. Data from Biraben (1975).

1450, the places recording frequent epidemics spread southwards along the major trade route of the Middle Ages: Lübeck–Hamburg–Magdeburg–Nuremberg–Augsburg (see Fig. 11.22).

The situation changed after 1450 and continued until 1670; the plague continued to erupt widely and epidemics persisted in the towns in the north but the major foci, with the highest number of outbreaks, were in the southwest, with Basel and Geneva now heading the lists, together with Augsburg and other towns on the trade route to Milan, Innsbruck and Venice. Augsburg experienced some 20 outbreaks of plague in the 16th century in which a total of some 60 000 died (Kohn, 1995). Basel, situated on the Rhine, was a large river port, Switzerland's only outlet to the sea and the terminus of the Rhine navigation; for many centuries the Mittlere Brücke in Basel was the only bridge on the Rhine. Basel suffered from a 15-month epidemic in 1610–11 when, out of a population of some 15 000, about 6000 contracted the disease, of which only approximately 3600 died. Were some of the population showing partial immunity or resistance because of previous exposure to the plague? The city also suffered in

Table 11.3. *Localities in Germany, Switzerland and Austria that recorded the largest number of outbreaks of plague in successive cohorts, 1348–1670*

Cohort					
1348–99	1400–49	1450–99	1500–49	1550–99	1600–70
Hamburg (10)	Hamburg (7)	Geneva (8)	Basel (12)	Basel (19)	Basel (15)
Lübeck (10)	Hanover (5)	Basel (7)	Augsburg (11)	Geneva (12)	Augsburg (14)
Bremen (8)	Nuremberg (5)	Brunswick (7)	Geneva (10)	Bremen (12)	Bremen (14)
Basel (8)	Bremen (5)	Nuremberg (6)	Dresden (9)	Augsburg (8)	Nuremberg (12)
Mayence (6)	Lübeck (4)	Hildesheim (5)	Wittenberg (8)	Brunswick (8)	Luneburg (12)
Cologne (4)	Basel (4)	Hanover (5)	Luneburg (7)	Dresden (7)	Hildesheim (12)
Magdeburg (4)	Vienna (3)	Cologne (4)	Cologne (5)	Nuremberg (5)	Dresden (11)
Wismar (4)	Nordhausen (3)	Luneburg (4)	Erfurt (5)	Hamburg (5)	Geneva (10)

The number of epidemics suffered by each locality in each cohort given in parentheses.
Data from Biraben (1975).

Switzerland's last epidemic in 1667–68 and a meeting was held in Bremgarten to which officials were invited to discuss how they might prevent the plague spreading from Basel (Kohn, 1995). There were major trade routes from Basel to Lyons and along the Rhine to Mainz, Cologne and to the port at Antwerp. Geneva also occupied a central position in Europe and because of its early adherence to the principles of the Protestant Reformation became, at the beginning of the 16th century, a refuge for the persecuted and a starting-point for missionaries. It was therefore crowded with immigrants and intending emigrants and was an ideal focus for plague outbreaks.

11.6 The Benelux countries

The Black Death arrived along the southern Belgian border in Hainaut and Flanders, and in Luxemburg in 1349, penetrating to Liège. It disappeared almost completely in the following year and thereafter, for 200 years, plague broke out only sporadically, usually in single localities (see Fig. 11.27) and confined to the French borders in Flanders, Hainaut, Luxemburg and the Ardennes. We infer that these minor epidemics came with infectives crossing the French border. Plague was reported from the port of Antwerp in 1511 and occurred there very regularly in the 16th century when it was a major commercial centre. Was the plague brought in from overseas via the port, or did it spread northwards from France or westwards from Germany? This pattern of outbreaks continued after 1550; Luxemburg was repeatedly the only locality affected. The plague spread more frequently into Holland and its port at Amsterdam after 1600.

11.7 Spread of the plague across Europe

The spread of haemorrhagic plague in France is central to understanding its behaviour after the Black Death because it is only in France that it was virtually pseudo-endemic for over 300 years from 1350 to 1670. We have traced its erratic appearance during this time in sections 11.2.4 and 11.2.5; the epidemics broke out in widely separated areas or towns and, in general, the major outbreaks in the southern, central and northern coastal regions correlated with one another. Figure 11.15 shows how the centroids moved in each successive 50-year period. The towns most frequently affected (see Table 11.1) were of larger size and sited on the main medieval trading routes or in the valleys of the Loire, Saône and Rhône. The ports on the Mediterranean coast were, surprisingly, less frequently affected and we

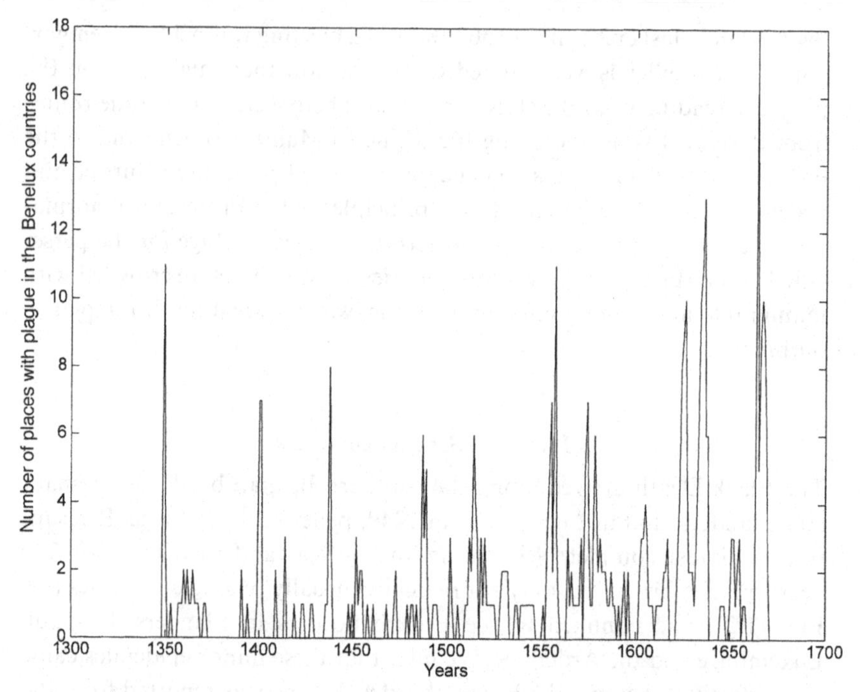

Fig. 11.27. Number of localities in Belgium, the Netherlands and Luxemburg in which plague epidemics were reported 1347–1688. Data from Biraben (1975).

conclude that the endemic status of the plague in France was only infrequently boosted by arrivals of infectives from overseas.

The Pyrenees to the southwest formed an effective barrier to the transmission of the plague and there is no evidence that the major epidemics in the separate metapopulation of the Iberian peninsula came over this mountain range; the infectives entered through the ports. There is no significant cross-correlation between the data-series for the Iberian peninsula (see Fig. 11.23) and the corresponding epidemics in southern or central France.

The behaviour of the plague in Italy superficially resembles that in Spain; it can be treated as a separate metapopulation and, although there were many minor epidemics, major outbreaks were few. The health authorities in the northern city states led the world in their control measures and were probably successful in containing or preventing some of the infections coming by sea from the Levant, North Africa, the Mediterranean islands and France. The Alps must have provided a partial barrier to the spread of the plague from the north but, nevertheless, study of the major epidemics suggests that some originated via this route, coming from France or the Holy Roman Empire in central Europe.

Spectral analysis of the data-series of the annual number of localities suffering from plague in Italy (see Fig. 11.16) shows that a 12-year oscillation was present from 1347 to 1447 and changed thereafter to a significant 25-year oscillation ($P = 0.005$). This medium-wavelength cycle cross-correlates significantly after filtering with a comparable oscillation in southern (ccf = +0.65) and central (ccf = +0.5) France. We do not suggest that outbreaks in Italy (where plague was epidemic) were a driven system; rather that, when plague was rampant in France, there was a greater chance of infectives coming to Italy and starting epidemics.

The plague, obviously, did not observe political boundaries and we have seen how epidemics broke out in Flanders and Luxemburg soon after the Black Death and contributed to the endemic status of the disease in France. Inspection of Fig. 11.26 shows that the plague was not endemic in the Holy Roman Empire before 1550, although there were repeated outbreaks along the main trade routes (see section 11.5 and Fig. 11.22) and we conclude that central Europe was repeatedly reinfected, probably usually from France, during this time. Plague was present more continuously in the Holy Roman Empire after 1550 and, for certain periods, approached an endemic situation. The data-series for both France (see Figs. 11.3 and 11.4) and Germany (see Fig. 11.26) show significant medium-wavelength oscillations that cross-correlate significantly. Cross-correlation studies suggest that the medium-wavelength oscillations in the frequency of epidemics in the metapopulation composed of Germany, Austria, Switzerland and Bohemia were initiated by comparable cycles in France, with infectives travelling long distances, usually along the different trade routes, and regularly bringing the plague into central Europe.

On this basis, middle Europe (present-day France, Germany, the Benelux countries, Switzerland and Austria) can be viewed as a vast metapopulation around which the plague was carried by apparently healthy, travelling infectives. We have, therefore, calculated the annual number of localities where plague was reported over this huge area, using the data given by Biraben (1975), and the results are shown in Fig. 11.28. The plague spread progressively through the years and became more and more firmly established, with a rising endemic level, culminating in the terrible outbreak which lasted from 1630 to 1637. The plague declined sharply thereafter and this is in marked contrast with the situation in England, where the major outbreak was in 1665–66. The outbreak of 1630–37 was driven, as usual, by France, where it appeared first, and it did not rise markedly in

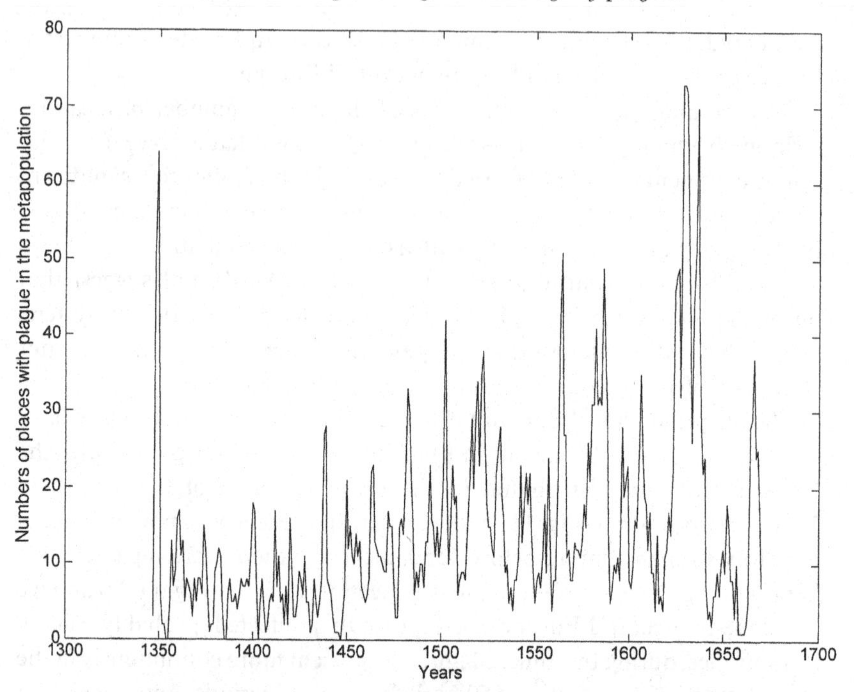

Fig. 11.28. Number of localities in the supermetapopulation composed of France, Germany, Austria, Switzerland and the Benelux countries in which plague epidemics were reported, 1347–1670. Spectral analysis reveals a significant oscillation ($P = 0.005$) of wavelength 21.6 years. Data from Biraben (1975).

Germany and the Benelux countries until 1634–35.

Spectral analysis of the data shown in Fig. 11.28 reveals a significant medium-wavelength oscillation of period 21.6 years ($P = 0.005$). When the cohorts are analysed separately, the following cycles were identified:

1347–1447	12.5 years	NS Mostly epidemics in France
1447–1547	12.6 years	NS
	22.2 years	NS
1547–1670	20.8 years	$P = 0.005$
	25.0 years	$P = 0.005$
	31.3 years	$P = 0.005$

NS, not significant.

It is evident that the medium-wavelength oscillation (see Fig. 11.29) changed from a weak 12.5-year cycle to a significant 21.6-year cycle after 1472. It is an open question as to what drove this oscillation and why it should have changed 125 years after the Black Death.

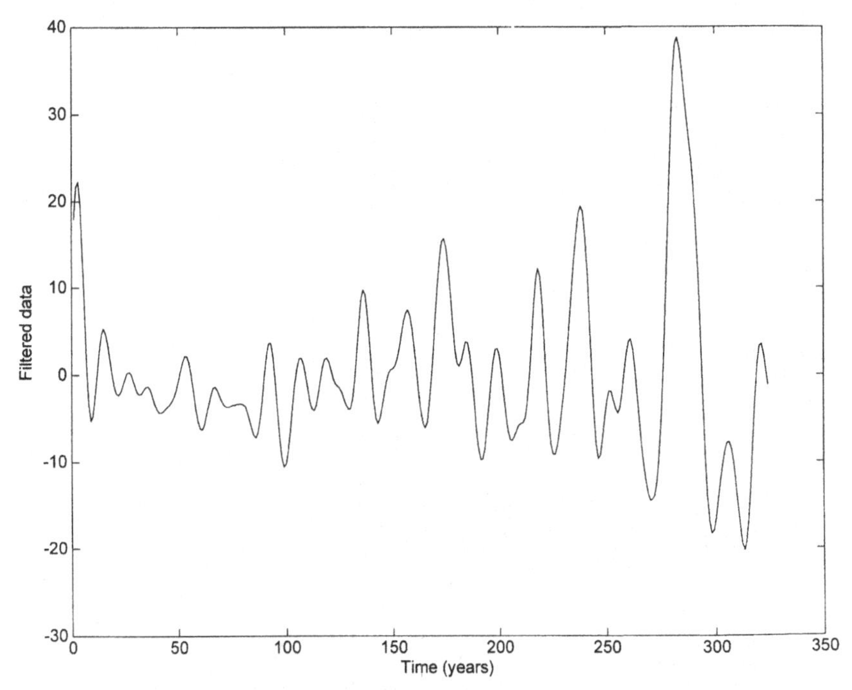

Fig. 11.29. The data-series of Fig. 11.28 filtered to reveal the medium-wavelength oscillation. Filter window = 16 to 30 years.

We have identified in this chapter the major trade routes in the Middle Ages running north–south (see Fig. 11.22) by which the plague was frequently spread within each country but, for the east–west transmission of the infection between France and Germany, the cities of Basel, Geneva, Bourg-en-Bresse, Strasburg and Luxemburg became key crossing-points (Fig. 11.22) and suffered from repeated epidemics (see Tables 11.1 and 11.3). Inspection of the data suggests that the plague frequently circulated between these key towns.

The British Isles were separated by sea from this huge metapopulation of northern and central Europe, so that the epidemics there were started by infectives coming through the ports: Newcastle-upon-Tyne, Hull, York (via the Humber and Ouse), the East Anglian river system, London and the many ports, large and small, on the Kent and south coasts. We have described the 14-year and 20- to 21-year, medium-wavelength oscillations in the data-series for the frequency of epidemics in Britain (see Fig. 9.13) in section 9.7.

The filtered medium-wavelength cycles for Britain and northern France (filter window = 10 to 25 years) for the period 1347 to 1667 cross-correlate

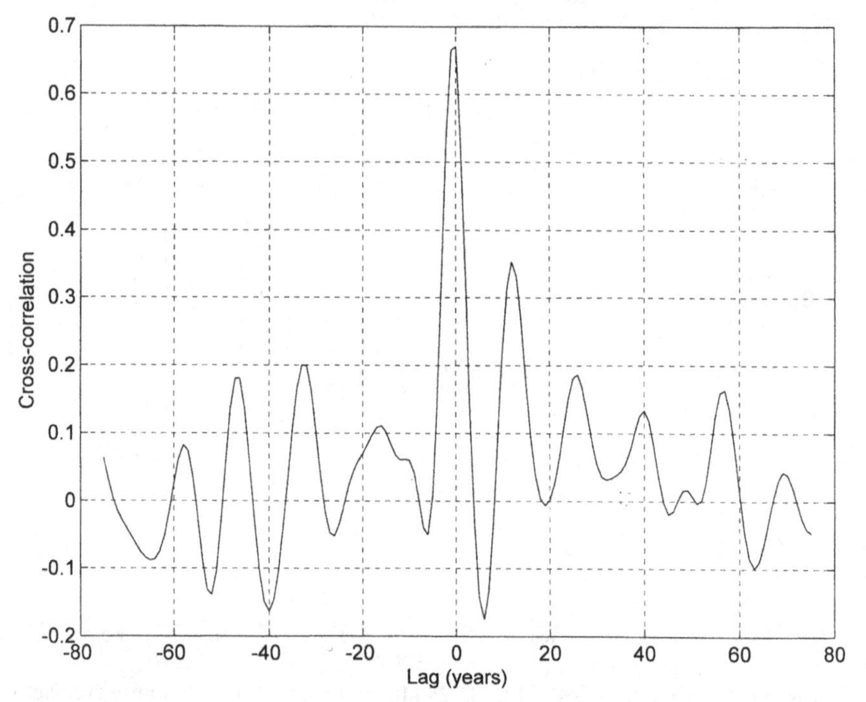

Fig. 11.30. Cross-correlation between the medium-wavelength oscillations in the number of localities in which plague was reported in Germany and Britain, 1347–1500. Filter window = 10 to 25 years. Ccf = +0.67.

moderately well, with a ccf of +0.36. This correlation was poor for the period before 1540, after which the major plagues emerged in the British Isles. This medium-wavelength cycle in British epidemics shows better correlation with the German series (see Fig. 11.26); overall the ccf is +0.46 but, in contrast with the French series, the significance is greater in earlier years, 1347–1500 (ccf = +0.67; Fig. 11.30). We conclude that, when plague was raging in northern France or in Germany, there was a much greater chance of apparently healthy infectives embarking on a voyage to England across the North Sea or the English Channel.

A study of the extensive list of places where plague outbreaks were reported in continental Europe (Biraben, 1975) shows, although it is not complete, a dramatic change in 1500. Prior to that date, for the 150 years after the Black Death, only very rarely were full-scale epidemics reported from the ports along the northwestern French, Belgian and Dutch coasts, from Brest to Amsterdam, and these would not have provided a ready supply of infectives. Epidemics were, of course, widespread in Paris and the towns in northern France during 1350–1500 so that infectives may have travelled on boats sailing out of the French channel ports. However, plague frequently broke out at Nantes during this time, although ships coming from there to England would have had to make a much longer voyage; they may have docked on the south coast, or in London, or possibly in the river complex of East Anglia. Perhaps more importantly, England and France were engaged in the Hundred Years War from 1337 to 1453, which was really a disjointed series of wars. England held a substantial territory in Gascony, around and to the south of Bordeaux, at the time of the Black Death and this area was greatly enlarged in 1360 after the treaty of Bretigny. Edward III, the Black Prince and John of Gaunt led six campaigns between them in the 14th century, moving all across France. Henry V led expeditions into northern France in 1415, 1417–20 and 1421–22. The English were eventually expelled from Gascogny in 1453 and then retained only Calais in France. The movement of troops and the supply of provisions to the territories held must have led to many infectives crossing the English Channel and sailing round from western France; plague was frequently reported at Bordeaux after 1394.

From 1353 onwards the ports of Hamburg and Bremen reported irregular epidemics, in perhaps 20 out of 150 years. Hamburg traded widely with the ports along the east coast of England and was presumably the source of some of the infections.

Figure 9.13 shows that 1500 marked a clear change in the pattern of the epidemics in England and, concomitantly, the ports along the French and Dutch coasts suffered from outbreaks of plague. There was rarely a year in which at least one port did not report an epidemic during 1500–1666 and Rouen was pre-eminent, particularly during the first half of the 16th century. Many of the boats plying over the English Channel and North Sea must frequently have been inadvertently bringing infectives from France, Antwerp, Amsterdam and Hamburg.

12

The plague at Marseilles, 1720–22: an outbreak of bubonic plague?

The epidemic that broke out in Marseilles in June 1720, some 50 years after haemorrhagic plague is believed to have disappeared, has been described and discussed in detail (Biraben, 1972, 1975; Bertrand, 1973), although the details of its origins, arrival at the port and early infections are not completely clear. These events are of critical importance in determining the characteristics of the infectious agent. The diary of events in 1720 are broadly as follows.

30 Jan.	The vessel, the *Grand St Antoine*, leaves Saida in Syria where plague was raging.
3 April	Turkish passengers embark at Tripoli, Libya.
5 April	Death of one of the Turks in the night.
27 April	Death of first sailor.
28 April	Death of second sailor.
4 May	Deaths of two sailors.
6 May	Death of the surgeon. Captain isolates the crew.
17 May	Arrival at Leghorn, Italy. Deaths of three sailors. Refused entry at port.
25 May	Arrival at Marseilles. The health commissioners impound the ship's merchandise and quarantine the passengers and crew in a lazaretto on an island in the middle of the port.
27 May	Death of a sailor who had been ill for 3 days.
31 May	Three ships from Saida arrive at Marseilles.
3 June	The *Grand St Antoine* and the three other ships are authorised to discharge their cargo in the roadstead.
12 June	One of the guards of the *Grand St Antoine* dies suddenly on board.

14 June	Passengers and crew are allowed to land. One passenger travels to Paris and two to Holland but there were no recorded cases of plague in these localities. The people of Marseilles buy contraband cloth from the sailors, so coming into contact with them.
21 June	First suspicious death in the town after an illness of 2 days – she was Marguerite Dauptaine, rue Bell-Table.
23 June	Death in the infirmary of the ship's boy from the *Grand St Antoine.*
24 June	Death in the infirmaries of three porters who had opened the bales of cotton from the *Grand St Antoine*; one of them had also opened the cargo from the other ships from Syria.
28 June	Death of the tailor Michel Crisp, place du Palais.
29 June	Death of the wife of M. Crisp.
3 July	Death in the infirmaries of three porters.
5–6 July	A secondhand clothier in the rue de l'Oratoire dies with his wife and children.
8 July	M. Bonche, a tailor, his wife and children die.
9–11 July	A boy aged 13 (surname Issalenc) living in the rue Jean-Galant dies in the presence of two doctors who diagnose the pest, report the case to the aldermen and have the whole family hospitalised. The daughter, who is a tailoress, dies. A neighbour in the rue Jean-Galant dies.
14 July	The outbreak spreads to rue de l'Echelle with the deaths of several neighbours. Dr Sicard declares it to be an outbreak of the pest whereas Dr Bauzon, who was asked by the aldermen for an expert opinion, maintains that these are malignant fevers, possibly caused by intestinal worms. [Evidently there was some doubt about the diagnosis.]
15–21 July	No new cases.
23 July	Fourteen deaths in rue de l'Echelle.
29 July	Great increase in the number of deaths and emergency measures are introduced. Nearly 10 000 inhabitants had already fled.

These events, baldly described above, cannot be fitted within a framework of an infectious disease with a long incubation period and we conclude that this was not an outbreak of haemorrhagic plague. It appears to have come, not from the Levant where the *Grand St Antoine* originally took its cargo on board, as seems to be generally supposed, but from Tripoli in

North Africa where the Turkish passengers embarked. The voyage from Syria to Libya had taken 9 weeks and was apparently trouble free. One of the passengers died 2 days after joining the ship, whereas the next death (the first sailor) did not occur until 22 days later. Was this an epidemic of bubonic plague? DNA sequences specific for *Yersinia pestis* have been recorded from the dental pulp of plague victims buried at Marseilles in 1722 (Drancourt *et al.*, 1998) and there is evidence from DNA studies of bubonic plague around Provence in the 14th and 16th centuries. The erratic and unpredictable sequence of deaths before the epidemic became established in one area of the town, described in the diary above, might suggest that one of the Turkish passengers who came aboard at Tripoli was infected and that he brought with him infected fleas and rats. When the rats on board were infected in turn and an epidemic was established, bubonic plague spread to the crew after an interval of 3 weeks. The outbreak spread ashore via the rats and porters, guards and possibly the crew of the *Grand St Antoine* and became established in an area of the town during late June–early July.

Subsequent events of this important plague at Marseilles are now briefly described, with the possibility that *Yersinia* was responsible kept in mind.

The disagreement between the doctors when initially diagnosing the disease suggests that the signs may have been different from the descriptions of haemorrhagic plague 50 years before. Most of those dying in the early stages of the outbreak had buboes and carbuncles (Bertrand, 1973). The faces of those dying presented different appearances: one pale, another livid, another yellow, another violet. Some complained of acute pains in the head and in all parts of the body, others were afflicted with severe vomiting or with violent swelling of the abdomen or with burning tumours.

On 1 August 1720 the doctors presented the following report of their visit to the sick of Marseilles:

> 1st . . . the body of a women of sixty years of age, dead after an illness of three days; on whom we found no marks of a pestilential disease on any part of the body. We afterwards visited, in another house, a women of thirty-five years of age, who had a bubo in the groin . . . As we did not, however, perceive any other symptom of a pestilential disease, we have reason to think that the tumour proceeded from a very different malady.
>
> 2dly . . . the body of a girl of twenty years of age . . . she was seized with a violent sickness and head-ache, accompanied with a general faintness, and died in thirty hours after her seizure, covered with livid purple spots, having the belly extremely distended, and of a purpule colour; and having discharged a great quantity of blood at the nose in a very liquefied and serous state . . . attacked with fever, pains in the head, and violent sickness . . . but in none did we find

symptoms which appeared to indicate the contagion.

3dly . . . a woman had died suddenly, four or five days before, suspected of having the plague. Her child of twelve years old had died this day, covered with livid purple spots, and with an excessive tension of the belly, and a swelling near the glands of the groin on the left side. She had been seized, according to the account of those about her, two days before with a violent nausea, and insupportable pains in the head. By the side of this body, on a wretched bed, was her father aged forty years lying down dressed; his face was livid, his eyes sunk and dying; he had been seized two days before with a violent head-ach and vomiting; he had a tumour on the groin . . . he was covered with livid purple spots, and his belly was extremely distended with very violent pains . . . a mother and daughter, both with the face livid and the eyes sunk, and with such an excessive faintness that they were scarcely able to open their eyes: the daughter had been ill two days with the most violent head-ache and nausea; neither of them had any tumours either on the groin or under the arms, nor had they any purple spots.

4thly . . . a girl of twenty years of age recently dead, covered with a livid purple, having been ill three days with violent head-ache and continued vomitings. In a little bed by her side was her brother, thirteen years old, who had been seized the day before with horrible pains in the head and violent efforts to vomit. His eyes were sparkling and inflamed; his tongue was dry and whitish, and his belly swelled, with excessive languor, and a considerable swelling near the groin on the right side, which occasioned him violent pain.

(Bertrand, 1973)

There is evidence from the accounts that the violence of the disease ameliorated in the later stages of the epidemic. Some of those infected suffered only a few days of fever without tumours, others had an eruption that soon regressed and thus a greater proportion of the patients in the autumn recovered. However, many of these experienced a fresh attack in the following spring and there is no doubt that people in this plague could be infected a second time, even at the height of the epidemic (Bertrand, 1973).

It is noteworthy that Roberts (1966), when reviewing the literature on plagues in England for the meeting of the Royal Society of Medicine to commemorate the Tercentenary of the Plague of London in 1665 related in 'a report of the 1720 plague of Marseilles, that the fishermen netted ten thousand dead animals in the harbour, and dragged the corpses out to sea. *Nowhere else has a reference to destruction of rats been found* [our italics]'.

The mortality was enormous; 10 000 people fled before the cordon sanitaire was established and some 50% of the 80 000 who remained died. This is probably a higher mortality than would be expected with bubonic plague, perhaps suggesting that pneumonic plague was responsible for a high proportion of the deaths (see Wu, 1936). The pattern of the outbreak is

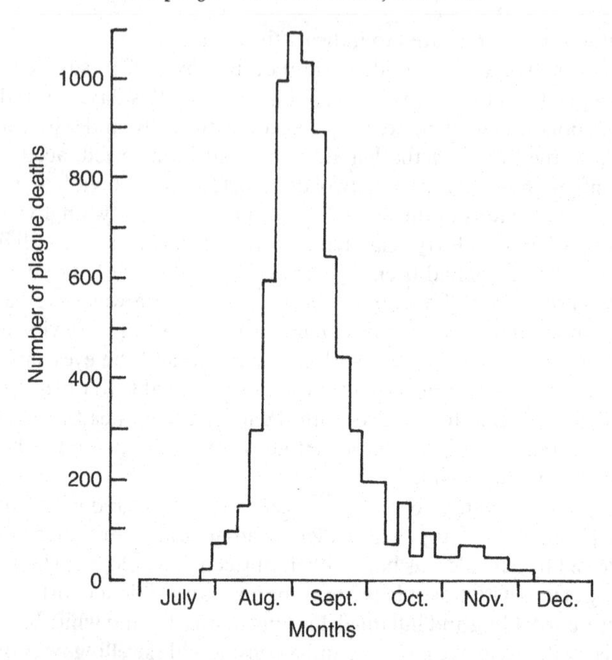

Fig. 12.1. Time-course of the epidemic at Marseilles in 1720: the number of reported plague deaths, July to December.

shown in Fig. 12.1: the deaths were largely confined to August and September 1720, a shorter time-scale than that typically recorded in a major outbreak of haemorrhagic plague. The mortality fell sharply during the autumn but a few cases were reported in the following year between 18 April and 19 August 1721, when the last case was reported and severe quarantine measures were introduced that lasted until 9 November (i.e. twice as long as the 40-day quarantine period usually decreed in England, France and Italy). No new cases were reported during the quarantine period, nor during the following 3 months. However, the plague began again on 3 February 1722; 260 persons were attacked, of which 194 died (75%) and draconian quarantine measures were reintroduced (Biraben, 1972). In previous epidemics throughout Europe, a 40-day quarantine period after the last death was accepted as safe because the disease did not reappear after this time. But human quarantine periods are not applicable to rodent and flea populations and this is good evidence for an outbreak of bubonic plague.

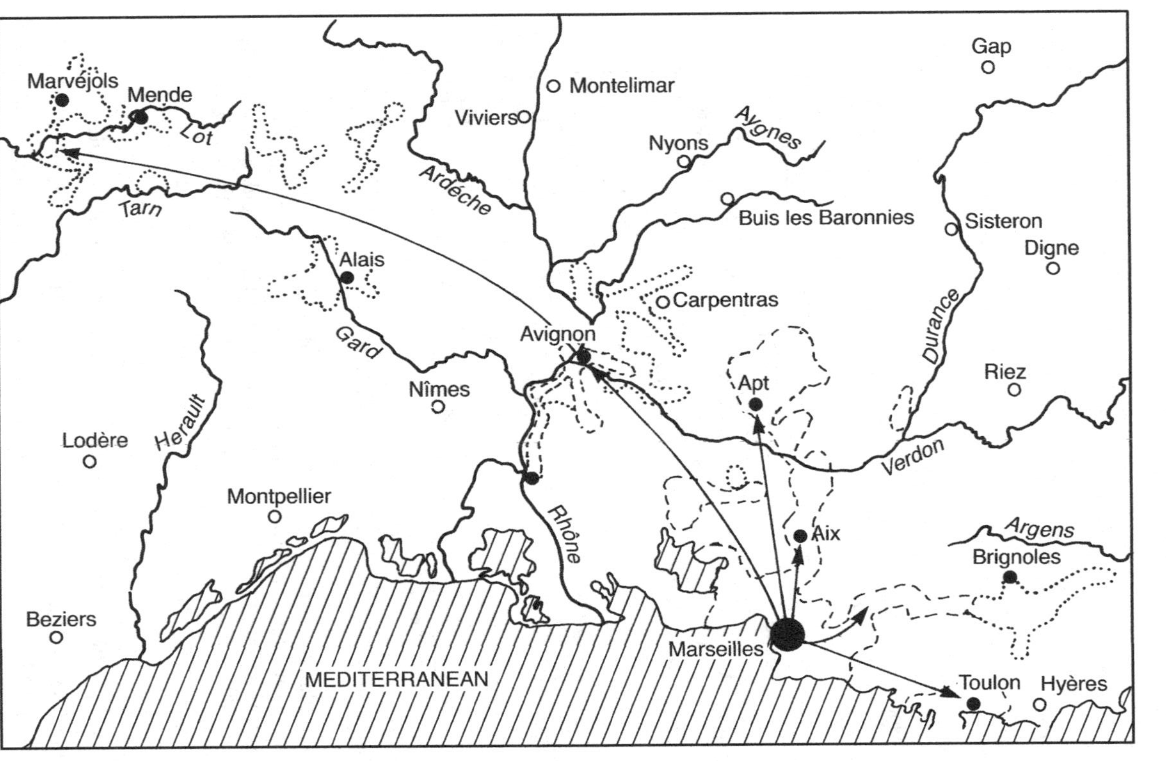

Fig. 12.2. Spread of the plague from Marseilles in 1720–22. Dashed and dotted lines represent approximate extent of the outbreak by 30 December 1720 and 30 December 1721, respectively. The main foci at Aix, Apt and Avignon (see Fig. 12.3) and at Mende and Marvéjols (see Fig. 12.4) are indicated. Populations: large closed circle, over 10 000; smaller closed circles, over 1000. Open circles, localities unaffected by plague. After Biraben (1975).

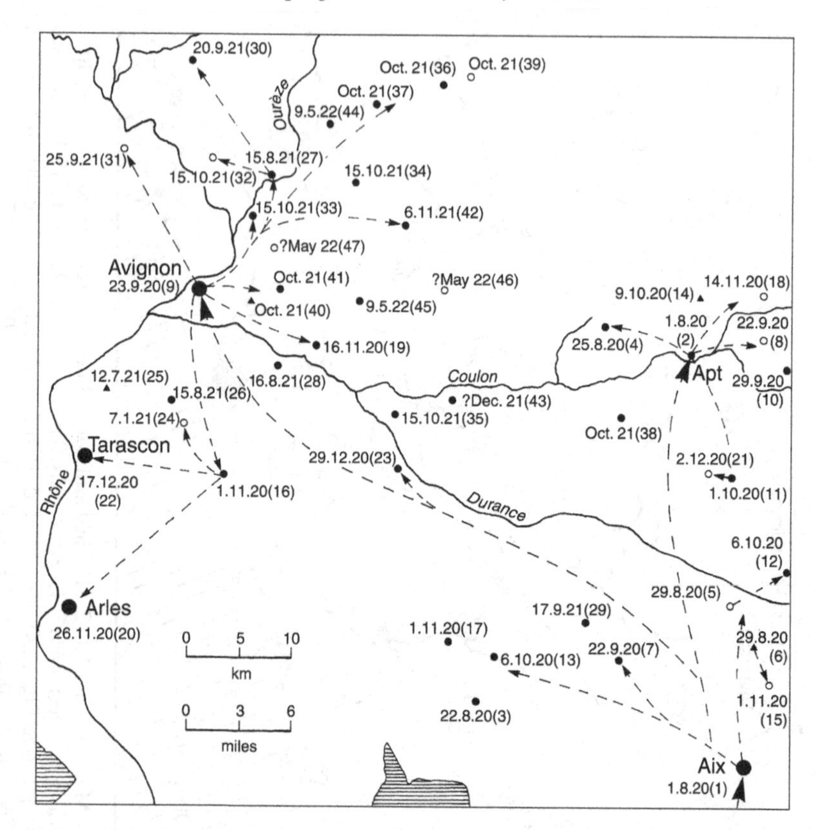

Fig. 12.3. Detailed reconstruction of the plague epidemic in the areas around Aix, Apt and Avignon. Dates indicate the first appearance of the plague in each of the localities which are numbered according to Table 12.1. Populations: closed circles, over 1000; open circles, 200–1000; closed triangles, less than 200. After Biraben (1975).

12.1 Spread of the plague from Marseilles to the countryside

Meanwhile, the plague had, soon after the start of the epidemic, spread outwards from Marseilles into the countryside of Provence, arriving simultaneously at Aix and Apt on 1 August 1720 (Figs. 12.2 and 12.3; Table 12.1). An inefficient blockade of Marseilles was instituted on 1 August and a decision was made on 4 August to install protective cordons, although these were not put in place until 20 August. From that time, people were not allowed to cross the barriers, extreme caution was exercised and merchandise was allowed to pass only at four specified points. In addition to the blockade round Marseilles, cordons sanitaires were established:

Table 12.1. *Sequential appearance of the plague in the area around Aix, Apt and Avignon*

Locality	Date of appearance of plague	Estimated population in 1720	Mortality %
1 Aix	1.8.1720	28 000	23.9
2 Apt	1.8.1720	4900	5.5
3 Lançon	22.8.1720	1700	4.8
4 Roussillon	25.8.1720	1000	13.8
5 Le-Puy-Sainte-Réparade	29.8.1720	800	5.4
6 Saint-Canadet	29.8.1720	100	34.0
7 Saint-Cannat	22.9.1720	1250	27.4
8 Caseneuve	22.9.1720	750	2.4
9 Avignon	23.9.1720	23 041	31.4
10 Saint-Martin-de-Castillon	29.9.1720	1250	26.1
11 Cucuron	1.10.1720	2700	30.8
12 Perthuis	6.10.1720	3000	12.1
13 Pélissanne	6.10.1720	2000	11.2
14 Villars-Brancas	9.10.1720	100	12.0
15 Venelles	1.11.1720	410	8.1
16 Saint-Rémy	1.11.1720	3000	33.6
17 Salon	1.11.1720	4200	21.7
18 Rustrel	14.11.1720	700	1.9
19 Caumont-sur-Durance	16.11.1720	1300	
20 Arles	26.11.1720	23 170	44.3
21 Vaugines	2.12.1720	250	13.2
22 Tarascon	17.12.1720	7000	3.0
23 Orgon	29.12.1720	1750	6.0
24 Maillane	7.1.1721	750	14.1
25 Boulbon	12.7.1721	60	31.7
26 Graveson	15.8.1721	1000	1.1
27 Bédarrides	15.8.1721	1400	22.0
28 Noves	16.8.1721	1228	14.5
29 Lambesc	17.9.1721	2500	
30 Orange	20.9.1721	5000	8.2
31 Saint-Genais	25.9.1721		
32 Châteauneuf-du-Pape	15.10.1721	900	
33 Sorgues	15.10.1721		
34 Monteux	15.10.1721	2600	
35 Cavaillon	15.10.1721	5000	
36 Caromb	Oct. 1721	3000	
37 Aubignan	Oct. 1721	1500	
38 Bonnieux	Oct. 1721	2900	
39 Crillon	Oct. 1721	650	
40 Montfavet	Oct. 1721	70	
41 Morières	Oct. 1721	1000	
42 Pernes	6.11.1721	3500	
43 Robion	?Dec. 1721	1200	

Table 12.1. *(cont.)*

Locality	Date of appearance of plague	Estimated population in 1720	Mortality %
44 Sarrians	9.5.1722	1750	
45 Le Thor	9.5.1722	2400	
46 Saumane	?May 1722	650	
47 Vedènes	?May 1722	800	10.1

See Fig. 12.3.
Data source: Biraben (1975).

(i) on the west bank of the Rhône; (ii) along lines in an easterly direction from the Rhône, north and south of Avignon and south of Arles; (iii) along the River Orb; and (iv) in a ring around the Languedoc area. The outbreak had not spread far before 20 August (Fig. 12.3; Table 12.1) but these cordons sanitaires were completely ineffective and the plague eventually spread to over 250 other localities in Provence and neighbouring districts (Fig. 12.2). This is in contrast with the situation at Eyam, 55 years previously, where none outside the cordon around the village was infected. We conclude that this difference in epidemiology is suggestive evidence of bubonic plague spread by rodents in Provence in 1720–22.

The times of appearance of the plague in the hamlets, villages and towns is listed sequentially in Table 12.1 and its initial geographical spread is shown in Fig. 12.3. In 2 months, by the end of September 1720, the epidemic had: (i) spread out radially in a cluster in the environs of Marseilles; (ii) appeared in Toulon and started to spread therefrom (Fig. 12.2); (iii) spread locally in clusters from its starting points at Aix and Apt; and (iv) appeared on 23 September at Avignon, 28 miles from Apt, where it persisted for 11 months. The epidemic then spread outwards from these foci so that over 50 localities were affected by the end of the year: (i) a tongue began to spread eastwards from Marseilles towards Brignoles (Fig. 12.2); (ii) Arles was first affected on 26 November 1720, the spread being from the environs of either Aix or Avignon; and (iii) plague appeared at its furthest westward point at the village of Auxillac in the Languedoc, presumably having travelled from Avignon, and 8 days later it appeared at nearby La Canourgue, where over 900 died (55% of the population) (see Fig. 12.4). The plague did not spread from here at first, but this outbreak acted later as the centre for an epidemic in the Mende district that began in summer 1721 and affected a cluster of 45 villages and hamlets (Table 12.2; Fig. 12.4).

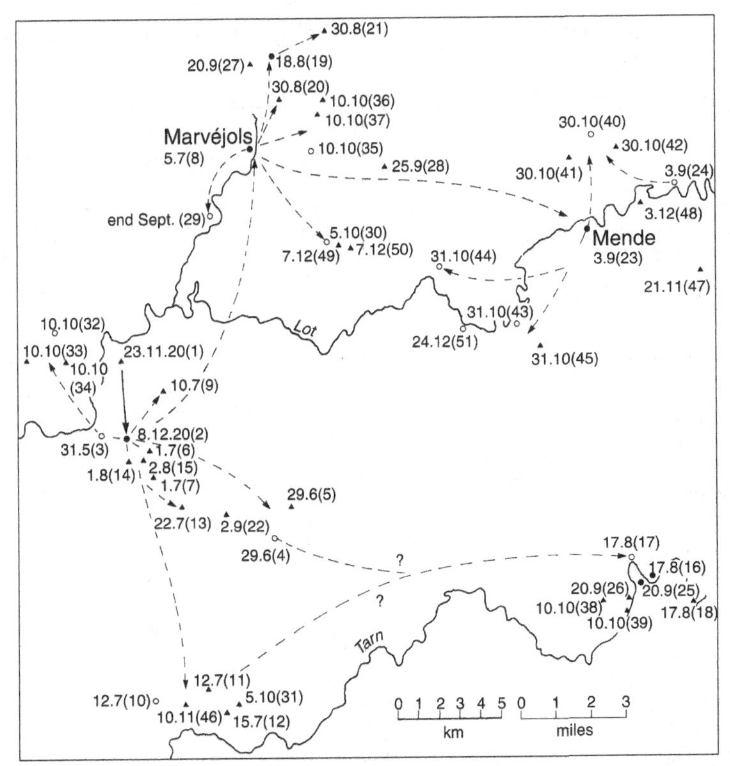

Fig. 12.4. Detailed reconstruction of the plague epidemic in the areas around Mende and Marvéjols. Each locality is numbered according to Table 12.2 and the dates indicate the first appearance of the plague. Populations: closed circles, over 1000; open circles, 200–1000; closed triangles, less than 200. Note the small size of the localities affected in this rural area. After Biraben (1975).

The epidemic continued to spread slowly from these foci to the east of the Rhône, affecting neighbouring villages in January 1721, but was markedly slower during February to the end of May 1721. Thereafter it exploded again during June to September 1721, again spreading to neighbouring villages and towns and establishing another focus in the region around Alais to the west of the Rhône.

This outbreak in the countryside around Marseilles to the east of the Rhône, therefore, lasted about 16 months although it persisted until summer 1722 in some localities in the Languedoc area. It spread a maximum distance from Marseilles of 135 miles. In general, its movement was gradual, travelling to nearby localities, but there were examples of apparently saltatory behaviour. This was a rural outbreak with the epidemic clustering in a great many localities in a relatively small area of the countryside. The

Table 12.2. *Sequential appearance of the plague in the rural areas around Mende and Marvéjols*

Locality	Date of appearance of plague	Estimated population in 1720	Mortality %
1 Corréjac	23.11.1720	109	61.5
2 La Canourgue	8.12.1720	1633	56.0–58.0
3 Banassac	31.5.1721	900	23.8
4 La Capelle	29.6.1721	200	43.5
5 Brunaves	29.6.1721		
6 Saint-Frézal	1.7.1721	110	40.0
7 Maleville	1.7.1721		
8 Marvéjols	5.7.1721	2756	
9 Cadoule	10.7.1721		
10 Saint-Georges-de-Levejac	12.7.1721	400	15.2
11 Serres	12.7.1721	40	
12 La Calcidouze	15.7.1721	<12	
13 La Masmontet	22.7.1721		
14 La Bastide	1.8.1721	<20	
15 Trémoulis	2.8.1721		
16 Ispagnac	17.8.1721	1800	11.8
17 Molines	17.8.1721		
18 Biesse	17.8.1721		
19 Saint-Léger-de-Peyre	18.8.1721	2000	19.0
20 Valadou	30.8.1721		
21 Rechiniac	30.8.1721		
22 Fontjulien	2.9.1721	87	
23 Mende	3.9.1721	3800	28.0
24 Badaroux	3.9.1721		
25 Quézac	20.9.1721	1150	7.0
26 Le Buisson	20.9.1721		
27 Les Gratoux	20.9.1721		
28 Gabriac	25.9.1721	130	6.9
29 Chirac	End Sept. 1721	850	1.3
30 Grézes	5.10.1721	400	41.5
31 Les Caynoux	5.10.1721		
32 Saint-Germain-du-Teil	10.10.1721		
33 Montagudet	10.10.1721		
34 Malbousquet	10.10.1721		
35 Montrodat	10.10.1721	500	32.0
36 Vimenet	10.10.1721		
37 Inosses	10.10.1721		
38 Le Mas-André	10.10.1721		
39 Le Chambonnet	10.10.1721		
40 Chastelnouvel	30.10.1721	200	22.0
41 Le Crouzet	30.10.1721		
42 Alteyrac	30.10.1721	225	
43 Balsiège	31.10.1721		
44 Barjac	31.10.1721		

Table 12.2. *(cont.)*

Locality	Date of appearance of plague	Estimated population in 1720	Mortality %
45 Lasfons	31.10.1721		
46 Le Masrouch	10.11.1721		
47 Lannuéjols	21.11.1721		
48 Les Bories	3.12.1721		
49 La Serre	7.12.1721		
50 Boudoux	7.12.1721		
51 Bramonas	24.12.1721		

See Fig. 12.4.
Data source: Biraben (1975).

plague struck indiscriminately at towns as well as at the tiniest hamlets (some with only four houses) so we conclude that it was not density dependent. Even so, some ancient and major centres that would have been on trading routes, such as Nîmes, Uzes, Montpellier and Carpentras, escaped completely (see Fig. 12.2). There was no pattern to the mortality: some localities, with populations varying from 400 to 26 000 recorded 40% to 70% mortalities; others with populations varying from 58 to 1200 recorded fewer than 12 deaths (see Table 12.2). There is little convincing evidence that the plague spread along the rivers, but it successfully crossed the Rhône and the Durance.

The patchy distribution, clustering in many of the villages and hamlets in a restricted area of the countryside is completely unlike the plague epidemics reported in France between 1347 and 1666, in which the major outbreaks in substantial towns were often widely separated, sometimes by hundreds of miles. The different mortalities experienced is noteworthy; only one person died from a population of 1200 in Pourrières and only six from 2000 in Gardane, whereas 313 died from 450 (70%) at Neoules.

We believe that the foregoing is good evidence for a localised spread of bubonic plague in Provence in 1720–21, although there are some surprising features. Some localities, both small and large, in both winter and summer, showed a mortality of over 50% of the population; the household death rate may have been markedly increased by pneumonic plague. The epidemic spread first from Marseilles northwards and erupted simultaneously on 1 August at Aix (24 000 inhabitants) and Apt (6000 inhabitants), distances of 15 and 39 miles, respectively, 41 days after its appearance at the seaport. This appears to be a short time to establish an infection of *Yersinia pestis* among the rodents in this northwards corridor (see Fig. 12.3). A possible explanation may lie in the account given by Bertrand (1973) of letters and papers that describe the plague as being present at Marseilles in 1719 and certainly before May 1720 when the *Grand St Antoine* docked. This observation was supported by a study of the mortuary registers, where several persons were registered as having died of the plague. In addition, Bertrand discovered that many men, women and children had shown symptoms of the disease. If these assertions were correct, bubonic plague would have had a much longer time to become established in the rodent populations of Marseilles and its environs before the arrival of the *Grand St Antoine*. The rural outbreak lasted, effectively, for about 2 years, before disappearing from the countryside forever. Presumably the rats were killed by *Yersinia pestis* and it was not possible to form a reservoir among local resistant species.

12.2 Spread of the plague at Aix and Apt

As we have said, the plague spread northwards from Marseilles early in the epidemic, appearing simultaneously on 1 August 1720 in Aix and Apt, both large-sized towns of some importance. The spread from these foci is shown in detail in Fig. 12.3, where the numbered locations are named in Table 12.1. Epidemics appeared at a number of sites, both large and small in the succeeding 2 months but thereafter dispersion was much slower and desultory. Plague then broke out at Avignon 7 weeks after its appearance at Aix (Fig. 12.3), perhaps being carried via the rat population from there or from Apt or because of direct transmission from Marseilles. There was limited spread of the plague from Avignon in 1720 before the winter and there were major epidemics only at Tarascon and Arles. However, in the summer and autumn of 1721 Avignon acted as a focus and epidemics broke out over a wide area; these continued in the spring of 1722. These outbreaks occurred mainly in villages and hamlets that were close to each other, a pattern that was completely different from the earlier epidemics of haemorrhagic plague in France.

12.3 Spread of the plague at Mende and Marvéjols

Plague arrived in this deeply rural area (see Fig. 12.4) at Corréjac (population 109) on 23 November 1720, probably having come from Avignon. There was little activity before winter set in (Table 12.2) but the epidemic became quite widespread in the summer and autumn of 1721: subfoci developed successively at La Canourgue, Marvéjols and Mende. Again, the majority of locations had a population of below 200 (Fig. 12.4; Table 12.2).

The epidemic of 1720–22 in Marseilles and its hinterland, some 50 years after the disappearance of haemorrhagic plague, is of particular interest because its spread has been reconstructed in such detail. The existence of *Y. pestis* DNA in the dental pulp of victims in Marseilles (Drancourt *et al.*, 1998) confirms the probability that this was an outbreak of bubonic plague. The characteristics of its spread through Mende and Marvéjols are completely different from those of haemorrhagic plague. There had been previous outbreaks of bubonic plague in this coastal region of southern France for hundreds of years. Raoult *et al.* (2000) have identified *Y. pestis* DNA in teeth from graves at Montpelier dated between the 13th century and the late 14th century; the victims did not necessarily die in the Black Death as these authors presume. Drancourt *et al.* (1998) have found similar evidence of bubonic plague in victims buried in 1590 at Lambesc, Provence.

13
Conclusions

13.1 The receptor for the entry of HIV

Information about the nature of the infectious agent in the Black Death has recently come from an unexpected source, namely the studies in molecular biology of acquired immunodeficiency syndrome (AIDS). The discovery that the *CCR5* gene product encodes a transmembrane G-protein-coupled receptor on macrophages and monocytes that serves as an entry port (or chemical doorway) for primary human immunodeficiency virus (HIV)-1 strains represented a major step forward in our understanding of this disease (Alkhatib *et al.*, 1996; Choe *et al.*, 1996; Deng *et al.*, 1996; Doranz *et al.*, 1996; Dragic *et al.*, 1996). A 32 base-pair deletion mutation that interrupts the coding region of the *CCR5* chemokine receptor locus on human chromosome 3p21 was subsequently described (Dean *et al.*, 1996; Liu *et al.*, 1996; Samson *et al.*, 1996); this *CCR5-Δ32* mutation, which causes a frameshift at amino acid residue 185 (Carrington *et al.*, 1997), leads to truncation and loss of the receptor on lymphoid cells so that homozygous individuals have nearly complete resistance to HIV-1 infection and heterozygotes for the mutation delay the onset of AIDS (Dean *et al.*, 1996; Huang *et al.*, 1996; Biti *et al.*, 1997; Michael *et al.*, 1997; O'Brien *et al.*, 1997; Theodorou *et al.*, 1997; Zimmerman *et al.*, 1997). Sixteen additional mutations in the coding region of the *CCR5* gene have now been identified (Carrington *et al.*, 1997).

The *CCR5-Δ32* deletion has a high allele frequency in several Caucasian populations (Dean *et al.*, 1996; Liu *et al.*, 1996; Samson *et al.*, 1996; Huang *et al.*, 1996; Michael *et al.*, 1997; Martinson *et al.*, 1997), and its rarity or absence in non-Caucasian populations led to speculation that the mutation occurred only once in the ancestry of the Caucasian ethnic group, subsequent to their continental isolation from African ancestors (Dean *et al.*, 1996; O'Brien & Dean, 1997). New mutations would have a very high likelihood of being lost within a few dozen generations and it is highly unlikely that a single *CCR5-Δ32* variant as a strictly neutral mutation did increase to modern frequencies across Europe/Asia by random genetic drift.

A recent survey of over 4000 individuals in 38 ethnic populations revealed a cline of *CCR5-Δ32* allele frequencies of 0% to 14% across Eurasia, whereas the variant is absent among native African, American Indian and East Asian ethnic groups. This might explain the rapid spread of HIV in sub-Saharan Africa, whereas possession of the *CCR5-Δ32* allele may have delayed its progress in Europe. The time of the origin of the *CCR5-Δ32* mutation has been estimated on the basis of the persistence of a common and likely ancestral three-locus haplotype retained in linkage disequilibrium. The age of the *CCR5-Δ32*-bearing haplotype has been computed by these methods to be approximately 700 years old (but with a wide range of 275–1875 years) and it has been suggested that this haplotype was driven upwards to the present-day frequencies of 5% to 15% by a historic, strong selective event, probably an enormous mortality mediated by a widespread epidemic of a pathogen that, like HIV-1, utilised *CCR5* for entry into lymphoid cells (Stephens *et al.*, 1998). The Black Death is an excellent candidate for such a catastrophic event and it has been suggested that the epidemic provided the strong selective pressure that drove up the *CCR5-Δ32* mutation some 650 years ago (Stephens *et al.*, 1998). However, Stephens *et al.* (1998) assumed, of course, that the Black Death was caused by the plague baccillus, *Yersinia pestis*, which carries a 70 kilobase plasmid that encodes an effector protein, Yop1, which enters macrophages, causing diminished immune defences. They inferred that this bacterial protein enters via the *CCR5* receptor.

Carrington *et al.* (1997) went further and concluded that the high predominance of codon-altering alleles among *CCR5* mutants is consistent with an adaptive accumulation of function-altering alleles for this gene, perhaps as a consequence of historic selective pressures. Mummidi *et al.* (1998) also drew attention to the complex array of polymorphisms and genetic determinants in the HIV–host interplay. Libert *et al.* (1998) concluded that most, if not all, Δ*CCR5* alleles originated from a single mutation event which probably took place 'a few thousand years ago' in northeastern Europe, which would pre-date the Black Death. They agreed that the high frequency of the *CCR5-Δ32* allele in Caucasian populations cannot be readily explained by random genetic drift and that a selective advantage is, or has been, associated with homo- or heterozygous carriers.

The Black Death may therefore have been a major factor in *a continuum of plague epidemics* that drove up the frequency of *CCR5* mutants (see also Rottman *et al.*, 1997; Samson *et al.*, 1997; Kirchhoff *et al.*, 1997; Simmons *et al.*, 1997; Madani *et al.*, 1998; Rizzuto *et al.*, 1998; Stephens *et al.*, 1998; Wang *et al.*, 1998; Zhang *et al.*, 1998; Hussain *et al.*, 1998).

Thus these recent studies of the receptor for the HIV virus (Stephens *et al.*, 1998), surprisingly, reveal new information about the probable etiology of the Black Death, the pandemic of 650 years ago. Since the *CCR5-Δ32* mutation provides protection against the entry of a virus (HIV-1) into lymphoid cells, the causative agent of the Great Pestilence was probably also viral and targeted macrophages and monocytes. The continuation of plague epidemics for the next 300 years probably maintained and boosted the *CCR5-Δ32* mutation and this continuing historic selection pressure also accounted for the appearance, accumulation and persistence of the additional mutations that have now been identified.

13.2 Was the same causative agent responsible for all the plagues in England from 1348 to 1666?

This is an impossible question to answer because we have no detailed information on the seasonal pattern of the deaths in the Great Pestilence and can rely only on the institutions to vacant benefices in the different dioceses that have been so carefully derived and collated by Shrewsbury (1970) (see Fig. 4.2). Indeed, until parish registers began to be kept in the mid-16th century we have little firm quantitative evidence of the pattern of any of the epidemics. Nevertheless, the seasonal pattern of the Black Death in 1349 in the dioceses of Hereford and Gloucester, Lichfield, Norwich, Lincoln and Ely show a remarkable similarity to the epidemics seen in towns and cities in the provinces in the 17th century in England; each outbreak lasted for the characteristic lengthy period of 9–12 months, with a clear peak in July–August.

The limited and largely anecdotal on-the-spot descriptions of the symptoms and autopsy reports of such plagues as the Black Death (section 4.2) and the subsequent epidemics in Italy (section 11.3.2), London (section 8.6.3) and Kendal (section 7.9), are broadly similar.

Whenever Shrewsbury described an epidemic that even he saw could not be bubonic plague because of the seasonality or other features such as a high mortality, he resorted to averring that it must have been an outbreak of typhus. Inspection of contemporaneous records suggests the exact reverse: it is evident that, certainly by the 16th century, they were able to identify accurately the pestilence and they could distinguish it from the English Sweating Sickness (section 6.1). They were required by law in Elizabethan times to record plague deaths in the parish registers and they certainly came to recognise the typical symptoms such as the tokens (section 8.6.3). Furthermore, typhus epidemics do not follow Reed and Frost dynamics because it is a disease with an arthropod vector.

There were probably many minor outbreaks in isolated families where the plague went unreported. Equally, there may have been many small and contained outbreaks that were faithfully recorded in the parish registers between 1560 and 1660 (e.g. Slaidburn, 1632; section 9.4) and would be of considerable interest; we should be very pleased to receive the details of such records from people searching the registers.

In conclusion, all the major plague epidemics exhibited Reed and Frost dynamics typical of an infectious disease with a long incubation period. All the results are consistent with latent and infectious periods of about 10–12 days and 25–27 days, respectively. We believe, therefore, that the same causative agent was probably responsible for all the plagues in England from 1348 to 1666.

However, there are features of the Black Death that distinguish it from the well-documented plagues after about 1560. In the diocese of Bath & Wells, the epidemic began in autumn 1348, was at its peak in mid-winter in January and was finished by April (see Fig. 4.2B). In the diocese of Salisbury, the epidemic was rampant through December to February and reached its peak in March (see Fig. 4.2A). Shrewsbury (1970) was unable to explain these mid-winter explosions as bubonic plague. This seasonal pattern is different from the usual behaviour of the plague in the 15th century in England, where the epidemic struggled to continue through the winter and infectivity between households was low (sections 13.4 and 13.5), although there are a few accounts of the plague peaking in winter in England in the 17th century, for example at Durham (section 9.5.1), Ottery St Mary (section 9.5.2) and Colchester (section 9.6.2). The Black Death arrived in England in summer 1348 and a major epidemic occurred whenever it reached a community, irrespective of season, albeit with a slow build-up because of the long incubation period. It arrived in the West Country in autumn 1348 and a winter epidemic ensued. All the evidence suggests, therefore, that the Black Death struck a completely naive metapopulation in England and infectivity was high irrespective of the time of year. As the years went by, with the seemingly interminable succession of plagues in the commonly infected foci, there would be changes in resistance (see the *CCR5-Δ32* receptor, section 13.1) and in the proportion of the population that were immune having recovered from an earlier epidemic. Social behaviour would change over 300 years, particularly during an outbreak and the rich learned to flee to the country at the first signs of an epidemic. The communities learned to recognise the first victims, information was exchanged as to where the plague was raging and quarantine measures were introduced. In addition it is possible that the infectious agent underwent mutations during the 300 years when plagues swept through the metapopulation. All these factors would interact to change the effective contact rate, the mortality and the spread of the disease.

13.3 Was *Yersinia pestis* the infectious agent in the plagues?

There is the possibility that some of the outbreaks in the 16th century that were confined to London (i.e. did not spread to the provinces) might have been bubonic plague caused by rats and blocked fleas that were brought into the port in the summer from overseas, spreading *Yersinia* to black rats in the crowded metropolis. There is no supporting evidence for this view and we believe that *Yersinia* was probably not responsible for any *significant* epidemics in England before the 20th century, although it probably broke out for short periods around the Mediterranean coast, as for example in Barcelona (section 11.4.2) and Marseilles (Chapter 12). It was certainly not the causative agent in the Black Death or in any other of the outbreaks in England where we have any detailed information.

We have drawn attention throughout this book to the many inconsistencies to the generally held view that *Yersinia* was responsible for the plagues in Britain between 1348 and 1666 and some of these may be briefly summarised as follows.

(i) The complex dynamics of bubonic plague, with fleas, rats, resistant rodents and, occasionally, humans all involved as hosts, impose severe constraints on the biology of the disease. The epidemics of bubonic plague are described in Chapter 3 and have completely different characteristics from those of the plague outbreaks in England and Europe; they do not follow Reed and Frost dynamics.

(ii) The foregoing conclusions were evident to many previous workers on English plagues but they asserted that the pestilence was spread by the pneumonic form of bubonic plague. This was impossible. Pneumonic plague develops and is infectious only in the later stages of the disease. It spreads by person-to-person infection and is a lethal form of bubonic plague; it can spread readily to others in the household, particularly those involved in the nursing. However, those dying in the terminal stages of haemorrhagic plague or pneumonic plague would have been too ill to move very far and it is impossible that they could have spread the epidemic over any distance.

(iii) The only rat in England during the period 1348–1666 was the black rat (*Rattus rattus*); the brown rat (*Rattus norvegicus*) did not spread from Russia and arrive in England until the early 18th century. The black rat does not move far from the warmth of human habitations or from warehouses in the seaports in temperate regions, although it spread inland more widely in the tropics. The status of the rat species during the plagues in England has been discussed in detail by Twigg (1984) and it is certain that there were no rodents, particularly in inland northern Britain, to carry bubonic plague through the supermetapopulation of Europe.

(iv) There is no evidence of any resistant rodent species available in England or Europe throughout the period 1348–1666 to establish a buffer epizootic through the metapopulation. Bubonic plague cannot spread without such a pre-established reservoir rodent population (Plague Research Commission, 1907a,b); a pre-infected rodent population throughout Europe would have been a prerequisite for the rapid spread of the Black Death.

(v) The life histories of the fleas and the rats are very sensitive to climatic conditions, particularly temperature. These are of lesser importance in subtropical regions, but are critical in temperate areas and it is impossible that fleas could have been breeding and black rats could have been active during the epidemics of the winter months in England and Scotland, nor in Greenland, Iceland and Norway during the Black Death and subsequent plagues. Shrewsbury (1970) was aware that the seasonality of some plague epidemics meant that it was impossible that *Yersinia* was the causative agent and so he assumed either that the winter 'must have been mild' or that it was really an outbreak of typhus.

(vi) The arrival of large steamships in the early years of the 20th century meant that, for the first time, large populations of infected rats and fleas could be transported rapidly by sea from the port warehouses of Asia to other subtropical countries where bubonic plague could be established. It is most unlikely that any infected rats could have survived the lengthy sea voyages to the northern latitudes of Greenland or Iceland in the boats available in the 14th and 15th centuries.

It is improbable that haemorrhagic plague was endemic in Britain before about 1600 and each epidemic was introduced from the Continent or Ireland. Some outbreaks were also spread by boats trading up the east coast at ports between London and Scotland. The inhabitants were well aware of the dangers of transmission by this means and, by the late 16th century, they were prohibiting from docking boats that came from ports where the plague was raging. It is *most* unlikely that all these introductions and reintroductions were of *Yersinia* carried by fleas, rats or humans. On the other hand, whereas it is difficult to envisage bubonic plague being spread by sea in this way, the transmission of an infectious disease with a long incubation period of about 32 days, spread person-to-person is readily understandable. The port authorities prohibited entry to any person showing symptoms of plague but this measure was largely ineffectual: a man could come aboard a ship on the Continent having contracted the plague, say, a week previously, he could then infect many of the crew during the voyage and all would have been apparently healthy and free of symptoms when they came ashore in London or any of the ports on the south or east coasts. Widespread tertiary infections would then begin in the crowded conditions around the docks.

(vii) Equally, bubonic plague could not have spread in the saltatory fashion seen in many of the epidemics in England and, particularly, in France when it jumped more than 30 miles in a few days with no intermediate outbreaks. In the outbreak at Malpas (section 9.3) it travelled nearly 200 miles in this way. The speed of this transmission is completely inconsistent with bubonic plague: the Plague Research Commission (1907a,b) gave an example where the epidemic in rats took 6 weeks to travel 300 feet; this slow diffusion is one of the most common characteristics of bubonic plague and we see this pattern in the hinterland of Marseilles in 1722 (Chapter 12). Bubonic plague in South Africa in 1899–1925, moved about 8–12 miles per year and this spread may have been aided by steamtrains (Twigg, 1984). The Black Death, in stark contrast, spread from the toe of Italy to northern Europe in 3 years, and, again, the rapidity of spread of haemorrhagic plague over long distances is explicable as the movement of apparently healthy infectives travelling on foot or by horse.

(viii) Wherever detailed information from the parish registers is available, as at Penrith and Eyam, analysis is consistent with an epidemic of an infectious disease that spread person-to-person with clearly defined characteristics. The build-up of the epidemic was slow, the consequence of a long incubation period, but thereafter it exploded and followed typical Reed and Frost dynamics.

(ix) *Yersinia* is a bacterium whereas the studies with the *CCR5* receptor suggest that the infectious agent in the plague was viral (see section 13.1). It is noteworthy that the *CCR5-Δ32* mutation, which is supposed to provide protection from bubonic plague, is not found in ethnic populations from eastern Asia, the area where the disease has been endemic for many years.

(x) Because bubonic plague is dependent on the biology and behaviour of fleas, rodents and humans, the outbreaks have complex dynamics and appear to be haphazard. In contrast, plague epidemics were sharply defined and predictable and all followed a standard pattern (see our analyses in different parishes in different years). Shrewsbury (1970) was aware of this marked discrepancy.

(xi) Mortality in outbreaks of bubonic plague is always relatively low whereas, where we have detailed information, the death toll when the pestilence struck a naive population in England or continental Europe was probably about 30% to 40%. Plague mortality was probably even higher during the Black Death, although we have no concrete evidence. In Venice in 1347–48 and Genoa in 1656–57, 60% of the population is estimated to have died; half the population of Milan died in an outbreak of plague in 1630, and perhaps half the population of Padua in 1405 and of Lyons in 1628–29; the death toll reached 30% in Venice in 1630–31 (Slack, 1988). Over the period October 1630 to August 1631, 1198 patients were admitted to the pest-houses in Pistoia in Tuscany and, of these, 607 (50%) died (Cipolla, 1981). Again, Shrewsbury (1970) was aware of this major discrepancy and was forced to the conclusion that the mortality in the Great Pestilence was greatly overestimated.

(xii) Endemic bubonic plague is essentially a rural disease because it is an infection of rodents. The Black Death, in contrast, struck indiscriminately in the countryside and in the towns. *Major* epidemics of later plagues were mostly, but not entirely, confined to the towns. In summary, haemorrhagic plague epidemics were density dependent whereas bubonic plague (as in the Mende area at Marseilles) was not.

(xiii) Since bubonic plague is a disease of rodents, the arrival of an outbreak is frequently presaged by rats dying in the streets and yet Shrewsbury (1970), who believed that *Yersinia* was the infectious agent in the pestilence, found only one mention in English writings on plague of rat and mouse mortality during an epidemic: in Leeds in 1645 there is a contemporaneous observation that 'in June the air was then very warm, and so infectious that dogs and cats, rats and mice died, also several birds in their flight over the town dropped dead' (Shrewsbury, 1970). Ell (1980) also commented that 'Historians have noted that contemporary accounts omit any mention of rat mortality'. Roberts (1966) recorded that 10 000 dead animals were found in the harbour during the plague (probably bubonic) at Marseilles in 1720 but 'nowhere else has a reference to destruction of rats been found'.

(xiv) Shrewsbury (1970) has calculated the population density for each English county during the Black Death and classified them as follows:

(a) The most densely populated: Norfolk, Bedfordshire, Suffolk, Northamptonshire and Leicestershire with over 100 persons to the square mile (range = 101 to 119).

(b) Medium densities, ranging from 99 (Rutland) down to 32 per square mile (Lancashire). Devonshire, Worcestershire, Shropshire and Yorkshire lay in the middle of this range with 50–60 persons to the square mile.

(c) Very low density ranging from 20 to 25 persons per square mile: Cumberland, Northumberland, Durham, Westmorland and Cheshire. Shrewsbury (1970) stated categorically that 'These densities are so low that it would have been biologically impossible for bubonic plague to have spread over any of these counties in the fourteenth century . . .'. He continued (and Shrewsbury firmly believed that the Black Death was an outbreak of bubonic plague): 'The epidemiology of bubonic plague renders it improbable that [*Yersinia*] *pestis* could have been distributed by rat-contacts as epizootic plague in any English county in 1348–9 having an average density of population of less than 60 persons to the square mile' and yet we have described in Chapters 7 and 9 how plague ravaged these counties of low density.

(xv) A 40-day quarantine period was first instituted in the city states of northern Italy in the late 14th century and this was gradually adopted throughout Europe and was maintained for the next 300 years until the plague disappeared. Presumably, this proved to be a tried and tested formula for handling *what they obviously believed to be an infectious disease.* It agrees well with our estimated period of about 37 days from infection to death. Obviously the quarantine period is applicable only for a directly transmitted infectious disease and would be completely inappropriate and ineffective for dealing with bubonic plague, a disease of rodents that was accidentally and erratically spread to humans, as was seen in the outbreak at Marseilles.

(xvi) Rat-borne bubonic plague rarely produces more than one case per household (Ell, 1980) and yet the analyses presented in the preceding chapters show that epidemics in England were characterised by a high household contact rate, particularly in the early stages of an epidemic, which compared with a low interhousehold rate. Ell (1980) described records from late medieval Italy showing that 96% of the families examined had multiple cases of plague per household and a high household effective contact rate was regarded as a diagnostic feature of plague by the city health authorities in Italy.

(xvii) Throughout the preceding chapters, we have emphasised the spatial components of the epidemiology of the plague in different metapopulations, both in England and in continental Europe at different times during its 300-year history (see summary in section 13.9). These varying patterns of spread are explicable as the interhuman transmission of an infectious disease with a long incubation period that permitted long distance movement of infectives. As transport improved and movement within a metapopulation became more general in the 16th and 17th centuries, so the epidemics became more numerous and widespread. Such a consistent saltatory movement and spread of the epidemics over 300 years would be impossible in bubonic plague, which is dependent on an epizootic in rats.

(xviii) The possibility of interhuman transmission of medieval plague in England and in continental Europe has been explored by several workers. Ell (1980) suggested that high mortality, the autumn seasonality, lack of evidence of rat mortality and multiple cases per household argue against rat-borne plague, whereas the seasonality plus the presentation of buboes rules out pneumonic spread. He added that few who visited the sick survived and family members abandoned one another for fear of infection. 'No text examined denied the danger of contact with plague victims' (i.e. that person-to-person transmission was possible). Biraben (1975) emphasised that plagues were strongly favoured by crowds of people, as in urban centres, markets, armies and processions and concluded that the spread was predominantly interhuman, although he suggested that this was via the agency of ectoparasites, the *human* flea and louse. Carmichael (1991) commented on the pestilence in 15th century Milan: 'Were these plagues actually contagious? If eye-witnesses insist that the identified cases of plague evidenced contagion, is there any reason we should doubt that precious testimony?' Cipolla (1981) said, 'it is difficult to believe that the same blocked flea kept jumping from one individual to the next, bringing the plague to three separate households' and he continued that the case in question seems to support the possibility of person-to-person transmission via the human flea.

(xix) Haemorrhagic plague disappeared in about 1670 but bubonic plague remained to the present day in its foci in Asia, and epidemics probably continued to break out in Mediterranean areas, including Marseilles in 1722. Furthermore, the brown rat arrived in Europe in the 18th century and would potentially have been a more effective host for bubonic plague.

(xx) The characteristic signs of haemorrhagic plague were the tokens but apparently these were not regular features of bubonic plague.

Perhaps the most telling point is that the accounts given in Chapters 4–11 form a consistent pattern and, taking them together and considering them objectively, it is impossible to conceive of the plagues not being a disease spread through contact with an infected person. Indeed, apart from the fact that the victims of both diseases presented with enlarged glands and subcutaneous swellings, it is difficult to suggest a more unlikely candidate than *Yersinia* as the infectious agent of haemorrhagic plague. In the 1720s, 60 years after the disappearance of the plague, London was swept by a series of epidemic fevers which were characterised by buboes and carbuncles (Creighton, 1894), confirming that these clinical signs were not exclusively diagnostic of bubonic plague.

13.4 Classification of the plague epidemics in the provinces in England

The variations in the patterns of the epidemics seen in populations in the provinces in England have been classified as follows:

Type (i): The standard plague epidemic seen throughout the period 1348–1666; it appeared in the spring, typically March or April, but sometimes in May, and developed slowly because of the long incubation period and then followed characteristic Reed and Frost dynamics with a peak in July–August followed by a decline through the autumn, with the last plague deaths in November or December. Typical duration of the epidemic: 8–9 months; typical mortality for a naive population: 30% to 50%.

A variant of a type (i) epidemic was at its peak in winter; it seems to have been recorded only during the Black Death and a few outbreaks in the provinces thereafter. Duration, 6–8 months; mortality, probably about 50%. Examples: dioceses of Salisbury and Bath & Wells, 1348–49; Durham, 1644; Ottery St. Mary, 1645.

Also included in this category are isolated, small-scale outbreaks in late spring or summer confined to one or two families or to a small hamlet. Epidemic spread restricted by density-dependent factors. Duration, about 2 months; mortality, probably under 20 persons.

Type (ii): The epidemic began with the arrival of an infective in late August or September who had probably travelled from a locality that was experiencing its type (i) summer peak. The outbreak developed slowly, as usual, and there was a small autumnal peak of plague burials but the spread of the infection was dramatically reduced by the cold weather of December. The epidemic broke out again in spring with the slow build-up and major peak in July–August as in type (i) epidemics. Typical duration of total outbreak: 14 months; typical total mortality of a naive population: 30% to 40%. Examples: Penrith, 1597–98; Eyam, 1665–66; Colyton, 1645–46; Chesterfield, 1586–87. A variant of this type of epidemic was seen at Shrewsbury in 1665–66 and at Colchester in 1665, which began in summer and died down in winter, as in type (i) epidemics, but reappeared in the following spring. Another variant was an epidemic with a small autumn peak but the infection died out completely over winter. Typical duration, 3–4 months; mortality, very low.

As we have shown, haemorrhagic plague was characterised by its long incubation period, namely a latent period of 10–12 days, an infectious period before the symptoms appeared of 20–22 days and about a 5-day period of showing symptoms. It is possible that the latent period in London was slightly shorter, with some people within the same household being infected after 9–10 days and possibly only 7 days after the primary was infected. The outbreak at Neston in 1665 (section 9.6.1; Fig. 9.11) establishes 35 days as the minimum duration of the latent and infectious periods and there is a 36-day gap in the outbreak at the parish of St Michael Bassishaw in London in 1641 (see Fig. 8.11). Our analyses of the London epidemics suggest that, exceptionally, the latent and infectious periods may have been extended beyond 37 days (up to 42 days or even longer) in cold weather in winter.

13.5 Seasonality of the epidemics

As we have said, seasonality apparently had little effect on the progress and spread of the Black Death either in England or continental Europe; a major epidemic followed whenever an infective arrived in a completely naive population, irrespective of the time of year. Peak mortality was recorded in every month, although the results suggest that the epidemic may have spread more readily and rapidly during the summer.

By the mid-16th century, when the parish registers began and we have more detailed information, it is evident that seasonality had a marked effect on the dynamics of the epidemics. In the autumn, in the provincial towns in England, although the effective household contact rate was high, the disease did not spread readily to other households and when transmission was achieved it was usually to relatives. We conclude that effective interhuman infection in autumn and winter was achieved only indoors and for an epidemic to continue it was imperative that the infection spread to other households. The hold of the epidemic through mid-winter was even more tenuous, with a very limited number of deaths even though the lengthy incubation period provided a long time during which each victim could achieve onward transmission. In complete contrast, interhousehold transmission was much higher in spring, generating a typical Reed and Frost epidemic with very heavy plague mortality in July–August. Furthermore, there seem to be fewer cases of multiple household deaths in summer, with more families recording only a single burial.

In the plague in London in 1665–66, Bell (1924) recorded that 6 February

> all men declared to be one of the coldest days they had ever experienced in England. In two separate months ice blocked the Thames, stopping the river traffic. But Plague occasionally showed its head even through the frost. There is Dr. Hodges' testimony that very few died that season, but he himself in January attended a case in which the Plague spot was apparent and the patient recovered. Josiah Westwood in the continuance of the frost also attended patients in whose condition he recognised Plague; they obtained a cure, 'the air then being so friendly to nature, and an enemy unto the Pestilence'. Other indications suggest that instances of Plague were not uncommon, but in the cold the disease never became severe, and the deadly symptoms afterwards so familiar were suppressed.

Three interacting factors could have contributed to this marked seasonal difference: human behaviour, resistance of the human host and sensitivity of the infectious agent to a combination of environmental factors such as temperature and humidity. These factors were probably the most important; the epidemics of many infectious diseases in the 18th and 19th centuries and even in the 20th century were strongly influenced by environmental and seasonal conditions. Measles epidemics in England between 1700 and 1785 were strongly correlated with low autumn temperatures; smallpox epidemics 1660–1800 in London were significantly correlated with low autumn rainfall; whooping cough epidemics in London 1720–50 were significantly associated with low autumn and winter temperatures; influenza epidemics usually occur in the winter months (Scott & Duncan, 1998). It seems that the causative agent of plague was most infectious, and droplet transmission person-to-person was most efficacious under warm conditions; in winter, this was achieved only indoors. In high summer, it might even have been warmer outside than within the houses.

A second, contributory factor to the differences in infectivity of the disease may have been seasonal changes in resistance associated with corresponding seasonal fluctuations in nutritive levels. In autumn and early winter, the poor in the population enjoyed relatively better nutrition following the harvest, although the overall diet in England from 1300 to 1700 was markedly deficient in many important respects (Scott *et al.*, 1997, 1998a,b; Scott & Duncan, 1999a,b,c, 2000). However, after the winter, in early spring, nutritive energy levels became progressively more compromised and some essential trace elements and vitamins fell below adequate levels, culminating in the well-known hungry season before the harvest. At this time of year, the poor in the community may have been more susceptible to infection by the plague. Conditions in England, during the period of the plague epidemics, may be compared with those in underdeveloped countries today where mortality from measles is accepted as being linked to poor nutrition and to vitamin A deficiency. Improved diet and vitamin A supplementation leads to a marked fall in the mortality of the disease (James, 1972; Barclay *et al.*, 1987; Berman, 1991; Duncan *et al.*, 1997; Scott & Duncan, 1998). Equally, in both London, 1700–1800, and Liverpool, 1850–1900, the endemic trend of whooping cough mortality correlated closely with wheat prices whilst the epidemics were strongly coherent with a short wavelength oscillation in the price of grains (Duncan *et al.*, 1996a, 1998; Scott & Duncan, 1998).

The third interacting factor that may have contributed to the differential infectivity of the plague was the seasonal behaviour of the rural population, who would have been more confined to their ill-heated houses during the cold winter months but would have emerged to work in the fields during the labour-intensive months of spring and summer. Not only would they establish more interhousehold contacts, but hard physical work on a calorie-deficient diet can only have exacerbated the physiological stresses and reduced resistance to disease. The arrival of the better weather in spring would also have promoted the movement of travellers within the metapopulation (see section 13.9), so bringing the infection to the rural communities. Social intercourse would be at its peak during the summer, not only during hay-making and harvest, but also with the long-established fairs at which people from all over the country assembled. A symptomless infective could spread the infection widely at such events and this was recognised by the authorities who cancelled the fairs, particularly those near London, once a plague epidemic was officially declared.

13.6 Density dependence of plague epidemics

We have little quantitative evidence concerning exactly how the Black Death struck at the different naive communities in the vast metapopulation of Europe but it appears that it attacked largely indiscriminately. Two hundred years later, after a succession of plagues, the dynamics of the epidemics were more formally established. In general, the outbreaks were density dependent, with the major epidemics confined to towns with some 1000 inhabitants or more. In villages or hamlets or in parishes where the population was scattered, the epidemics rarely exploded and the mortality was usually confined to a few dozen households and we have given examples in Chapter 9. This is the normal pattern of infectious diseases spread person-to-person (smallpox and measles are examples) where a sufficient population density is necessary to establish an epidemic (for a discussion of the mathematical modelling of the dynamics of the epidemics of infectious diseases, see Scott & Duncan, 1998). Of course, there were exceptions and Eyam was apparently below the critical size and the pestilence was only tenuously maintained through the winter, but it may have been a tightly knit community; with only a small proportion of the inhabitants fleeing when the plague came and with the establishment of the cordon sanitaire, there was probably a sufficiently high effective concentration of people to produce an explosive epidemic which followed Reed and Frost dynamics. The enormous conurbation of London represented the other end of the spectrum and large-scale epidemics were readily established and spread rapidly, dependent on the movements of symptomless infectives through the city.

13.7 Endemic versus epidemic plague

It is instructive to trace the evolution of smallpox epidemics (another lethal and infectious disease) in London, the dynamics of which were determined by density-dependent constraints and are described in section 2.9 and shown in Fig. 2.6. Throughout the period 1650–1800, this infectious disease was endemic, with superimposed regular epidemics, the periodicity of which changed from 4 years to 3 years and, finally, to 2 years (1750–1800). The frequency of the epidemics was determined by population density (section 2.9) and overall nutritional levels, which modified the susceptibility to the disease. Except, perhaps, for the 17th century, these dynamics of smallpox do not correspond with the evolving pattern of plague epidemics either in London or in the metapopulation: plague was certainly not endemic between 1350 and 1600 and there is no discernible regularity in the periodicity of the epidemics.

Figure 2.6 illustrates the climax of the dynamics of smallpox in a large and densely populated city during 1750–1800 but the situation elsewhere in the metapopulation of England was very different and we have classified the patterns of smallpox epidemics that can be identified, integrating them with the population dynamics. The following categories are arranged in order of increasing density but probably form a continuum. For the purpose of this classification of lethal smallpox epidemics in England, the story is assumed to begin in about 1630 when a more lethal form of smallpox is believed to have emerged (Razzell, 1977).

Category 1. Low density; small, scattered population. No epidemics; if an outbreak of smallpox occurs the disease does not spread and explode into a full epidemic because of the low density. Hence, many of the inhabitants remain as susceptible and would be at risk if they migrated to cities where smallpox was endemic. Compare with the mortality during the plague epidemics of apprentices and servants who migrated to London in search of work.

Category 2. Larger parishes, probably with a slowly rising population size and density. Large, but very sporadic smallpox epidemics. The interepidemic interval varied widely in the parish from 3 to 30 years. Probably many susceptibles escaped infection. An undriven system.

Category 3. Generally larger towns, perhaps reaching a critical density only during the population boom post-1750, with a regular supply of infectives coming to the community. Regular, clear smallpox epidemics with a frequency of 7 or 6 years; the interepidemic interval is again governed by the time taken to build up a pool of susceptibles that was of sufficient density for an epidemic to explode.

Category 4. Typically rural market towns with a higher population size and density and a regular supply of smallpox infectives coming into or through the parish. Epidemics on a regular 5-yearly basis with no infections in the interepidemic years. This category represents the climax epidemic situation for rural England. The interepidemic interval is again governed by the length of time taken to build up the necessary pool of susceptibles by new births, whereupon an epidemic will explode. Most of the other inhabitants will be immune from previous infections.

Category 5. High population density and size, as in cities and major conurbations, but smallpox probably still not fully endemic. Epidemics at 4- or 3-year intervals.

Category 6. Endemic smallpox in cities of high density; 4- and 3-year epidemics, driven by climatic factors. See Fig. 2.6.

Category 7. Endemic smallpox in large cities with 2-year epidemics driven by climatic factors in large cities. See Fig. 2.6.

The foregoing illustrates how the dynamics of a 'standard' infectious disease in England can be analysed over a 200-year period. The pattern of the smallpox epidemics was dependent on the density and size of the population and on the presence of a continuous source of infectives spreading from cities such as York, Chester and London where the disease became endemic. Although the incubation period for smallpox was 10–15 days, the infectious period was about 7 days so that an infective had time to travel out into the provinces but only a limited period during which he or she could effect transmission to susceptibles who were usually children, the adults being immune.

Although the plague epidemics were also largely density dependent, the disease was not endemic in rural England (unlike France) so that there was no regular and continuous supply of infectives in the metapopulation. Each epidemic burnt out completely (Reed and Frost dynamics), although it might take 2–3 years for the infection to be eliminated from the metapopulation. For another epidemic to be initiated, an infective had to arrive at one of the ports from overseas, usually from continental Europe, but also from Ireland. This failure to establish an endemic infection in Britain was the keynote of the pestilence in this island metapopulation. Possibly the cold weather of winter contributed to the final elimination of the infection in each epidemic. As a consequence, the plagues in England for 250 years were critically dependent on reintroductions of the infectious agent, either directly or indirectly, from an endemic source outside the metapopulation, namely central Europe.

By the end of the 16th century and during the 17th century, the records suggest that plague was almost endemic in London, either because the environmental conditions (including population size and density), or the dynamics of the disease had changed, or because there was now a steady stream of infectives coming into the ports. The plague was now increasing its spread, persistence and ferocity all over Europe. As a consequence, the pattern of the epidemics in London changed, as we suggest in section 2.9, to an undriven system that followed simple SEIR dynamics with decaying epidemics.

During the 17th century, there seem to have been minor or major epidemics in almost every year somewhere in the provinces of England. This may be because of repeated introductions of the disease via the many ports and river systems coupled with a regular traffic in infectives coming from the near-endemic conditions in London. The epidemics in each locality burnt out completely but could plague be regarded as pseudo-endemic in the metapopulation considered as a whole after 1600?

The status of the plague in France is in complete contrast with that in the British Isles because it became pseudo-endemic there soon after the Black Death had finished. It was not truly endemic and present everywhere, like smallpox in London (see above), but an epidemic was reported from a medium-sized town somewhere in almost every year, so that it never died out completely. The epidemics in each town usually lasted only 1–2 years but they acted as foci from which infectives could travel considerable distances along the major trading routes (see Fig. 11.22), so re-establishing the epidemics and also spreading plague to other metapopulations. We conclude that conditions in France were uniquely suitable for the establishment of this pseudo-endemic status of the plague which continuously cycled round the metapopulation: (i) the warmer temperatures, particularly in winter, and the relatively high humidity facilitated the persistence and spread of the infection; (ii) the metapopulation was large enough for the plague to circulate round between the large towns; (iii) there was a good communications network; and (iv) there were open borders allowing the ingress of infectives.

The other metapopulations lacked some of these factors and so the plague never or, only lately, became endemic.

Iberian peninsula: summers too hot and dry; poor internal communications; infectives arrived by sea.

Italy: summers too hot and dry; physically smaller; partially effective quarantine at the ports.
Holy Roman Empire: low winter temperatures.
England: low winter temperatures; infectives could come only by sea.

13.8 How were plagues initiated?

The arrival of the Black Death at Messina in Sicily in 1347 on the Genoese galleys coming from the Crimea is described in sections 4.2 and 4.3. The accounts are somewhat contradictory and confused but, in brief, the voyage would have lasted 4 or 5 weeks (Twigg, 1984), the sailors coming ashore were apparently healthy, there were no cases of plague on board and yet the infection appeared in its most deadly form a day or two after arrival. As explained in section 4.2, bubonic plague could not have arrived in this way: symptomless carriers could not have voyaged for over a month, nor could they have immediately initiated an outbreak by pneumonic contact; infected rats would have died during the voyage; infected rats and fleas could not have established a widespread epidemic in so short a time.

However, the story of the Genoese galleys is also inconsistent with a 'standard' infectious disease spread person-to-person. A long period before the appearance of symptoms would allow an infective to survive the voyage and also to infect fellow crew members but it would be a long time before the secondary infections appeared in Sicily. For a disease with a short incubation period, all the crew would have been dead before arrival, although an epidemic could have been established quickly after docking.

We conclude that the story of the galleys is a biological impossibility and that their arrival probably coincided with the early stages of an epidemic that had already begun.

Nevertheless, it is noteworthy that the Black Death began in Italian ports and so was probably brought by an infective (or infectives) from an overseas focus some time before the Genoese galleys arrived. Since the incubation period is believed to be about 37 days, and the infectives were probably not showing signs of the disease on arrival, the voyage would have taken less than 1 month. The Great Pestilence most probably originated in the Levant or from the ports of North Africa.

There appears to be no firm evidence that transmission of the plague was ever achieved via bundles of cloth, clothing or wigs. The stories of Lady Howard in Cumberland in 1625 taking infection from a new gown from London and dying on the same day that she received it (Creighton, 1894) and the box of cloth opened at Eyam (section 10.1) are probably coincidences. Consequently, epidemics of all scales, the European pandemic of the Black Death, the introduction of the infectious agent into Britain for each outbreak, the initiation of each major epidemic in towns and cities in provincial England or continental Europe, or the start of the pestilence in a village, always began with the arrival of an infective who was presumably usually male. Each infective may have travelled some distance from the focus where he was infected.

Each outbreak of plague in Britain was re-started by infective(s) coming by boat to the seaports or up the river systems that fed them. The Port of London received ships from continental Europe as well as from further afield; the ports of the east coast, including Scotland, and the south coast traded predominantly with the Continent; Chester and Pembrokeshire traded with Ireland. Coastal trade, particularly up the east coast, had an important role in transporting infectives and in disseminating the plague.

Further transmission over land was achieved by the movement of infectives on foot or horseback who initiated new epidemics. In many of the outbreaks of the pestilence in England that we have described in earlier chapters (Carlisle, Bridekirk, Eyam, York, Penrith) the incoming stranger has been identified. In summer in the 16th and 17th centuries, the initiating infective may have needed only to mingle with the populace in the market and to have infected one person and so started the epidemic, which then progressed remorselessly. If he attended one of the big fairs in summer, he may have infected several people who then returned to their home towns and villages so that several separate epidemics were initiated simultaneously. However, in autumn and winter, with the apparently lower infectivity of the plague, it was probably usually necessary for the incoming infective to stay in one of the houses if transmission were to be achieved and most of the initial infections were within that family.

13.9 Spread of the plague through the metapopulation

The Black Death spread through both continental Europe and England as a rolling succession of waves, travelling in summer, on average, about 1.5 miles per day, being quickest during the first 6 months of 1348 as it spread through southern and central France. Movement at that time would have been slow and limited because most people would have travelled on foot, but infected travellers steadily spread the disease forward on a wide front and the period of each advancing wave would be approximately related to the incubation period. Behind the advancing front, each epidemic burnt itself out over several months, leaving a devastated population.

After the Black Death, the island metapopulation of England was no longer naive and since each epidemic died out completely, a fresh outbreak, as we have seen, was critically dependent on new infectives coming from overseas. Some of the epidemics were initiated in London in this way via an entry through the docks. The metropolis grew steadily during 1350–1665 and the pattern of the epidemics within the city developed accordingly: the plague spread in waves from its point of initiation and, it has been claimed, had largely died down once the adjacent areas were infected. We conclude that (i) people moved around freely in their local area in London, less frequently to adjoining areas and rarely further afield, and (ii) the plague was gathering momentum in adjacent areas because of the long incubation period before it was recognised.

The epidemics spread out from London in a radial fashion, probably mainly because of people fleeing, and the suburbs to the north and south frequently became infected; its erratic movement from London 22 miles northwards to Ware in 1603 is described by Bradley (1977). The magnitude of the outbreak in each locality was probably density dependent. It became understood that, for safety, it was necessary to ride out of London for a day, about 20 miles. The plague also spread up and down the Thames, which was an important channel of communication, indeed people fled from the plague by barge.

Infections brought in through other ports to smaller cities and towns exhibited 9-month epidemics with normal Reed and Frost dynamics and the plague then also spread inland via the river systems and radially to adjacent communities (see Fig. 13.1). The ports of the Wash in East Anglia, notably Boston and King's Lynn, were of particular importance for communications by sea and as gateways for the river systems. By the time of the Black Death the galleys of the Mediterranean and the cogs of northern Europe were important and exporting a variety of goods, as shown in Table 13.1.

The epidemics also spread linearly from the outbreaks in London and the provincial towns, moving progressively along the recognised trade routes, brought by apparently healthy infectives (see Fig. 13.1). The north-east corridor between Newcastle and York was a regular transmission channel. The Tudor Royal Post system was established with staging points set about 10 miles apart between which letters were carried by postboys; about 5 miles per hour was an average speed but this would be only 1 mile per hour over difficult terrain (Campbell-Kease, 1989). However, this would have been an effective way of spreading an infection.

The plague also moved in a saltatory (and apparently erratic) fashion,

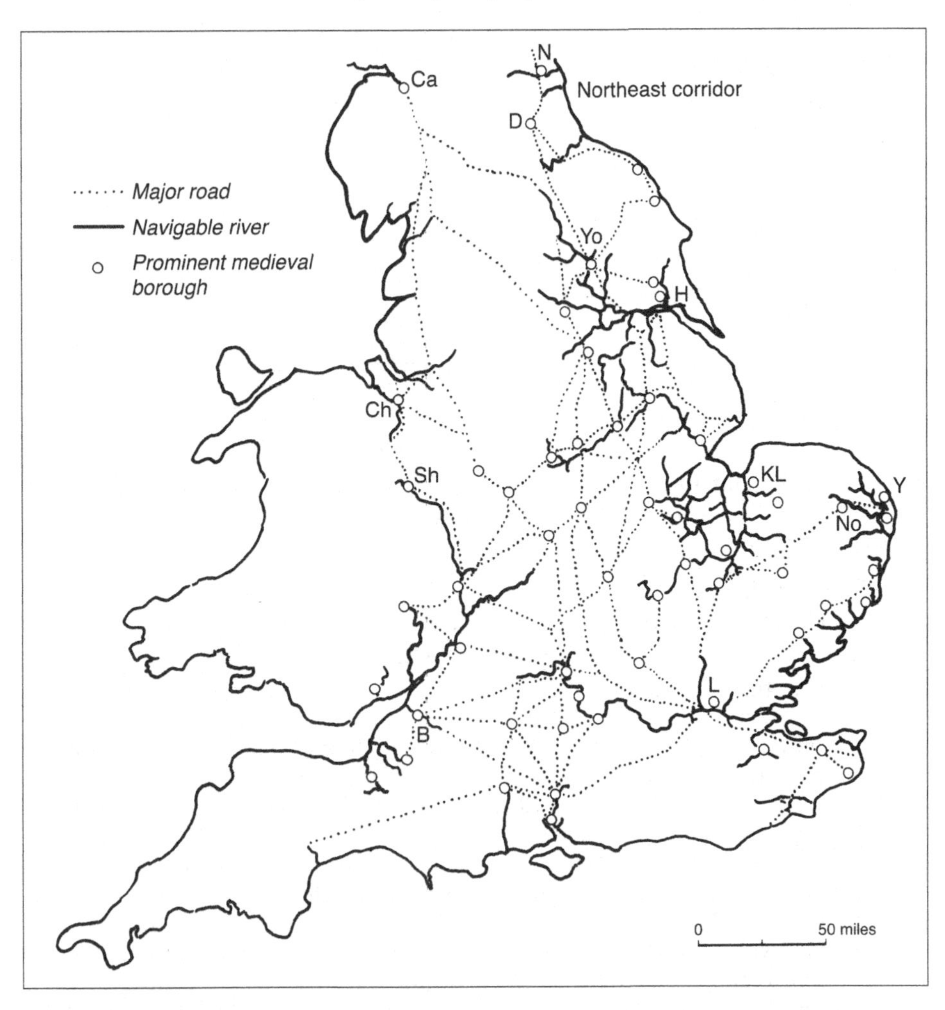

Fig. 13.1. The transportation system of medieval England. The major towns were linked by road and river. Note (i) the northeast corridor where plague was readily transmitted between Newcastle and York, (ii) the river systems of East Anglia, which brought both trade and the plague into the country, and (iii) the paucity of roads in Wales where there was little plague. Abbreviations: B, Bristol; Ca, Carlisle; Ch, Chester; D, Durham; H, Hull; KL, King's Lynn; L, London; N, Newcastle; No, Norwich; Sh, Shrewsbury; Y, Yarmouth; Yo, York. After Hindle (1993).

appearing many miles away from the point of infection in both England and France. This behaviour was possible because of the long incubation period of the disease; an example is the outbreak at Malpas in 1625, which was brought almost 200 miles by a member of the family from a plague in London (see section 9.3).

Table 13.1. *Imports and exports via the ports of Boston and King's Lynn in East Anglia from 1300*

Trading centre	Imports	Exports
Iceland	Fish	Cloth, corn
Bergen	Fish, timber, furs	Cloth, corn
The Baltic	Fish, oil, furs, wax	Cloth
Germany	Fish	Cloth
The Low Countries and Calais	Cloth, linen, madder	Wool, cloth
France	Wine, woad	Cloth, corn, fish

Source: King (1988).

The great fairs which were held annually were probably a major means of disseminating the plague in England (section 13.8) and possibly also a source for bringing the infection into the country. This was recognised by the authorities in the later years and fairs were cancelled once a plague was officially declared to have broken out. The major fairs lay south and east of a line from Exeter to York and were situated on or near the navigable rivers linking them to seaports or in places where there was an easy overland haul from the nearest port.

The great fair at Sturbridge [in East Anglia], which opened on 18 September and lasted for three weeks, was the most important of the English fairs and was of sufficient repute to attract merchants from many European trade centres. 'You can be present at the great Sturbridge fair and there see Venetian glass, Bruges linen, Spanish iron, Norwegian tar, Hanse fur, Cornish tin and Cretan wine, all for sale in the half of a square mile which was occupied for three whole weeks' . . . during Sturbridge Fair, Blakeney, Colchester, King's Lynn, and perhaps Norwich, were filled with foreign vessels. The ships that brought the merchandise of the Levant from Venice, and the other commodities from overseas . . .'

(Shrewsbury, 1970)

September 18th was probably too late for the initiation of a major, type (i) epidemic in that year but there were over a dozen other fairs and the Great Pestilence is said to have erupted in London at the time of the St Bartholomew fair.

The open market in rural English towns catered primarily for local demand and dealt in relatively small transactions; the private trader, by contrast, was essentially a traveller and an individualist. He traversed alone the southern heaths and northern moors in search of livestock, corn, or wool; if necessary he was prepared to pluck up his roots and plant anew in other cities and villages; he was ready to forsake home and family in search of individual reward. In the inns, where he bought and sold, he met others of his own kind, and there he discussed with them the burning issues of the time (Thirsk, 1967) and, more importantly for our study, he must have provided the ideal means for the transmission of the plague, talking at length with the locals *indoors*.

The medieval wool trade had the greatest impact on long-range movement within the metapopulation of England, the handling and transportation of the fleeces and the finished goods proceeding throughout the year on a large scale. Sheep were usually sheared in June after washing, which may have been done by hired workmen who, like the shearers, moved round the country. After sale, the wool was packed and delivered to the buyer either immediately or at some later date. The clothiers and middlemen themselves frequently came with transport to fetch wool from a grower's house, sometimes in instalments, although carriers were also employed. Leaden Hall in the City of London, where the staplers (wool merchants) sold much of their wool, was not only a market but was also the largest wool warehouse in England.

The middlemen either stored their wool for a number of months or sold them almost immediately. One large Shropshire dealer, for example, sold wool to a Shepton Mallet clothier at Shrewsbury and later sent his servant to deliver the wool at Bristol. In the mid-16th century, wealthy Coggeshall clothiers regularly travelled to Shropshire every June to arrange with middlemen for a supply of March wool.

In the 16th and 17th centuries a very large volume of wool was sold through the weekly public markets and, in manufacturing areas and wool-growing districts, some towns held a special market for the sale of wool and yarn. It was estimated that 40–50 horse packs of wool a week were sold at Halifax in 1577 (Bowden, 1971).

London was the centre of commerce as well as being the epicentre of plague mortality in England. Henry VII systematically assisted the wool merchants in their trade with the Netherlands (Taylor & Morris, 1939) and, during the 1540s, the volume of cloth exports through London doubled; vast fortunes were made by the Merchant Adventurers whose convoys of little ships sailed across to Antwerp twice yearly in May and November (Sheppard, 1998). It is clear that several major epidemics in London came via this route (for example in 1603; see Shrewsbury, 1970).

The possible effects of trade on the spread of the disease was tested by multivariate analysis of deaths from the pestilence in London in 1578–1649 versus commodity prices, wheat, wool, straw, hides, dairy products and hay. Only high wool prices ($P = 0.009$) and low prices for dairy products

($P = 0.02$) proved to be significantly associated with high plague mortality in London (interaction, $P = 0.02$), possibly suggesting that epidemics were favoured by the greater movement of people and of shipping during times of high wool prices.

We have shown in Chapter 11 how the plague was moved around the vast metapopulation of central Europe via the complex network of trade routes shown in Fig. 11.22, by sea, along river systems and by road. By parasitising healthy travellers, it crossed alpine passes into northern Italy; it travelled enormous distances inland by the river systems; it crossed the Mediterranean to Italy and Spain and the Atlantic to Portugal and the northern Spanish coast; it crossed the Channel and North Sea to England, Scotland and Iceland and the Baltic to Norway.

13.10 Resistance, immunity and virulence

After the plagues ceased in England by about 1670, smallpox became the most feared lethal infectious disease, having apparently mutated from a mild form early in the 17th century. Once recovered, people were immune to the smallpox virus for a long time so that each new epidemic swept with high infectivity through only the children who had been born in the interim, of whom about 20% died. This is a typical picture of a population that had suffered from a disease in its milder and more severe forms for over 100 years. In contrast, only six persons survived out of an estimated population of 200 when smallpox was introduced to the naive community on the Island of Foula (Shetland Islands) in 1720 (Razzell, 1977). The mortality of such a highly infectious disease in a given population changes subtly with time, particularly because of the development of resistance; breeding was confined to those members of the community who recovered from a childhood infection. Smallpox epidemics were density dependent and did not develop in small or low-density communities. Constant migration from such villages produced a mix of semi-resistant and non-resistant individuals in nearby towns where smallpox epidemics occurred.

It is impossible now to determine either the proportion of the population of England that was exposed to the Black Death in 1348–50 or the proportion of these who contracted the disease or the proportion that died. Overall mortality was clearly very high and perhaps 50% of the population of England died. The scattered reports suggest that the household contact in summer was very high indeed and probably almost all that contracted the plague died. In Venice in 1347–48, 60% of the population is estimated to have died (Slack, 1988).

It is generally concluded that the Black Death struck at a completely naive population, so causing the dreadful mortality, but is this necessarily true? Twigg (1984) has reviewed visitations of pestilences in England between AD 500 and the Black Death and it is clear that none of these outbreaks were bubonic plague (Shrewsbury, 1970; Twigg, 1984). *Pestis flava* (the Yellow Plague) was rife in the 6th century and was clearly severe with high mortality. It reappeared in the 7th century and continued for many years; in AD 664 it appeared in southern England and spread to Northumbria, with great mortality, and to Ireland. It is noteworthy that the plague was active during the summer months. This epidemic was widespread over Europe and contemporaneous accounts described it in Italy, Gaul and Spain (Twigg, 1984). There were many further epidemics in England between AD 667 and 1348. Were these forerunners of the Black Death that forced up the *CCR5-Δ32* and other mutations, so producing a small resistant fraction of the population? More accurate information is available by the late 16th century and it is evident that many people were in close contact indoors with infectives but did not contract the disease; examples are the vicar at Eyam whose wife died in his arms and the much-feared pest-nurses who supervised the victims when they were dying and who escaped infection, although examination of the pattern of an epidemic suggests that patients may have been *less* infectious in the terminal stages when they were displaying symptoms.

During the 17th century, there is evidence that the mortality and relative severity of the plagues in London were declining slightly (Sunderland, 1972; Benedictow, 1987). Plagues in Florence after 1424 killed many fewer people than they did in the 14th century (Carmichael, 1986), suggesting a greater resistance in the population or a decrease in the virulence of the causative agent.

It has been suggested that more people recovered from the plague in the cold weather in winter (sections 8.1, 8.6.1 and 13.5) than in the warmer weather of summer.

A few contracted the disease and recovered and were then believed to be immune. There is evidence that the pestilence became less virulent during the later stages of the 1665 epidemic in London, with more people recovering from the infection (section 8.6.4). This is in agreement with the observations of the outbreak at Bologna, Italy, in 1630 where 'at the beginning and the height of this dreadful scourge very few infected survived. Then by the beginning of July it seemed as if the sharp ferocity of [death's] scythe lost something of its edge, though very many still died'. This is at variance with the 1575 epidemic at Palermo, Sicily, where the 'greatest part of the patients recovered' at the beginning whereas 'the disease became more cruel day by day', and with the 1632–33 epidemic at Poggibousi, Tuscany, 'while earlier some of those who became sick recovered, now all die and in very short time' (Cipolla, 1981).

The registers at Penrith record a plague there in 1554 but there is no other evidence concerning its magnitude nor of another outbreak in the northwest. This is believed to be the only possible plague at Penrith before the major epidemic of 1597–98. Eyam, apparently, did not suffer from the pestilence before the outbreak in 1666, although it might have been attacked in the Black Death because it has been said that nearly every village in Derbyshire suffered except that Tissington (about 35 miles from Eyam) escaped from infection (Porteous, 1962). Mortality in Penrith and Eyam was of the order of 40% of the population, which can be compared with a value of about 12% to 15% (Slack, 1988) in the epidemics in London in the 17th century. Plague was virtually endemic in London at this time, with major outbreaks at irregular intervals, albeit with a relatively low *percentage* mortality. We have analysed the sequential plagues during the 17th century in London parishes (Chapters 6 and 8) and it is evident that mortality was often modest and, in the later outbreaks, deaths were restricted largely to one per household, perhaps suggesting that the other occupants were resistant or immune. Furthermore, those that died were frequently young people; for example, apprentices (who may have been susceptible immigrants) and infants (who were susceptible being born since the last plague). There is, therefore, some circumstantial evidence that the population in London was developing some immunity during the 17th century. Having suffered from this continuous series of epidemics, in addition to a low grumbling endemic level of mortality, it might be expected that the population of London was largely resistant and that mortality levels even lower than 15% would have been recorded. However, each epidemic or famine crisis brought fresh waves of immigrants from the countryside, particularly young people, the apprentices and maidservants, who would not be resistant and we see that the complex dynamics of the plague were inextricably bound up with the demography of the different communities within the metapopulation.

13.11 Medium-wavelength oscillation in the spread of the plague

We have shown that a strong medium-wavelength oscillation is detectable in the data-series of the number of towns reporting plague epidemics in each year in France (Fig. 11.4). This cycle emerged after 1436 and had a wavelength of 22–25 years ($P < 0.01$). France was the epicentre for the plague in Europe and this medium-wavelength oscillation correlated with cycles in the Holy Roman Empire and Italy. We suggest that when the epidemics were widespread in France, at the peaks of the oscillation, there was a greater chance of infectives travelling to other metapopulations and starting new outbreaks. This had a knock-on effect: the medium-wavelength cycle of plague deaths in England was then driven by the cycle in Germany.

But what was the cause of the original, highly statistically significant, oscillation in France that acted as a driver for the spread of plague through other parts of the supermetapopulation? It does not appear to correlate well with external factors and was probably not exogenous. We suggest the following explanation for this endogenous oscillation. France, throughout the age of plagues, was composed of about 400 discrete populations (i.e. medium- and large-sized towns) that were clearly separated from one another and suffered regularly from epidemics. Thus plague circulated round the metapopulation and was carried by travelling infectives along the major and minor trade routes (Fig. 11.22). Populations at the cross-roads on the major routes (by road and river) were the most commonly affected and probably, with repeated infections over the years, these communities may have built up an immunity (section 13.10). The period of the oscillation, 22–25 years, may represent a generation effect. Major epidemics were probably density dependent (section 13.6) and it may have taken this length of time to replace the individuals in each community with a new generation of susceptibles, so imposing an endogenous oscillation on the plague dynamics as it circulated round the metapopulation of France, the wavelength of which was determined by the approximate generation time.

In summary, this may be an example of an endogenous oscillation in France that drove (and interacted with) a corresponding exogenous oscillation in the Holy Roman Empire and Italy (for a discussion of population oscillations, see section 13.17).

13.12 Symptoms

People in England were well able to recognise the plague, certainly by the 16th century, even far out in the Provinces; the symptoms must have been readily recognisable, although in Italy in the mid-15th century the physicians went to considerable lengths to ensure a correct diagnosis and there may have been more than one lethal infectious disease extant there. Over a period of 300 years and a geographical spread across continental Europe with an enormous number of people infected, the causative agent of plague (probably a virus) may have undergone some mutations, but inspection of the historical records suggests that it remained the same recognisable disease. When the ministers and their clerks in England were required by law to record deaths from pestilence, the evidence suggests that they were accurate and that they were able to identify the plague from the start of an outbreak. Not so the scholars of the 20th century, who repeatedly designated an epidemic as typhus when the epidemiology clearly did not fit with that of bubonic plague; conversely they frequently averred that a crisis mortality in mid-summer was the result of bubonic plague, even when there was no evidence of an infectious disease that spread through households nor any record of pestilence in the church registers.

We have relied, where possible, on quoting contemporaneous accounts of the symptoms of plague from the Black Death to the last major epidemic of 1665–66 and a clear thread runs through them, although Slack (1985) has stated that the symptoms of the plague were variable. The patient probably exhibited the symptoms for 4–5 days before death but the range was from 2–11 days. The tokens (or God's tokens), large subcutaneous spots, often red and able to change colour between orange and black and between blue and purple, were regarded as diagnostic and often appeared on the chest but also on the throat, legs and arms (Slack, 1985; see section 8.6.3). They were regarded as evidence of subcutaneous bleeding and were so-called because they usually betokened approaching death (Roberts, 1966). Defoe (1722) wrote:

> Many persons never perceived that they were infected till they found, to their unspeakable surprise, the tokens come out upon them, after which they seldom lived six hours; for those spots they called the tokens were really gangrenous spots, or mortified flesh, in small knobs as broad as a little silver penny, and hard as a piece of callus or horn, so that when the disease was come up to that length, there was nothing could follow but certain death, and yet, as I said, they knew nothing of their being infected, nor found themselves so much as out of order, till those mortal marks were upon them.

In addition, various swellings were characteristic: carbuncles, blains and, of course, the buboes, which were swollen lymph glands in the neck, armpits and groins. Shrewsbury (1970) recorded that Barrough stated in 1634 that those cases in which buboes did not appear were the most dangerous. There was an accompanying fever, continual vomiting and prolonged bleeding from the nose. The physicians in Milan in 1468 regarded blood-tinged urine, in addition to the red (or black or violet) signs, as evidence of plague. The authorities in Italy, when deciding whether to declare an official plague outbreak, were apparently guided in most cases by the pattern of spread and lethality of the disease.

The description of the findings of an autopsy, given in section 8.6.3, shows conclusively that, in patients who died, these outward signs were accompanied by extensive internal necrosis in which 'no organ was found to be free from changes'; the stomach contained a black liquid. Comparable degradative and necrotic changes were found in the autopsies in Rome and Naples in 1656–57 (section 11.3.2). These signs are all indicative of a severe haemorrhagic illness.

13.13 Public health measures and the significance of the period of quarantine

People evidently believed at the time that the plague was an infectious disease spread through contact with an infected person, although they may have considered that transmission was also possible via contaminated clothing or bedding, probably because this would have apparently accounted for the big gaps between the death of one victim and the appearance of symptoms in the next. Preventative and control measures became progressively more strict and sophisticated; in particular, towns throughout Europe regularly exchanged information concerning where plague epidemics were established and all visitors coming from an area with an outbreak were banned. Likewise, ships from an afflicted port trading up the east coast in England were prohibited from docking and from transferring goods by barge to an inland destination, as for example from Hull to York. These measures must have been of real value in limiting the spread of the epidemics because it would have prevented the arrival of apparently healthy infectives.

Less effective, as we can see with the benefit of hindsight, although perhaps conferring marginal benefits on the rest of the community, was the practice of incarcerating victims in pest-houses and shutting up homes where any of the occupants were displaying the diagnostic symptoms – the infection would have already been passed on and the victims were probably less infectious by the terminal stages. It would certainly have exacerbated the effective household contact rate. Disinfection and burning of bedding was practised quite extensively and, in Venice, this was enforced by law in 1576; both the plague victims and their clothing and blankets were considered infectious (Ell, 1980). These procedures were probably of little use.

The Italian City States led in the development of epidemic surveillance, and maritime quarantine was devised in 1377 in the Venetian colony of Ragusa. At first, it was 30 days but this was soon extended and regularised to a 40-day quarantine period, which was strictly maintained and adhered to throughout Europe for 300 years. Its purpose in northern Italy was to prevent the importation of the disease rather than to isolate individuals who were already ill (Carmichael, 1997). People in Genoa in 1652 who had been in close and direct contact with infected people or merchandise were put in complete isolation for 40 days or more, plus a further period of isolation described as convalescence. The port of Genoa followed strict quarantine procedures in 1652:

(i) Vessels from England, if they come directly without touching at infected or suspected places, and with clean bills, are allowed entry after a few days; first, however, goods and merchandise are sent to the pesthouse where they are purified for 20 days, and if they touch any of the above [infected] places they must observe complete quarantine.
(ii) Vessels coming from ports uninfected, but under suspicion, are subject to quarantine for 30 or 35 days according to the suspicion held, but nevertheless the goods are sent immediately to the pesthouse.
(iii) If perchance any deaths occur or if anyone falls sick during the voyage or during the time the quarantine is being observed, the quarantine is to be extended for 50 or 60 days according to the danger and circumstances; the people and the goods are to be sent to the pesthouse.
(iv) Vessels from the Levant are quarantined for 30, 35, 40 days according to information received and if they come with a clean bill; the goods at the pesthouse are purified for the same length of time (Cipolla, 1981).

When an epidemic was subsiding in Florence, as in 1630, it was customary for the health authorities to declare a general quarantine in which as many people as possible were locked up in their homes for a period of 40 days, thus reducing human intercourse to a minimum. This procedure was supposed to terminate the epidemic more quickly (Cipolla, 1981). Regulations introduced during an outbreak of plague in Paris in 1668 forbade river traffic with Rouen, all suspect goods being unloaded at Mantes and remaining in quarantine for 40 days (Trout, 1973).

A quarantine period of 40 days was established in London and the provinces (see numerous references in the foregoing chapters) during the reign of Henry VIII and was determined presumably by experience; a similar period since the last death seems to have been adjudged adequate to declare the end of an epidemic. This value is directly comparable with our estimate of 37 days (with an 8% margin for error) between the point of infection and death. Since the 40-day quarantine period was accepted throughout the age of plagues, it is suggestive evidence that the same causative agent was responsible for all of this time (see section 13.2).

The cordons sanitaires established at Eyam and Penrith and other places were probably effective in containing the epidemic within the community, although the more wealthy citizens probably fled at the first signs of an outbreak. However, the extensive and efficient cordons sanitaires set up around Marseilles in 1720–22 were completely ineffective in preventing the spread of the epidemic – circumstantial evidence in favour of this being an outbreak of bubonic plague because the movement of rats would not be constrained by these measures (see Chapter 12).

13.14 Why did the plague disappear?

As we have seen, the plague in England, and probably in northern continental Europe also, persisted through the winter only with difficulty. As a consequence, even when plague was superficially endemic in England in the 17th century, it was probably dependent on introductions of infectives from overseas for it to be perpetuated. Plague quickly fizzled out in England after 1666 because it could not be maintained through winter and there were no further introductions from continental Europe.

Infectives ceased to arrive in England from overseas because the plague also disappeared abruptly in France and the Low Countries after 1670. The following, possibly interacting, factors may have contributed to this state of affairs:

(i) Winter conditions, even in southern France, may have constrained the plague and eliminated its endemic status.

(ii) It has been suggested (Appleby, 1980; Carmichael, 1997) that the disappearance of the plague corresponded with a period of low sunspot activity, globally cooler temperatures, the advance of glaciers and the early stages of the Little Ice Age (Lamb, 1977). Pfister (1980) has analysed the most significant effects during the Little Ice Age: the month of March reflects the climatic change most persistently and distinctly. Cooling began in 1560 and March remained cold and wintry together with a prevailing drought, suggesting that the month was frequently dominated by northerly winds and blocking anticyclones. The dominant impression of the three spring months is one of coldness and drought, with the coldest decades being the 1640s and

1690s. The second feature that has been identified by Pfister is that the month of June became wet and cool.

There is an exception to the general rule of an increasingly cold climate during the Little Ice Age: the 1630s were clearly warmer in marked contrast with the following decade, the 1640s, which was the coldest period during the age of plagues (Pfister, 1980). These findings may be compared with the severity of the plague in continental Europe where it was at its most rampant, with widespread epidemics, in France (see Fig. 11.3) and Germany (see Fig. 11.26) in the 1630s, which contrast sharply with the 1640s when plague grumbled on at a lower level in continental Europe.

It is possible, therefore, that factors contributing to the elimination of the plague in Europe were not only the colder winters, but also the cold weather conditions in March persisting through to June that may have reduced the ability of the plague to re-establish itself after the winter.

(iii) Introduction of fresh infectives into southern France, probably from the Levant, ceased.

(iv) The causative agent mutated to a more benign form, as the haemolytic streptococcus of scarlet fever is believed to have changed in England in the early 20th century (Duncan *et al.*, 1996b).

(v) An improvement in the general level of nutrition, particularly in southern France, may have contributed to increased resistance and hence to reduced infectivity, comparable with the changes in whooping cough lethality in England in the 19th century which were clearly linked to improved nutrition (Scott & Duncan, 1998). Corn prices in France were dropping sharply after 1662 (see Fig. 11.7; Baulant, 1968), which would have improved the level of nutrition for the bulk of the population. In contrast, 1628–29 were years of widespread famine in northern Italy with unusually high prices of grain; this period immediately preceded the greatest outbreak of plague in continental Europe (Cipolla, 1981).

The plague was not unique in suddenly appearing and disappearing. Many zoonoses have appeared during the 20th century and have then lain dormant; the English Sweating Sickness disappeared after the fifth epidemic.

13.15 What was the causative agent of haemorrhagic plague?

It is unlikely that we shall ever be able to answer this question with certainty, but the foregoing suggests that the disease may have been some form of viral haemorrhagic fever. Sensationalised accounts of dying patients have fuelled intensive public fascination with filoviruses and highlighted the global threat of emerging and re-emerging infectious diseases. Filoviruses are the prototypical emerging pathogens: they cause a haemorrhagic disease of high case fatality associated with explosive outbreaks caused by person-to-person transmission, have no known treatment, occur unpredictably, and have an unknown reservoir (Khan *et al.*, 1998). The term viral haemorrhagic fever is a clinical and imprecise definition for several different diseases (Le-Guenno, 1997) and has been described as any of a diverse group of virus diseases that are characterised by sudden onset, fever, aching, bleeding from internal organs, petechiae (spots resulting from the effusion of blood under the skin) and shock. A number of distinct haemorrhagic fever viruses are known, based on a shared ability to induce haemorrhage by poorly understood mechanisms that typically involve the formation of blood clots, known as disseminated intravascular coagulation (Ramanathan & Taylor, 1997).

Filoviruses are the main causative agents of haemorrhagic fever in humans (Feldmann *et al.*, 1996b) but these infections can also be caused by arenaviruses, flaviviruses and bunyaviruses. Filoviruses are enveloped, non-segmented negative-stranded RNA viruses (Palese *et al.*, 1996) and, once inside an infected cell, the virus transcribes its single strand of RNA and replicates by the synthesis of an antisense positive RNA strand that serves as a template for additional viral genome. As the infection progresses, the cytoplasm of the cell develops prominent inclusion bodies that contain the viral nucleocapsid; the virus then assembles and buds off the host cell, obtaining its lipoprotein coat from the plasma membrane of the cell.

Filoviruses cause a number of serious diseases in humans today, including Ebola and Marburg. Ebola infection is arguably the most gruesome way to die; it begins with a sudden fever and kills by the liquefaction of the internal organs. There are four known variants of the Ebola filovirus (Zaire, Sudan, Ivory Coast and Reston strains) together with Marbug, a genetically related monkey virus. Since filoviruses reproduce as RNA they do not have the ability to check each copy (unlike the retrovirus HIV where the RNA is converted into DNA inside the host cell before the virus reproduces) so that the error rate is one million-fold greater than that of DNA-based systems (Sanchez *et al.*, 1996; Takada *et al.*, 1997). Another incredible aspect of the biology of the filovirus is the speed with which it replicates in its host (Folks, 1998).

Marburg and Ebola viruses are believed to be evolving at similar rates that are 100 times slower than those of retroviruses and human influenza virus. The divergence time between Marburg and Ebola has been estimated to be more than several hundred years ago and most of the nucleotide substitutions were transitions and synonymous for Marburg, suggesting that purifying selection has operated during its evolution (Suzuki & Gojobori, 1997). However, the genetic sequences of two isolates of Ebola-Zaire from epidemics 600 miles apart and separated by 18 years proved to be virtually identical and it appears that something is restraining the natural tendency of filoviruses towards genetic divergence. Sanchez *et al.* (1996) suggested that natural selection pressures have favoured the survival of the original strain over any mutants. Filoviruses may also work by the suppression of the immune system and Ebola victims usually die without evidence of an effective immune response (Folks, 1998); Ebola-Zaire virus was found to suppress the basal expression of the major histocompatability complex class I family of proteins and inhibited the induction of multiple genes by alpha interferon (IFN-alpha) and IFN-gamma. The events that lead to the blockage of IFN signalling are believed to be critical for Ebola virus-induced immunosuppression and to play a role in the pathogenesis of the infection (Harcourt *et al.*, 1999).

It is not yet known in which animals Ebola lies dormant during the long stretches between human outbreaks, although hundreds of African species have been screened in search of a reservoir. The high degree of similarity between the 1976 and 1995 Ebola-Zaire strains suggests that the reservoir is the same in both locations and that the animal is either widespread in Zaire or else is a migratory species. Bats were initially regarded as likely candidates because they could distribute Ebola over a wide area of Africa, but this theory is now discounted. It has been noted that the geographical distribution of different types of the Ebola virus coincided with the distributions of different small mammals that live at the forest margins. Ebola RNA has been found in three mouse and one shrew species (MacKenzie, 1999).

Transmission of Ebola is usually by some form of direct contact with the carrier or host species through the skin and its secretions, but an airborne-route is considered a possibility. The symptoms appear after an incubation period of 2–21 days, finally resulting in fulminant shock and death (Peters *et al.*, 1996; Feldmann & Klenk, 1996). Schnittler & Feldmann (1998) proposed the following hypothetical model of Ebola haemorrhagic fever. The filovirus enters through minute lesions of the skin and mucosae and obtains direct access to the vascular system or indirect access via the lymphatic system (Fig. 13.2), after which virus particles are transported to the lymph nodes where macrophages, known to be the primary sites for

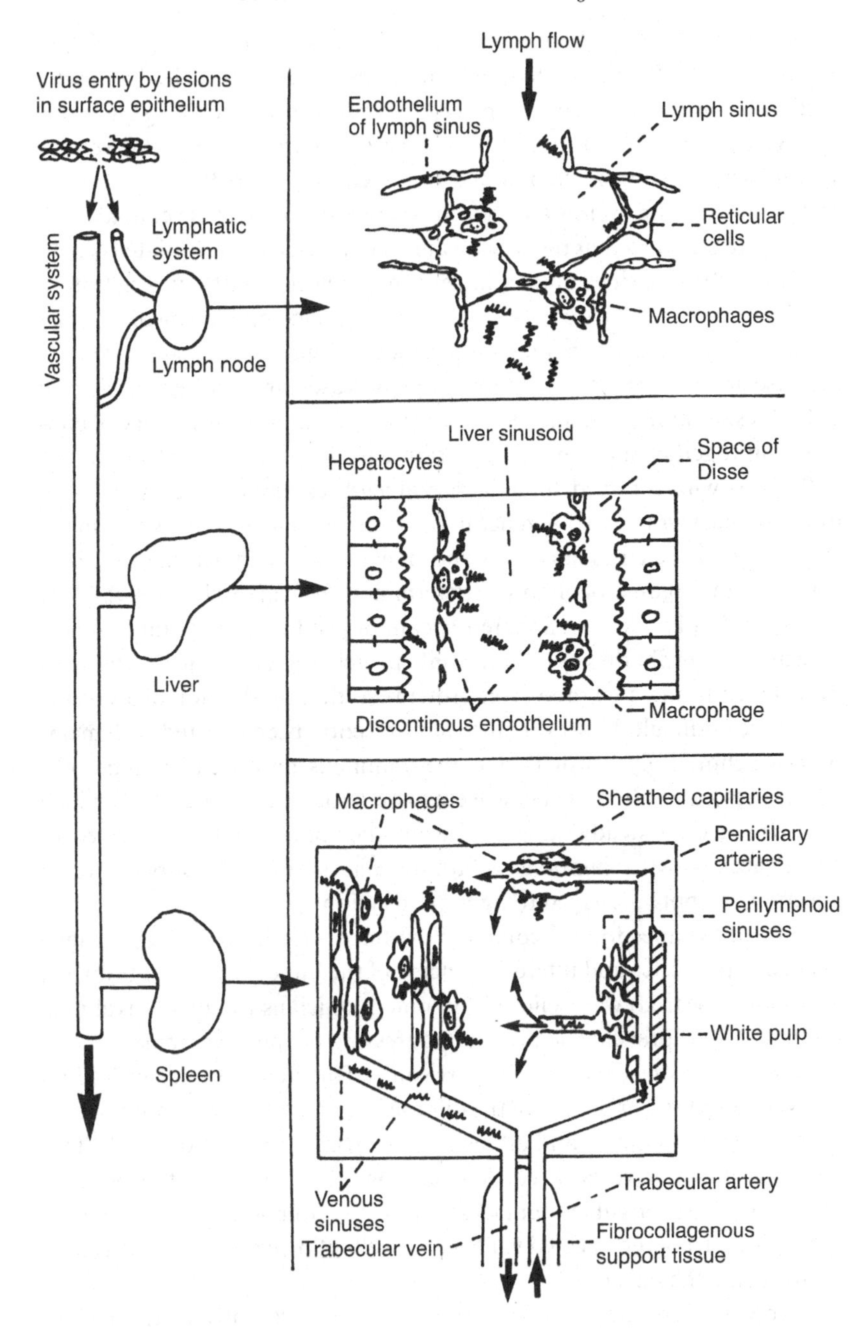

Fig. 13.2. Hypothetical model of Ebola and Marburg haemorrhagic fever. After Schnittler & Feldmann (1998).

virus replication (Geisbert *et al.*, 1992; Feldmann *et al.*, 1996a; Ryabchikova *et al.*, 1996; Zaki & Goldsmith, 1998), are bathed by lymphatic fluid. After primary infection, replication continues in secondary and tertiary lymph nodes, resulting in the release of particles into the venous system. The subsequent step of infection seems to be mediated by macrophages in the liver sinuses and spleen where they are in close contact with circulating blood so that replication and activation may be initiated quickly in these organs, since macrophages can be infected without penetration of cellular or tissue barriers (Fig. 13.2). Macrophage-derived mediators act on multiple organs, but mainly on the endothelium, which responds by an increase in permeability, dysregulation of vascular tone, expression of cell-adhesion molecules and the development of a procoagulatory phenotype that together substantially contribute to the occurrence of shock.

Patients who survived showed several features that distinguished them from non-survivors. No difference in the viral antigen load, measured in the serum, was detected in survivors and non-survivors, indicating that host defence rather than inoculum size determines survival (Nabel, 1999). Most victims of Ebola infection apparently cannot produce significant numbers of protective antibodies against the only protein on the surface of the virus (called membrane-anchored glycoprotein), so that production of a vaccine has proved difficult. However, mice have recently been injected with membrane-anchored glycoprotein and these animals produced immune cells that expressed antibodies. These were fused with mouse cancer cells which then produced a steady supply of antibodies that could be collected to make vaccines. Mice immunised with these antibodies 24 hours before an injection of Ebola managed to fight off infection.

Some of the anecdotal accounts of treatments for haemorrhagic plague may be of relevance and interest. Lancing of the buboes (presumably when they first appeared) was believed by some physicians to aid survival and, presumably, to prevent the disease progressing to the development of the tokens. A surgeon serving in the garrison at Dunster in Somerset bled all sick soldiers at the first signs of the disease in 1645, taking blood until 'they were like to drop down' and all his patients recovered (section 9.5.2). This bleeding would have been carried out some 30 days after the point of infection. Was it possible that lacing swollen lymph nodes or the removal of a major part of the infected white blood cells allowed the immune system to overcome the infection?

Crimea-Congo haemorrhagic fever re-emerged recently in the United Arab Emirates, with 11 cases admitted to hospital between June 1994 and January 1995. Symptoms on admission were high fever, vomiting, diarrhoea and haemorrhagic signs. Eight patients died at a mean of 6.8 days after admission (Schwarz *et al.*, 1997; Hassanein *et al.*, 1997).

We do not suggest that haemorrhagic plague in Europe during 1348–1670 was because of Ebola or any of the present-day viral haemorrhagic fevers that have been identified, but the close similarities with their symptoms and pathology described above suggest that a filovirus may have been the causative agent.

13.16 Co-existence of two plagues

Two distinct plagues existed together for hundreds of years: haemorrhagic plague in Europe from 1347 to 1670 and bubonic plague predominantly in central and eastern Asia but which also broken out sporadically and erratically along the Mediterranean coastline during the age of plagues. Traces of *Yersinia pestis* DNA have recently been reported in the dental pulp of skeletons of people who died in the area around Provence in the 14th and 16th centuries as well as, of course, at Marseilles in 1720. The frequent and irregular minor pests of Italy, and some of the epidemics at Barcelona and along the Spanish Mediterranean coast may also have been serious outbreaks of bubonic plague.

Bubonic plague spread in the 20th century to many subtropical regions but the epidemics never became permanently established in Europe. We know why this is. The climatic conditions are unsuitable over much of the continent and there are no resident resistant rodent species.

Why, on the other hand, did haemorrhagic plague apparently not appear in central Asia or sub-Saharan Africa? We suggest that a number of factors may have contributed to this. Firstly, the appropriate trade routes were much longer than in Europe and infectives may not have survived the journey. Secondly, trading was less intensive when compared with the internal routes in Europe, so that fewer potential infectives would be passing to and from Asia. Thirdly, the climate may have been too hot and dry for ready droplet transmission.

13.17 Population recovery after the mortality crisis of a plague epidemic

Historians have long been puzzled by the speed with which populations and metapopulations recovered after a major plague epidemic, particularly the Black Death (section 4.7). As we have seen, the outbreaks were density dependent and the *major* epidemics in the 16th and 17th centuries were confined to the towns, where the mortality was frequently between 30% and 50%.

In this section, we analyse the events at Penrith after the epidemic of 1597–98 (see Chapter 5). The community had suffered from famine and hardship in the preceding year so that during the period 1596–98 about 50% of the population died – a severe mortality crisis. During the 40 years preceding the crisis the *mean* annual number of baptisms and burials at Penrith were equal at 60 events each, showing that the population was in steady-state and under density-dependent control. The dynamics of this population can be represented by a conventional matrix model that shows that such a system would be sensitive to a perturbation, such as a mortality crisis, when it would respond by generating oscillations in the number of annual births and deaths. These oscillations are described as endogenous because they are dependent on the properties of the system, in contrast with exogenous cycles, which are driven by external factors (Duncan *et al.*, 1992; Scott & Duncan, 1998).

The recovery of the population of Penrith after the mortality crisis is remarkable: for example, baptisms in 1598 during the epidemic fell to 27, as would be expected, but had returned to 56 and 67 in 1599 and 1600, respectively. *Mean* annual baptisms and burials returned to their pre-epidemic levels very quickly (both equal to 60 per annum) and steady-state conditions continued for the next 150 years before a population boom began in 1750. Annual baptisms at Penrith for the period after the epidemic are shown in Fig. 13.3 and, although the population was in steady state for 150 years, oscillations in the basal level are detectable by eye. Spectral analysis (see section 2.8) shows that the wavelength of this endogenous oscillation triggered by the plague was 44 years ($P = 0.005$) and it is shown after filtering in Fig. 13.4.

A comparable 44-year, endogenous oscillation is detectable by spectral analysis in the burial data-series and this cross-correlates strongly (ccf = +0.74; Fig. 13.5) with the baptism series at almost zero lag, i.e. these long-wavelength oscillations were in synchrony (Scott & Duncan, 1998).

When a 30% mortality crisis is applied to the computer-based matrix model described above, it also responds by generating *synchronous* long wavelength, endogenous oscillations in births and deaths that are superimposed on their steady-state levels.

We have used the matrix model to explore which features of the population dynamics control the wavelength of these oscillations and, in brief, if the steady-state conditions are not governed by density-dependent feedback, the model predicts that the oscillations would have a wavelength equal to the mean of the fertility function (effectively the mean age at which a woman had her median child) and would decay. However, if the popula-

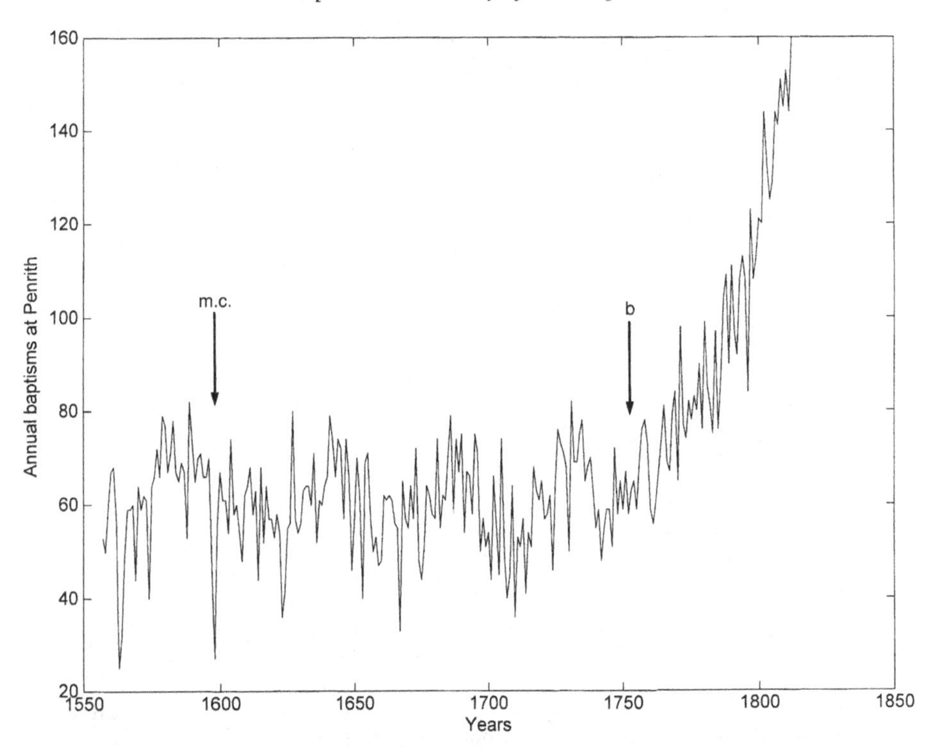

Fig. 13.3. Annual baptisms at Penrith, Cumbria, England, 1557–1812. Three cycles of the long-wavelength oscillation during the steady-state period and the marked rising trend after 1750 can be seen. Abbreviations: m.c., mortality crisis resulting from the combined effects of the famine in 1596 and the plague of 1597–98; b, start of the population boom and the end of the steady-state period.

tion dynamics are governed by feedback, the wavelength is greater, so that, with a mean of the fertility function of 28 to 30 years, the predicted wavelength of the synchronous cycles in births and deaths is 44 years. These observed long-wavelength endogenous cycles reflect the interaction of the demographic parameters and the dynamics of the population, i.e. the fertility function and the feedback (Duncan *et al.*, 1992, 1994c; Scott & Duncan, 1998, 1999b).

The population at Penrith was living under density-dependent constraints in the 16th and 17th centuries: there were no ecological niches available and if the population rose above a critical level there were insufficient food and other resources, and so mortality, particularly among the infants, rose inexorably, maintaining steady-state conditions. The mortality crisis of 1596–98 removed all constraints and the available ecological niches (jobs, trades, land-holdings) were quickly filled by immigration from

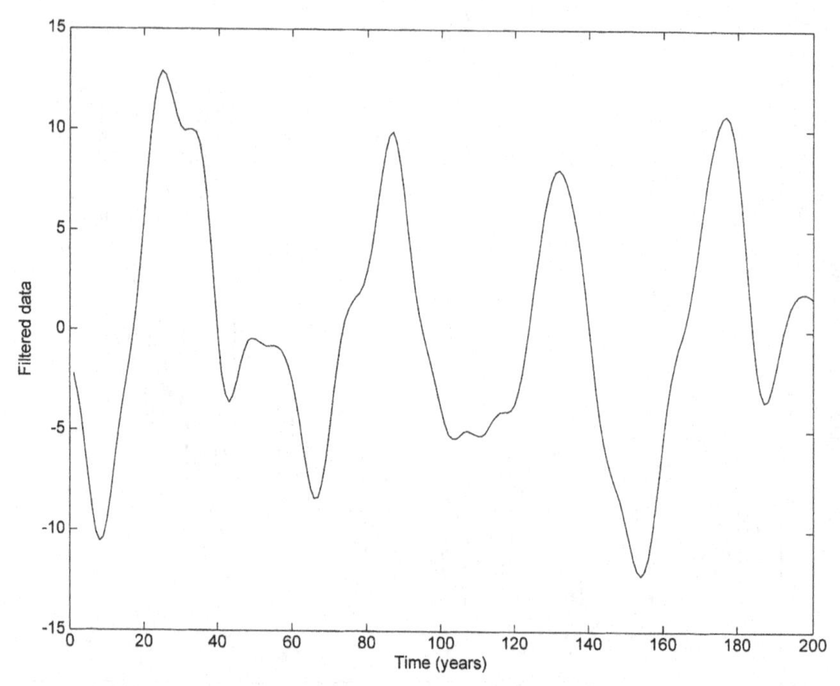

Fig. 13.4. Baptism series at Penrith, 1557–1757 (see Fig. 13.3) filtered to reveal the long-wavelength oscillation. Filter window = 20 to 100 years.

the surrounding parishes that had escaped the plague so that a surge in births and deaths followed. Once the population had reached a critical size at the limits of the density-dependent constraints, there was a compensatory rise in emigration (Duncan *et al.*, 1994c) and increased mortality, so that the endogenous oscillations were maintained. These persisted at Penrith until 1750 when the economic and farming conditions changed and a population boom began (see Fig. 13.3). Thus the demographic effects of the mortality crisis were detectable for the following 150 years.

Similar long-wavelength oscillations in synchronous baptisms and burials can be detected by time-series analysis after the plague of 1604 in York and after the plague of 1665–66 in London (Scott & Duncan, 1998).

Major epidemics in the 16th and 17th centuries in England (and in continental Europe) were confined to the towns and cities, whereas the surrounding villages and smaller towns largely escaped. Immigration from the countryside, particularly from the younger people, speedily filled the gaps left by the mortality thereby assisting in the demographic recovery. London took in a very large number of servants and apprentices from as

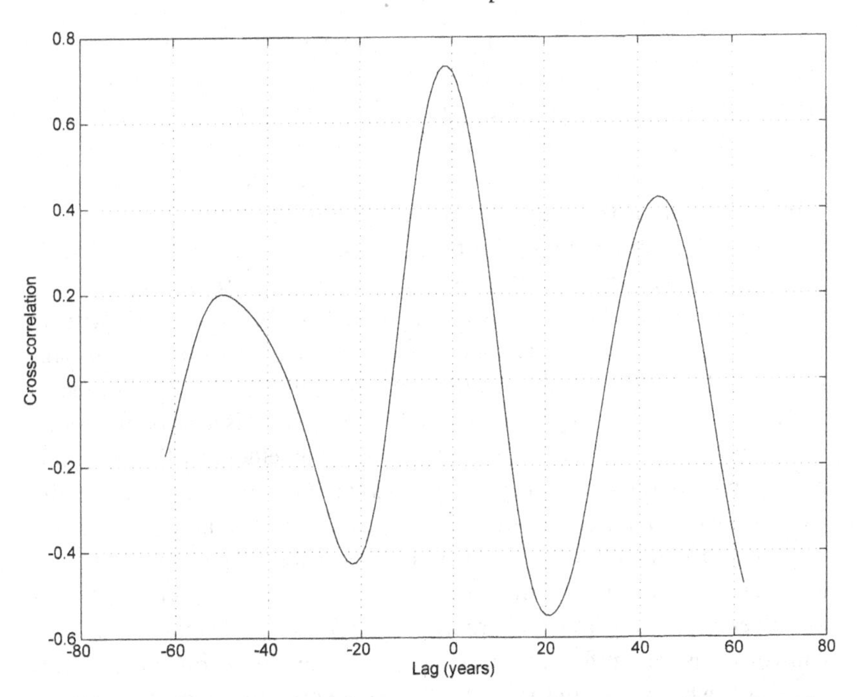

Fig. 13.5. Cross-correlation between the burial and baptism series at Penrith, 1625–1750. Filter window = 40 to 50 years.

far afield as York (Galley, 1998) and these would have added to the pool of susceptibles, probably being neither resistant nor immune. Table 8.1 shows that many of them died in the plague epidemics, leaving more gaps to be filled by the next wave of susceptible young immigrants. We have shown in this section, therefore, how the epidemiology of the plague interacted with the demography, dynamics and social behaviour of the population.

13.18 Postscript

We have used two main approaches in our study of the biology of haemorrhagic plague. Firstly, we have analysed parish records in detail and have traced the spread of the infection through and between families in identified populations, simulating as far as is possible 400 years after the event, the work of a modern epidemiologist. The characteristics of the disease can be elucidated in this way. Secondly, we have examined the different spread of the disease through different metapopulations. These two approaches together present haemorrhagic plague in a new light.

Obviously, it is impossible to do justice to the wealth of information about plagues in Europe but we believe that we have established a completely new perspective on which a proper understanding of the epidemiology and demography can be built. The key lies with our study of pseudo-endemic plague in the metapopulation of France; if the cycle of epidemics here had been broken, the plague would probably have disappeared. Previously, the different patterns of spread of the plague within individual communities and within metapopulations and across the seas were inexplicable when the epidemics were attributed to *Yersinia pestis* with its complex biology, being dependent on rats, fleas and resistant rodents, and some scholars soon abandoned their attempts to explain the seasonality and propagation of the epidemics. But bubonic plague is now banished to a very minor role indeed (as in the outbreak at Marseilles in 1720) and, once it is accepted that haemorrhagic plague was an infectious disease (probably viral) spread person-to-person, the biology of the plague epidemics immediately become obvious. The key to the success of the infectious agent for over 300 years lay in the lengthy incubation period, which allowed infectives a long time to move around over substantial distances and made control measures difficult. Nevertheless, the health authorities in Italy, even with what we would regard today as a very limited medical knowledge, quickly identified the important features of the disease and established sensible control measures in the 14th century, including the all-important 40-day quarantine period. They were most concerned with the importation of the disease via the ports.

A full understanding of the plagues in bygone days may be of importance to epidemiology in the 21st century because, as we relate in Chapter 1, there is the continuing fear that other lethal infectious diseases will emerge at any time from their animal hosts. The HIV and Spanish influenza pandemics of the 20th century are grim warnings.

Now that we have a new integrated picture of the plague epidemics, in which epidemiology, biology and demography are combined, it will be possible to continue with the analyses and to make rapid progress. Topics that could be pursued include:

(i) Is there any evidence to substantiate the belief that a growing proportion of the population, particularly in London by the 17th century, was composed of resistant or immune individuals? Resistance might be the result of the *CCR5-Δ32* deletion and immunity might be conferred by previous exposure.
(ii) Is there evidence, other than the anecdotal, that the disease decreased

in virulence during the course of an epidemic?

(iii) Does the apparent difference in infectivity in summer and winter provide information of the nature of the causative agent?

(iv) Where an epidemic struck a well-defined, medium-sized town with full parish records, it would be of interest to reconstruct the course of the outbreak as we have done at Penrith and Eyam and also to follow the subsequent population dynamics as described in section 13.17 and in Scott & Duncan (1998). We have studied the 404 parishes of rural England described by Wrigley & Schofield (1981) by time-series analysis and there are many mortality crises that apparently have the hallmarks of haemorrhagic plague that are not listed by Shrewsbury (1970). Possible towns for initial study in Leicestershire alone are Husband's Bosworth, Kibworth Beauchamp, Hinckley, Castle Donnington, Ashby de la Zouche and Bottesford.

(v) Biraben (1975) has listed the towns in France that suffered from the plague in each year from the Black Death until 1670 and, if it were possible to determine the start and finish of the epidemics in these foci, the probable movements of the infectives cycling round the metapopulation could be determined.

(vi) Details of minor outbreaks of plague in England, hidden away in parish registers and specifically designated as the pest could be reported and analysed. Nil returns would also be relevant and gradually the detailed pattern of spread of an epidemic during the period 1565–1665 would be established.

Identification of a lethal disease that was prevalent over 300 years ago is difficult and it is only by the accretion and integration of a large body of evidence that we can hope to discern the epidemiology and biology of this terrible disease that, arguably, had the greatest demographic impact on Europe.

References

Alkhatib, G., Combadiere, C., Broder, C.C., Feng, Y., Kennedy, P.E., Murphy, P.M., *et al.* (1996). CCCKR5: a RANTES, MIP-1α, MIP-1β receptor as a fusion cofactor for macrophage-tropic HIV-1. *Science*, **272**, 1955–1958.

Anderson, R.M. & May, R.M. (1982a). Coevolution of hosts and parasites. *Parasitology*, **85**, 411–426.

Anderson, R.M. & May, R.M. (1982b). Directly transmitted infectious diseases: control by vaccination. *Science*, **215**, 1053–1060.

Anderson, R.M. & May, R.M. (1985). Vaccination and herd immunity to infectious diseases. *Nature*, **318**, 323–329.

Anderson, R.M. & May, R.M. (1991). *Infectious Diseases of Humans.* Oxford: Oxford University Press.

Annotation (1924). Reservoir of plague in South Africa. *British Medical Journal*, **1**, 875.

Appleby, A.B. (1973). Disease or famine? Mortality in Cumberland and Westmorland 1580–1640. *Economic History Review*, **26**, 403–432.

Appleby, A.B. (1975). Agrarian capitalism or seigneurial reaction? The northwest of England, 1500–1700. *American Historical Review* **80**, 574–594.

Appleby, A.B. (1980). Epidemics and famine in the Little Ice Age. *Journal of Interdisciplinary History*, **10**, 643–663.

Axon, W.E.A. (1894). Chronological notes on the visitations of plague in Lancashire and Cheshire. *Transactions of the Lancashire and Cheshire Antiquarian Society*, **12**, 52–99.

Bacot, A.W. & Martin, C.J. (1914). Observations on the mechanism of the transmission of plague by fleas. *Journal of Hygiene*, **13**, 423–439.

Bailey, N.T.J. (1975). *The Mathematical Theory of Infectious Diseases and its Applications.* London: Griffin.

Barclay, A.J.G., Foster, A. & Sommer, A. (1987). Vitamin supplements and mortality related to measles: a randomised clinical trial. *British Medical Journal*, **294**, 294–296.

Barnes, H. (1891). Visitations of the plague in Cumberland and Westmorland. *Transactions of the Cumberland and Westmorland Antiquarian and Archaeological Society*, **11**, 158–186.

Bartlett, M.S. (1957). Measles periodicity and community size. *Journal of the Royal Statistical Society*, **120**, 48–70.

Bartlett, M.S. (1960). The critical community size for measles in the U.S. *Journal of the Royal Statistical Society*, **123**, 38–44.

Batho, G.R. (1964). The plague at Eyam: a tercentenary re-evaluation. *Derbyshire Archaeological Journal*, **84**, 81–91.

Baulant, M. (1968). Le prix des grains à Paris de 1431 à 1788. *Annales: Économies, Sociétés, Civilisations*, **23**, 520–540.

Bell, W.G. (1924). *The Great Plague in London in 1665*. London: John Lane, The Bodley Head Ltd.

Benedictow, O.J. (1987). Morbidity in historical plague epidemics. *Population Studies*, **41**, 401–431.

Benenson, A.S. (1990). *Control of Communicable Diseases in Man*. 15th Edition. New York: American Public Health Association.

Bercovier, H., Mollaret, H.H., Alonso, J.M., Brault, J., Fanning, G.R., Steigerwalt, A.G., *et al.* (1980). Intra- and interspecies relatedness of *Yersinia pestis* by DNA hybridization and its relationship to *Yersinia pseudotuberculosis*. *Current Microbiology*, **4**, 225–229.

Berman, S. (1991). Epidemiology of acute respiratory infections in children in developing countries. *Review of Infectious Diseases*, **13**, S454–S462.

Bertrand, J.B. (1973). *A Historical Relation of the Plague at Marseilles in the Year 1720*. Brookefield, VT: Gregg International Publishers Limited.

Biraben, J.-N. (1972). Certain demographic characteristics of the plague epidemic in France, 1720–1722. In *Population and Social Change*, ed. D.V. Glass & R. Revelle, pp. 233–241. London: Edward Arnold.

Biraben, J.N. (1975). *Les Hommes et la Peste en France et dans les Pays Européens et Méditerranéens*. Tome I. *La peste dans l'histoire*. Paris: Mouton & Co and École des Hautes Études en Sciences Sociales.

Biraben, J.N. (1976). *Les Hommes et la Peste en France et dans les Pays Européens et Méditerranéens*. Tome II. *Les hommes face a la peste*. Paris: Mouton & Co and École des Hautes Études en Sciences Sociales.

Biti, R., French, R., Young, J., Bennetts, B. & Stewart, G. (1997). HIV–1 infection in an individual homozygous for the CCR5 deletion allele. *Nature Medicine*, **3**, 252–253.

Bolin, I., Norlander, L. & Wolf-Watz, H. (1982). Temperature-inducible outer membrane protein of *Yersinia pseudotuberculosis* and *Yersinia enterocolitica* is associated with the virulence plasmid. *Infection and Immunity*, **37**, 506–512.

Bolker, B. & Grenfell, B. (1993). Chaos and biological complexity in measles dynamics. *Philosophical Transactions of the Royal Society B*, **251**, 75–81.

Bolker, B. & Grenfell, B. (1995). Space, persistence and dynamics of measles epidemics. *Philosophical Transactions of the Royal Society B*, **348**, 309–320.

Bolton, J. (1996). The world upside down. Plague as an agent of economic and social change. In *The Black Death in England*, ed. W.M. Ormrod & P.G. Lindley, pp. 17–78. Stamford: Paul Watkins.

Bowden, P.J. (1967). Agricultural prices, farm profits, and rents. In *The Agrarian History of England and Wales*, ed. J. Thirsk, vol. IV, *1500–1640*, pp. 593–695. Cambridge: Cambridge University Press.

Bowden, P.J. (1971). *The Wool Trade in Tudor and Stuart England*. London: Frank Cass.

Bowden, P.J. (1985). Agricultural prices, wages, farm profits and rents. In *The Agrarian History of England and Wales*, ed. J. Thirsk, vol. V, *1640–1750*, pp. 1–117. Cambridge: Cambridge University Press.

Boyce, N. (1999). Out of Africa, into the Bronx. *New Scientist*, no. 2206, 13.

Bradley, L. (1977). The most famous of all English plagues: a detailed analysis of the plague at Eyam 1665–6. In *The Plague Reconsidered*, pp. 63–94. Stafford:

Hourdsprint.
Bradshaw, F. (1907). Social and economic history. III. The Black Death. In *The Victoria History of the County of Durham*, vol. II, ed. W. Page, p. 210. London: Constable.
Bridges, C.B., Katz, J.M., Seto, W.H., Chan, P.K.S., Tsang, D., Ho, W., *et al.* (2000). Risk of influenza A (H5N1) infection among health care workers exposed to patients with influenza A (H5N1), Hong Kong. *Journal of Infectious Diseases*, **181**, 344–348.
Campbell-Kease, J. (1989). *A Companion to Local History Research*. London: A. & C. Black Ltd.
Carmichael, A. (1986). *Plague and the Poor in Renaissance Florence*. Cambridge: Cambridge University Press.
Carmichael, A.G. (1991). Contagion theory and contagion practice in fifteenth-century Milan. *Renaissance Quarterly*, **44**, 213–256.
Carmichael, A.G. (1997). Bubonic plague: the Black Death. In *Plague, Pox and Pestilence*, ed. K.F. Kiple, pp. 60–67. London: Weidenfeld & Nicholson.
Carrington, M., Kissner, T., Gerrard, B., Ivanov, S., O'Brien, S.J. & Dean, M. (1997). Novel alleles of the chemokine-receptor gene *CCR5*. *American Journal of Human Genetics*, **61**, 1261–1267.
Chambers, J.D. (1972). *Population, Economy and Society in Pre-industrial England*. Oxford: Oxford University Press.
Chandler, A.C. & Read, C.P. (1961). *Introduction to Parasitology*. 10th Edition. New York: John Wiley and Sons.
Choe, H., Farzan, M., Sun, Y., Sullivan, N., Rollins, B., Ponath, P.D., *et al.* (1996). The β-chemokine receptors CCR3 and CCR5 facilitate infection by primary HIV-1 isolates. *Cell*, **85**, 1135–1148.
Christie, A.B. (1969). *Infectious Diseases: Epidemiology and Clinical Practice*. 3rd Edition. Edinburgh: Churchill Livingstone.
Cipolla, C.M. (1981). *Fighting the Plague in Seventeenth-Century Italy*. Wisconsin: University of Wisconsin Press.
Clapham, J. (1949). *A Concise Economic History of Britain*. Cambridge: Cambridge University Press.
Clark, P., Gaskin, K. & Wilson, A. (1989). *Population Estimates of English Small Towns 1550–1851*. Centre for Urban History, University of Leicester, Working Paper No. 3.
Cliff, A. (1995). Incorporating spatial components into models of epidemic spread. In *Epidemic Models: Their Structure and Relation to Data*, ed. D. Mollison, pp. 119–149. Cambridge: Cambridge University Press.
Cliff, A.D. & Haggett, P. (1988). *Atlas of Disease Distributions: Analytic Approaches to Epidemiological Data*. Oxford: Blackwell Reference.
Clifford, J. (1989). *Eyam Plague 1665–1666*. Sheffield: The Print Centre.
Clifford, J.G. & Clifford, F. (1993). *Eyam Parish Register 1630–1700*, vol. XXI. Chesterfield: Derbyshire Record Society.
Coale, A.J. & Demeny, P. (1966). *Regional Model Life Tables and Stable Populations*. Princeton, NJ: Princeton University Press.
Comford, M.E. (1907). Religious houses. In *The Victoria History of the County of Durham*, vol. II, ed. W. Page, p. 99. London: Constable.
Cornelis, G.R. & Wold-Wulz, H. (1997). The Yersinia Yop virulon: a bacterial system for subverting eurkaryotic cells. *Molecular Microbiology*, **23**, 861–867.
Cox, J.C. (1910). *The Parish Registers of England*. London: Methuen.
Creighton, C. (1894). *History of Epidemics in Britain*. Cambridge: Cambridge

University Press.
Davis, D.E. (1986). The scarcity of rats and the Black Death: an ecological history. *Journal of Interdisciplinary History*, **XVI**, 455–470.
Dean, M., Carrington, M., Winkler, C., Huttley, G.A., Smith, M.W., Allikmets, R., *et al.* (1996). Genetic restriction of HIV-1 infection and progression to AIDS by a deletion allele of the CKR5 structural gene. *Science*, **273**, 1856–1862.
Deaux, G. (1969). *The Black Death 1347*. New York: Weybright and Talley.
Defoe, D. (1722). *Diary of a Plague Year*. London: Everyman's Library.
Deng, H., Liu, R., Ellmeier, W., Choe, S., Unutmaz, D., Burkhart, M., *et al.* (1996). Identification of a major co-receptor for primary isolates of HIV-1. *Nature*, **381**, 661–666.
Dietz, K. (1975). In *Epidemiology*, ed. D. Ludwig & K.L. Cooke, pp. 104–121. Philadelphia: Society for Industrial and Applied Mathematics.
Doranz, B.J., Rucker, J., Yi, Y., Smyth, R.J., Samson, M., Peiper, S.C., *et al.* (1996). A dual-tropic primary HIV-1 isolate that uses fusin and the β-chemokine receptors CKR-5, CKR-3, and CKR-2b as fusion cofactors. *Cell*, **85**, 1149–1158.
Dragic, T., Litwin, V., Allaway, G.P., Martin, S.R., Huang, Y., Nagashima, K.A., *et al.* (1996). HIV-1 entry into CD4+ cells is mediated by the chemokine receptor CC-CKR5. *Nature*, **381**, 667–673.
Drancourt, M., Aboudharam, G., Signoli, M., Dutour, O. & Raoult, D. (1998). Detection of 400-year-old *Yersinia pestis* DNA in human dental pulp: an approach to the diagnosis of ancient septicemia. *Proceedings of the National Academy of Science*, **95**, 12 637–12 640.
Drummond, J.C. & Wilbraham, A. (1991). *The Englishman's Food*. London: Pimlico.
Duncan, S.R., Scott, S. & Duncan, C.J. (1992). Time series analysis of oscillations in a model population: the effects of plague, pestilence and famine. *Journal of Theoretical Biology*, **158**, 293–311.
Duncan, S.R., Scott, S. & Duncan, C.J. (1993a). An hypothesis for the periodicity of smallpox epidemics as revealed by time series analysis. *Journal of Theoretical Biology*, **160**, 231–248.
Duncan, S.R., Scott, S. & Duncan, C.J. (1993b). The dynamics of smallpox epidemics in Britian 1550–1800. *Demography*, **30**, 405–423.
Duncan, S.R., Scott, S. & Duncan, C.J. (1994a). Modelling the different smallpox epidemics in England. *Philosophical Transactions of the Royal Society B*, **346**, 407–419.
Duncan, S.R., Scott, S. & Duncan C.J. (1994b). Smallpox epidemics in cities in the 17th and 18th centuries. *Journal of Interdisciplinary History*, **25**, 255–271.
Duncan, S.R., Scott, S. & Duncan, C.J. (1994c). Determination of a feedback vector that generates a non-decaying oscillation in a model population. *Journal of Theoretical Biology*, **167**, 67–71.
Duncan, C.J., Duncan, S.R. & Scott, S. (1996a). Whooping cough epidemics in London 1701–1812: infection dynamics, seasonal forcing and malnutritional effects. *Proceedings of the Royal Society B*, **263**, 445–450.
Duncan, C.J., Duncan, S.R. & Scott, S. (1996b). The dynamics of scarlet fever epidemics in England and Wales in the 19th Century. *Epidemiology & Infection*, **117**, 493–499.
Duncan, C.J., Duncan, S.R. & Scott, S. (1997). The dynamics of measles epidemics. *Theoretical Population Biology*, **52**, 155–163.
Duncan, C.J., Duncan, S.R. & Scott, S. (1998). The effects of population density

and malnutrition on the dynamics of whooping cough. *Epidemiology & Infection*, **121**, 325–334.

Duncan, S.R., Scott, S. & Duncan, C.J. (1999). A demographic model of measles epidemics. *European Journal of Population*, **15**, 185–198.

Dyer, A. (1997). The English Sweating Sickness of 1551: an epidemic anatomized. *Medical History*, **41**, 362–84.

Ell, S.R. (1980). Interhuman transmission of medieval plague. *Bulletin of the History of Medicine*, **54**, 497–510.

Eversley, D.E.C. (1965). Epidemiology as social history. In *History of Epidemics in Britain*, vol. 1, *From A.D. 664 to the Great Plague*, ed. C. Creighton, pp. 3–39. London: Frank Cass & Co. Ltd.

Farrar, S. (2000). Bug that bears the mark of death. *The Times Higher Education Supplement*, no. 1431, 14 April, 20–21.

Fauman, E.B. & Saper, M.A. (1996). Structure and function of the protein tyrosine phosphatases. *Trends in Biochemical Sciences*, **21**, 413–417.

Feldmann, H. & Klenk, H.D. (1996). Marburg and Ebola viruses. *Advances in Virus Research*, **47**, 1–52.

Feldmann, H., Bugany, H., Mahner, F., Klenk, H.D., Drenckhahn, D. & Schnittler, H.J. (1996a). Filovirus-induced endothelial leakage triggered by infected monocytes/macrophages. *Journal of Virology*, **70**, 2208–2214.

Feldmann, H., Slenczka, W. & Klenk, H.D. (1996b). Emerging and reemerging of filoviruses. *Archives of Virology*, **S11**, 77–100.

Fine, P.E.M. (1993). Herd immunity: history, theory, practice. *Epidemiologic Review*, **15**, 265–304.

Folks, T. (1998). Ebola takes a punch. *Nature Medicine*, **4**, 16–17.

Furness, W. (1894). *The History of Penrith from the Earliest Period to the Present Time by Ewanian (William Furness)*. Penrith: William Furness.

Galley, C. (1994). A never-ending succession of epidemics? Mortality in early-modern York. *Social History of Medicine*, **7**, 29–57.

Galley, C. (1995). A model of early-modern urban demography. *Economic History Review*, **48**, 448–469.

Galley, C. (1998). *The Demography of Early Modern Towns: York in the Sixteenth and Seventeenth Centuries*. Liverpool: Liverpool University Press.

Garrett, L. (1995). *The Coming Plague. Newly Emerging Diseases in a World out of Balance*. London: Virago Press.

Geisbert, T.W., Jahrling, P.B., Hanes, M.A. & Zack, P.M. (1992). Association of Ebola-related Reston virus particles and antigen with tissue lesions of monkeys imported to the United States. *Journal of Comparative Pathology*, **106**, 137–152.

Gisecke, J. (1994). *Modern Infectious Disease Epidemiology*. Bristol: J.W. Arrowsmith Ltd.

Gordon, J.E. & LeRiche, H. (1950). The epidemiologic method applied to nutrition. *American Journal of the Medical Sciences*, **219**, 321–345.

Gottfried, R.S. (1978). *Epidemic Disease in Fifteenth Century England. The Medical Response and the Demographic Consequences*. Leicester: Leicester University Press.

Gottfried, R.S. (1983). *The Black Death*. New York: The Free Press.

Grenfell, B.T. (1992). Chance and chaos in measles epidemics. *Journal of the Royal Statistical Society B*, **54**, 383–398.

Guan, K. & Dixon, J.E. (1990). Protein tyrosine phosphatase activity of an essential virulence determinant in *Yersinia*. *Science*, **249**, 553–556.

Halloran, M.E. (1998). Concepts of infectious disease epidemiology. In *Modern Epidemiology*, 2nd Edition, ed. J. Rothman & S. Greenland, pp. 529–554. Philadelphia: Lippincott-Raven.

Hankin, E.H. (1905). On the epidemiology of the plague. *Journal of Hygiene*, **5**, 48–83.

Harcourt, B.H., Sanchez, A. & Offermann, M.K. (1999). Ebola virus selectively inhibits responses to interferons, but not to interleukin-1beta, in endothelial cells. *Journal of Virology*, **73**, 3491–3496.

Harvey, G. (1769). *The City Remembrancer: Being Historical Narratives of the Great Plague at London, 1665; Great Fire, 1666; and Great Storm, 1703*, vol. I. London: W. Nicoll.

Hassanein, K.M., el-Azazy, O.M. & Yousef, H.M. (1997). Detection of Crimean-Congo haemorrhagic fever virus antibodies in humans and imported livestock in Saudi Arabia. *Transactions of the Royal Society for Tropical Medicine and Hygiene*, **91**, 536–537.

Hindle, B.R. (1993). *Roads, Tracks and Their Interpretation*. London: Batsford.

Hinnebusch, B.T., Perry, R.D. & Schwan, T.G. (1996). Role of the *Yersinia pestis* hemin storage (hms) locus in the transmission of plague by fleas. *Science*, **273**, 367–370.

Hirshleifer, J. (1996). *Disaster and Recovery: The Black Death in Western Europe*. Santa Monica: Rand Corporation.

Hirst, L.F. (1952). Plague. In *British Encyclopaedia of Medical Practice*, 2nd Edition, vol. 9, pp. 144–145. London: Butterworths.

Hollingsworth, M.F. & Hollingsworth, T.H. (1971). Plague mortality by age and sex in the Parish of St. Botolph's without Bishopsgate, London 1603. *Population Studies*, **25**, 131–146.

Horrox, R. (1994). *The Black Death*. Manchester: Manchester University Press.

Howells, J. (1985). Haverfordwest and the plague, 1652. *Welsh History Review*, **12**, 411–419.

Howson, W.G. (1961). Plague, poverty and population in parts of North-West England 1580–1720. *Transactions of the Historic Society of Lancashire and Cheshire*, **112**, 29–55.

Huang, Y., Paxton, W.A., Wolinsky, S.M., Neumann, A.U., Zhang, L., He, T., *et al.* (1996). The role of a mutant CCR5 allele in HIV-1 transmission and disease progression. *Nature Medicine*, **2**, 1240–1243.

Hughes, J. (1971). The plague at Carlisle 1597/8. *Transactions of the Cumberland and Westmorland Antiquarian and Archaeological Society*, **81**, 52–63.

Husbands, C. (1984). Hearth Tax figures and the assessment of poverty in the seventeenth-century economy. In *The Hearth Tax: Problems and Possibilities*, ed. N. Alldridge, pp. 45–58. Hull: Humberside College of Higher Education.

Hussain, S., Goila, R., Shahi, S. & Banerjea, A. (1998). First report of a healthy Indian heterozygous for delta 32 mutant of HIV-1 co-receptor-*CCR5* gene. *Gene*, **207**, 141–147.

Irving, W. (1935). *Historic Penrith. Rambles in Ancient Byways*. Published privately as a local pamphlet.

Isberg, R. R. & Falkow, S. (1985). A single genetic locus encoded by *Yersinia pseudotuberculosis* permits invasion of cultured animal cells by *Escherichia coli* K-12. *Nature*, **317**, 262–264.

James, J.W. (1972). Longitudinal study of the morbidity of diarrhoeal and respiratory infections in malnourished children. *American Journal of Clinical Nutrition*, **25**, 690–694.

Khan, A.S., Sanchez, A. & Pflieger, A.K. (1998). Filoviral haemorrhagic fevers. *British Medical Bulletin*, **54**, 675–692.
King, E. (1988). *Medieval England.* London: Guild Publishing.
Kiple, K.F. & Ornelas, K.C. (1997). Sweating sickness: an English mystery. In *Plague, Pox and Pestilence*, ed. K.F. Kiple, pp. 156–159. London: Weidenfeld & Nicolson.
Kirchhoff, F., Pohlmann, S., Hamacher, M., Means, R.E., Kraus, T., Uberla, K., *et al.* (1997). Simian immunodeficiency virus variants with differential T-cell and macrophage trophism use *CCR5* and an unidentified cofactor expressed in CEMx174 cells for efficient entry. *Journal of Virology*, **71**, 6509–6516.
Kitch, M.J. (1986). Capital and kingdom: migration to later Stuart London. In *London 1500–1700. The Making to the Metropolis*, ed. A.L. Beier & R. Finlay, pp. 224–251. London: Longman.
Kohn, G.C. (1995). *The Wordsworth Encyclopedia of Plague and Pestilence.* New York: Facts on File.
Ladurie, E.L.R. & Baulant, M. (1980). Grape harvests from the fifteenth through the nineteenth centuries. *Journal of Interdisciplinary History*, **10**, 839–849.
Lamb, H.H. (1977). *Climate History and the Future.* Princeton, NJ: Princeton University Press.
Landers, J. (1993). *Death and the Metropolis.* Cambridge: Cambridge University Press.
Langer, W.L. (1964). The Black Death. *Scientific American*, **210**, 114–121.
Langmuir, A.D., Worthen, T.D., Solomon, J., Ray, C.G. & Petersen, E. (1985). The Thucydides syndrome: a new hypothesis for the cause of the plague of Athens. *New England Journal of Medicine*, **313**, 1027–1030.
Last, J.M. (ed.) (1995). *A Dictionary of Epidemiology*, 3rd Edition. New York: Oxford University Press.
Law, A. (1908). Social and economic history. In *The Victoria History of the County of Lancaster*, vol. II, ed. W. Farrer & J. Brownbill, p. 285. London: Constable.
Leasor, J. (1962). *The Plague and The Fire.* London: George Allen & Unwin Ltd.
Le-Guenno, B. (1997). Haemorrhagic fevers and ecological perturbations. *Archives of Virology Supplement*, **13**, 191–199.
Lenski, R.E. (1988). Evolution of plague virulence. *Nature*, **334**, 473–474.
LeRiche, W.H. & Milner, J. (1971). *Epidemiology as Medical Ecology.* London: Churchill Livingstone.
Levin, S. & Pimentel, D. (1981). Selection of intermediate rates of increase in parasite host systems. *American Naturalist*, **117**, 308–315.
Libert, F., Cochaux, P., Beckman, G., Samson, M., Aksenova, M., Cao, A., *et al.* (1998). The delta CCR5 mutation conferring protection against HIV-1 in Caucasian populations has a single and recent origin in Northeastern Europe. *Human Molecular Genetics*, **7**, 399–406.
Lilienfeld, D.E. & Stolley, P.D. (1994). *Foundations of Epidemiology*, 3rd Edition. Oxford: Oxford University Press.
Lindley, P. (1996). The Black Death and English art. A debate and some assumptions. In *The Black Death in England*, pp. 125–146. Stamford: Paul Watkins.
Liston, W.G. (1924). The Milroy Lectures. The Plague. *British Medical Journal*, **1**, 900–997.
Littman, R.J. & Littman, M.L. (1969). The Athenian plague: smallpox. *Transactions of the American Philological Association*, **100**, 261–275.

Liu, R., Paxton, W.A., Choe, S., Ceradini, D., Martin, S.R., Horuk, R., *et al.* (1996). Homozygous defect in HIV-1 coreceptor accounts for resistance of some multiply-exposed individuals to HIV–1 infection. *Cell*, **86**, 367–377.

Livi-Bacci (1997). *A Concise History of World Population.* Cambridge: Cambridge University Press.

London, W.P. & Yorke, J.A. (1973). Recurrent outbreaks of measles, chickenpox and mumps. I. Seasonal variation in contact rates. *American Journal of Epidemiology*, **98**, 453–468.

Longrigg, J. (1980). The great plague of Athens. *History of Science*, **18**, 209–225.

Lunn, J. (1937). The Black Death in the Bishop's Registers. Ph.D. thesis, Cambridge University.

MacKenzie, D. (1999). Bats off the hook. *New Scientist*, **164**, 16.

Madani, N., Kozak, S.L., Kavanaugh, M.P. & Kabat, D. (1998). gp120 envelope glycoproteins of human immunodeficiency viruses. *Proceedings of the National Academy of Sciences, USA*, **95**, 8005–8010.

Maia, J. de O.C. (1952). Some mathematical developments on the epidemic theory formulated by Reed and Frost. *Human Biology*, **24**, 167–200.

Mann, J.M. (1995). Preface. In *The Coming Plague. Newly Emerging Diseases in a World Out of Balance*, pp. xi–xii. London: Virago Press.

Martinson, J.J., Chapman, N.H., Rees, D.C., Lui, Y.-T. & Clegg, J.B. (1997). Global distribution of the CCR5 gene 32-base pair deletion. *Nature Genetics*, **16**, 100–102.

Matheson, C. (1939). A survey of the status of *Rattus rattus* and its subspecies in the seaports of Great Britain and Ireland. *Journal of Animal Ecology*, **8**, 76–93.

May, R.M. & Anderson, R.M. (1983). Parasite–host co-evolution. In *Coevolution*, ed. D.J. Futuyma & M. Slatkin, pp. 186–206. Sunderland, MA: Sinauer Associates.

McNeill , W.H. (1977). *Plagues and Peoples.* Oxford: Blackwell.

Michael, N.L., Chang, G., Louie, L.G., Mascola, J.R., Dondero, D., Birx, D.L., *et al.* (1997). The role of viral phenotype and CCR-5 gene defects in HIV-1 transmission and disease progression. *Nature Medicine*, **3**, 338–340.

Miller, E. (1961). Mediaeval York. *The Victoria History of the City of York*, vol. III, ed. P.M. Tillott, pp. 84–85. London: Constable.

Mills, S.D., Boland, A., Sory, M.P., Van De Smissen, P., Kerbourch, C., Finlay, B.B., *et al.* (1997). *Yersinia enterocolitica* induces apoptosis in macrophages by a process requiring functional type III secretion and translocation mechanisms and involving YopP, presumably acting as an effector protein. *Proceedings of the National Academy of Sciences, USA*, **94**, 12 638–12 643.

Morens, D.M. & Littman, R.J. (1972). Epidemiology of the plague of Athens. *Transactions of the American Philological Association*, **122**, 271–304.

Morens, D.M. & Littman, R.J. (1994). The Thucydides syndrome reconsidered: new thoughts on the plague of Athens. *American Journal of Epidemiology*, **140**, 621–627.

Morris, C. (1977). Plagues in Britain. In *The Plague Reconsidered*, pp. 37–47. Stafford: Hourdsprint.

Morris, J.N. (1957). *Use of Epidemiology*, 1st Edition. London: E. & S. Livingstone Ltd.

Mummidi, S., Ahuja, S.S., Gonzalez, E., Anderson, S.A., Santiago, E.N., Stephen, K.T., *et al.* (1998). Genealogy of the *CCR5* locus and chemokine system gene variants associated with altered rates of HIV-1 disease progression. *Nature*

Medicine, **4**, 786–793.

Nabel, G.J. (1999). Surviving Ebola virus infection. *Nature Medicine*, **5**, 373–374.

Nicholson, J. & Burn, R. (1777). *The History and Antiquities of Westmorland and Cumberland*. London: Strachan and Cadell.

Nohl, J. (1926). *The Black Death. A Chronicle of the Plague*. London: George Allen & Unwin Ltd.

O'Brien, S.J. & Dean, M. (1997). In search of AIDS-resistance genes. *Scientific American*, **277**, 44–51.

O'Brien, T., Winkler, C., Dean, M., Nelson, J.A.E., Carrington, M., Michael, N.L., *et al.* (1997). HIV-1 infection in a man homozygous for CCR5-Δ32. *Lancet*, **349**, 1219.

Olshansky, S.J., Carnes, B., Rogers, R.G. & Smith, L. (1997). Infectious diseases. New and ancient threats to world health. *Population Bulletin*, **52**, 2–54.

Olsen, L.F. & Schaffer, W.M. (1990). Chaos versus noisy periodicity: alternative hypotheses for childhood epidemics. *Science*, **249**, 499–504.

Olson, P.E., Hames, C.S., Benenson, A.S. & Genovese, E.N. (1996). The Thucydides syndrome: Ebola déjà vu (or Ebola reemergent?). *Emerging Infectious Diseases*, **2**, 155–156.

Ormrod, M. (1996). The politics of pestilence. Government in England after the Black Death. In *The Black Death in England*, pp. 147–181. Stamford: Paul Watkins.

Ormrod, W.M. & Lindley, P.G. (1996). *The Black Death in England*. Stamford: Paul Watkins.

Oxford, J.S. (2000). Viral clues in frozen brain cells. *The Times Higher Education Supplement*, 21 January 2000, 28.

Page, D.L. (1953). Thucydides' description of the great plague. *Classical Quarterly*, **47**, 97–119.

Palese, P., Zheng, H., Engelhardt, O.G., Pleschka, S. & Garcia Sastre, A. (1996). Negative-strand RNA viruses: genetic engineering and applications. *Proceedings of the National Academy of Sciences, USA*, **93**, 11 354–11 358.

Pepys, S. (1665/1906). *Diary of Samuel Pepys*. London: J.M. Dent & Sons Ltd.

Peters, C.J., Sanchez, A., Rollin, P.E., Ksiazek, T.G. & Murphy, F.A. (1996). Filoviridae: Marburg and Ebola viruses. In *Fields' Virology*, 3rd Edition, ed. B.N. Fields, D.M. Knipe & P.M. Howley, pp. 1161–1176. Philadelphia: Raven Press.

Pfister, C. (1980). The Little Ice Age: thermal and wetness indices for Central Europe. *Journal of Interdisciplinary History*, **10**, 665–696.

Plague Research Commission (1907a). The epidemiological observations made by the Commission in Bombay City. *Journal of Hygiene*, **7**, 724–798.

Plague Research Commission (1907b). Observations made in four villages in the neighbourhood of Bombay. *Journal of Hygiene*, **7**, 799–873.

Pollitzer, R. (1954). *Plague*. Geneva: World Health Organization Monography Series No. 22.

Poole, J.C.F. & Holladay, A.J. (1979). Thucydides and the plague of Athens. *Classical Quarterly*, **29**, 282–300.

Porteous, C. (1962). *The Well-Dressing Guide*. Hanley: Wood, Mitchell & Co. Ltd.

Portnoy, D.A., Wolf-Watz, H., Bolin, I., Beeder, A.B. & Falkow, S. (1984). Characterization of common virulence plasmids in *Yersinia* species and their role in the expression of outer membrane proteins. *Infection and Immunity*, **43**, 108–114.

Race, P. (1995). Some further consideration of the plague in Eyam, 1665/6. *Local*

Population Studies, **54**, 56–57.
Ramanathan, C.S. & Taylor, E.W. (1997). Computational genomic analysis of hemorrhagic fever viruses. Viral selenoproteins as a potential factor in pathogenesis. *Biological Trace Element Research*, **56**, 93–106.
Raoult, D., Aboudharam, G., Crubézy, E., Larrouy, G., Ludes, B. & Drancourt, M. (2000). Molecular identification by 'suicide PCR' of *Yersinia pestis* as the agent of Medieval Black Death. *Proceedings of the National Academy of Sciences, USA*, **97**, 12800–12803.
Rappaport, S. (1989). *Worlds Within Worlds: Structures of Life in Sixteenth-Century London.* Cambridge: Cambridge University Press.
Rasheed, J.K., Arko, R.J., Feeley, J.C., Chandler, F.W., Thornsberry, C., Gibson, R.J., *et al.* (1985). Acquired ability of *Staphylococcus aureus* to produce toxic shock-associated protein and resulting illness in a rabbit model. *Infection and Immunity*, **47**, 598–604.
Razzell, P. (1977). *The Conquest of Smallpox: The Impact of Inoculation on Smallpox Mortality in Eighteenth-Century, Britain.* Sussex: Caliban.
Retief, F.P. & Cilliers, L. (1998). The epidemic of Athens, 430–426 BC. *South African Medical Journal*, **88**, 50–53.
Richards, R. (1947). *Old Cheshire Churches: A Survey of their History, Fabric and Furniture, with Records of the Older Monuments.* London: Batsford.
Rizzuto, C.D., Wyatt, R., Hernandez-Ramos, N., Sun, Y., Kwong, P.D., Hendrickson, W.A., *et al.* (1998). A conserved HIV gp120 glycoprotein structure involved in chemokine receptor binding. *Science*, **280**, 1949–1953.
Roberts, J.I. (1935). The endemicity of plague in East Africa. *East African Medical Journal*, **12**, 200–219.
Roberts, R.S. (1966). The place of plague in English history. *Proceedings of the Royal Society of Medicine*, **59**, 101–116.
Rosqvist, R., Skurnik, M. & Wolf-Watz, H. (1988). Increased virulence of *Yersinia pseudotuberculosis* by two independent mutations. *Nature*, **334**, 522–525.
Rottman, J.B., Ganley, K.P., Williams, K., Wu, L., Mackay, C.R. & Ringler, D.J. (1997). Cellular localization of the chemokine receptor *CCR5*. Correlation to cellular targets of HIV-1 infection. *American Journal of Patholology*, **151**, 1341–1351.
Russell, J.C. (1948). *British Medieval Population.* Albuquerque: University of New Mexico Press.
Russell, J.C. (1968). That Earlier Plague. *Demography*, **5**, 174–184.
Ryabchikova, E., Kolesnikova, L., Smolina, M., Tkachev, V., Pereboeva, L., Baranova, S., *et al.* (1996). Ebola virus infection in guinea pigs: presumable role of granulomatous inflammation in pathogenesis. *Archives of Virology*, **141**, 909–921.
Sakata, T. (1982). Some original materials concerning the plague at York in 1604. *Osaka Gakuin Review of Economics and Business*, **8**, 59–109.
Samson, M., Libert, F., Doranz, B.J., Rucker, J., Liesnard, C., Farber, C.-M., *et al.* (1996). Resistance to HIV-1 infection in caucasian individuals bearing mutant alleles of the CCR-5 chemokine receptor gene. *Nature*, **382**, 722–725.
Samson, M., LaRosa, G., Libert, F., Paindavoine, P., Detheux, M., Vassart, G., *et al.* (1997). The second extracellular loop of *CCR5* is the major determinant of ligand specificity. *Journal of Biological Chemistry*, **272**, 24 934–24 941.
Sanchez, A., Trappier, S.G., Mahy, B.W.J., Peters, C.J. & Nichol, S.T. (1996). The viron glycoproteins of Ebola viruses are encoded in two reading frames and are expressed through transcriptional editing. *Proceedings of the National*

Academy of Sciences, USA, **93**, 3602–3607.
Savage, W.G. & Fitzgerald, D.A. (1900). A case of plague from a clinical and pathological point of view. *British Medical Journal*, 1232–1236.
Schevill, F. (ed.) (1928). *The First Century of Italian Humanism.* New York: F.S. Crofts and Co.
Schnittler, H.J. & Feldmann, H. (1998). Marburg and Ebola hemorrhagic fevers: does the primary course of infection depend on the accessibility of organ-specific macrophages? *Clinical Infectious Diseases*, **27**, 404–406.
Schofield, R. (1977). An anatomy of an epidemic: Colyton November 1645–November 1646. In *The Plague Reconsidered*, pp. 95–126. Stafford: Hourdsprint.
Schubert, H.L., Fauman, E.B., Stuckey, J.A., Dixon, J.E. & Saper, M.A. (1995). A ligand-induced conformational change in the *Yersinia* protein tyrosine phosphatase. *Protein Science*, **4**, 1904–1913.
Schwarz, T.F., Nsanze, H. & Ameen, A.M. (1997). Clinical features of Crimean-Congo haemorrhagic fever in the United Arab Emirates. *Infection*, **25**, 364–367.
Scott, S. (1995). Demographic study of Penrith, Cumberland, 1557–1812, with particular reference to famine, plague and smallpox. Ph.D. thesis, University of Liverpool.
Scott, S. & Duncan, C.J. (1993). Smallpox epidemics at Penrith in the 17th and 18th centuries. *Transactions of the Cumberland and Westmorland Antiquarian and Archaeological Society*, **93**, 155–160.
Scott, S. & Duncan, C.J. (1997). The mortality crisis of 1623 in North-West England. *Local Population Studies*, **58**, 14–25.
Scott, S. & Duncan, C.J. (1998). *Human Demography and Disease.* Cambridge: Cambridge University Press.
Scott, S. & Duncan, C.J. (1999a). Malnutrition, pregnancy and infant mortality: a biometric model. *Journal of Interdisciplinary History*, **30**, 37–60.
Scott, S. & Duncan, C.J. (1999b). Characteristics of population cycles in pre-industrial England. *Local Population Studies*, **62**, 70–76.
Scott, S. & Duncan, C.J. (1999c). Reproductive strategies and sex-biased investment – the role of breast-feeding and wet-nursing. *Human Nature*, **10**, 85–108.
Scott, S. & Duncan, C.J. (2000). Interacting effects of nutrition and social class differentials on fertility and infant mortality in a pre-industrial population. *Population Studies*, **54**, 71–87.
Scott, S., Duncan, S.R. & Duncan, C.J. (1995). Infant mortality and famine: a study in historical epidemiology. *Journal of Epidemiology & Community Health*, **49**, 245–252.
Scott, S., Duncan, C.J. & Duncan, S.R. (1996). The plague at Penrith, Cumbria 1597/8: its causes, biology and consequences. *Annals of Human Biology*, **23**, 1–21.
Scott, S., Duncan, S.R. & Duncan, C.J. (1997). Critical effects of malnutrition during pregnancy. *Local Population Studies*, **58**, 62–65.
Scott, S., Duncan, S.R. & Duncan, C.J. (1998a). The interacting effects of prices and weather on population cycles in a preindustrial community. *Journal of Biosocial Science*, **30**, 15–32.
Scott, S., Duncan, S.R. & Duncan, C.J. (1998b). The origins, interactions and causes of cycles in grain prices in England. *Agricultural History Review*, **40**, 1–14.

Seal, S.C. (1969). Epidemiological studies of plague in India. I. The present position. *Bulletin of the World Health Organization*, **23,** 283–292.

Sellers, M. (1913). Social and economic history. In *The Victoria History of the County of York*, vol. III, ed. W. Page, p. 442. London: Constable.

Sharif, M. (1951). Spread of plague in southern and central divisions of Bombay Province and plague endemic centres in the Indo-Pakistan subcontinent. *Bulletin of the World Health Organization*, **4**, 75–109.

Sharpe France, R. (1939). A history of plague in Lancashire. *Transactions of the Historic Society of Lancashire and Cheshire*, **90**, 1–175.

Sheppard, F. (1998). *London A History*. Oxford: Oxford University Press.

Shrewsbury, J.F.D. (1950). The plague at Athens. *Bulletin of the History of Medicine*, **24**, 1–25.

Shrewsbury, J.F.D. (1970). *A History of Bubonic Plague in the British Isles*. Cambridge: Cambridge University Press.

Shumway, R.H. (1988). *Applied Statistical Time Series Analysis*. Wisconsin: Prentice Hall International Editions.

Simmons, G., Clapham, P.R., Picard, L., Offord, R.E., Rosenkilde, M.M., Schwartz, T.W., *et al.* (1997). Potent inhibition of HIV-1 infectivity in macrophages and lymphocytes by a novel CCR5 antagonist. *Science*, **276**, 276–279.

Slack, P. (1977a). Introduction. In *The Plague Reconsidered*, pp. 5–10. Stafford: Hourdsprint.

Slack, P. (1977b). The local incidence of epidemic disease: the case of Bristol 1540–1650. In *The Plague Reconsidered*, pp. 49–62. Stafford: Hourdsprint.

Slack, P. (1985). *The Impact of Plague in Tudor and Stuart England*. London: Routledge and Kegan Paul.

Slack, P. (1988). Responses to plague in early modern Europe: the implications of public health. *Social Research*, **55**, 433–453.

Stephens, J.C., Reich, D.E., Goldstein, D.B., Shin, H.D., Smith, M.W., Carrington, M., *et al.* (1998). Dating the origin of the CCR5-Δ32 AIDS-resistance allele by the coalescence of haplotypes. *American Journal of Human Genetics*, **62,** 1507–1515.

Stewart, J. (1928). *Nestorian Missionary Enterprise. The Story of a Church on Fire*. Edinburgh: T. & T. Clark.

Sunderland, I. (1972). When was the Great Plague? Mortality in London, 1563–1665. In *Population and Social Change*, ed. D.V. Glass & R. Revelle, pp. 287–320. London: Edward Arnold.

Suzuki, Y. & Gojobori, T. (1997). The origin and evolution of Ebola and Marburg viruses. *Molecular Biology and Evolution*, **14**, 800–806.

Takada, A., Robison, C., Goto, H., Sanchez, A., Murti, K.G., Whitt, M.A., *et al.* (1997). A system for functional analysis of Ebola virus glycoprotein. *Proceedings of the National Academy of Sciences, USA*, **94**, 14 764–14 769.

Taviner, M., Thwaites, G. & Gant, V. (1998). The English Sweating Sickness, 1485–1551: a viral pulmonary disease? *Medical History*, **42,** 96–98.

Taylor, G. & Morris, J.A. (1939). *A Sketch-Map History of Britain and Europe 1485–1783*. London: George G. Harrap & Company Ltd.

Theodorou, I., Meyer, L., Magierowska, M., Katlama, C., Rouzioux, C. and the Seroco Study Group. (1997). HIV-1 infection in an individual homozygous for *CCR5-Δ32*. *Lancet*, **349**, 1219–1220.

Thirsk, J. (1967). The farming regions of England. In *The Agrarian History of England and Wales*, ed. J. Thirsk, vol. IV, pp. 1–112. Cambridge: Cambridge

University Press.
Thompson, A.H. (1914). The pestilences of the fourteenth century in the Diocese of York. *Archaeological Journal*, **LXXI**, 97–154.
Tidd, C.W., Olsen, L.F. & Schaffer, W.M. (1993). The case for chaos in childhood epidemics. II. Predicting historical epidemics from mathematical models. *Proceedings of the Royal Society B*, **254**, 257–273.
Trevelyan, G.M. (1945). *English Social History*. London: Longmans.
Trout, A.P. (1973). The municipality of Parish confronts the plague of 1668. *Medical History*, **17**, 418–423.
Twigg, G. (1978). The role of rodents in plague transmission: a worldwide review. *Mammal Review*, **8**, 77–110.
Twigg, G. (1984). *The Black Death: A Biological Reappraisal*. London: Batsford Academic.
Twigg, G. (1989). The Black Death in England: an epidemiological dilemma. *Maladie et Société*, siècles XII–XVII, 75–98.
Twigg, G. (1993). Plague in London: spatial and temporal aspects of mortality. In *Epidemic Disease in London*, ed. J.A.I. Champion, pp. 1–17. London: University of London Centre for Metropolitan History.
Underwood, E.A. (1965). Charles Creighton, the man and his work. In *History of Epidemics in Britain*, vol. 1. *From A.D. 664 to the Great Plague*, ed. C. Creighton, pp. 43–135. London: Frank Cass & Co. Ltd.
Vernadsky, G. (1953). *The Mongols and Russia*. New Haven, CT: Yale University Press.
Walker, J. (1860). *The History of Penrith from the Earlier Period to the Present Time*. Penrith: Hodgson Printers.
Wang, W.K., Dudek, T., Zhao, Y.J., Brumblay, H.G., Essex, M. & Lee, T.H. (1998). *CCR5* coreceptor utilization involves a highly conserved arginine residue of HIV type 1 gp120. *Proceedings of the National Academy of Sciences, USA*, **95**, 5740–5745.
Watson, J. (1992). *Lakeland Towns*. Milnthorpe, Cumbria: Cicerone Press.
White, A. (ed). (1993). *A History of Lancaster 1193–1993*. Keele: Ryburn Publishing Keele University Press.
Willan, T.S. (1983). Plague in perspective: the case of Manchester in 1605. In *Transactions of the Historic Society of Lancashire and Cheshire*, ed. J.I. Kermode & C.B. Phillips, vol. 132, pp. 29–40. Gloucester: Alan Sutton Publishing Ltd.
Wilson, E.M. (1975). Richard Leake's plague sermons, 1599. *Transactions of the Cumberland and Westmorland Antiquarian and Archaeological Society*, **75**, 150–173.
Wood, W. (1865). *The History and Antiquities of Eyam*, 4th Edition. London: Bell and Daldy.
Wrigley, E.A. (ed.) (1966). *Introduction to English Historical Demography*. London: Weidenfeld and Nicholson.
Wrigley, E.A. & Schofield, R.S. (1981). *Population History of England and Wales, 1541–1871. A Reconstruction*. London: Edward Arnold.
Wu, L.-T. (1926). *A Treatise on Pneumonic Plague*. Geneva: League of Nations Health Organization.
Wu, L.-T., Chun, J.W.H., Pollitzer, R. & Wu, C.Y. (1936). *Plague: A Manual for Medical and Public Health Workers*. Shanghai: Mercury Press.
Wylie, J.A.H. & Collier, L.H. (1981). The English Sweating Sickness (Sudor Anglicus): A reappraisal. *Journal of the History of Medicine and Allied*

Sciences, **36**, 425–445.
Yorke, J.A. & London, W.P. (1973). Recurrent outbreaks of measles, chickenpox and mumps. II. Systematic differences in contact rates and stochastic effects. *American Journal of Epidemiology*, **98**, 469–482.
Yorke, J.A., Nathanson, N., Pianigiani, G. & Martin, J. (1979). Seasonality and the requirements for perpetuation and eradication of viruses in populations. *American Journal of Epidemiology*, **109**, 103–123.
Zaki, S.R. & Goldsmith, C.S. (1999). Pathologic feature of filovirus infections in humans. *Current Topics in Microbiology and Immunology*, **235**, 97–116.
Zhang, L., He, T., Talal, A., Wang, G., Frankel, S.S. & Ho, D.D. (1998). In vivo distribution of the human immunodeficiency virus simian immunodeficiency virus coreceptors: *CXCR4*, *CCR3*, and *CCR5*. *Journal of Virology*, **72**, 5035–5045.
Ziegler, P. (1969). *The Black Death.* London: Collins.
Zimmerman, P.A., Buckler-White, A., Alkhatib, G., Spalding, T., Kubofcik, J., Combadiere, C., *et al.* (1997). Inherited resistance to HIV-1 conferred by an inactivating mutation in CC chemokine receptor 5-studies in populations with contrasting clinical phenotypes, defined racial background, and quantified risk. *Molecular Medicine*, **3**, 23–26.

Index

Numbers in *italics* refer to pages with figures and tables